Heat Transfer in Engineering

Heat Transfer in Engineering

Editors

Moo-Yeon Lee
Jae-Hyeong Seo

MDPI • Basel • Beijing • Wuhan • Barcelona • Belgrade • Manchester • Tokyo • Cluj • Tianjin

Editors

Moo-Yeon Lee
Mechanical Engineering
Dong-A University
Busan
Korea, South

Jae-Hyeong Seo
Thermal Management R&D
Center, Convergence Technology
R&D Division
Korea Automotive Technology
Institute
Cheonan-Si
Korea, South

Editorial Office
MDPI
St. Alban-Anlage 66
4052 Basel, Switzerland

This is a reprint of articles from the Special Issue published online in the open access journal *Symmetry* (ISSN 2073-8994) (available at: www.mdpi.com/journal/symmetry/special_issues/Heat_Transfer_Engineering).

For citation purposes, cite each article independently as indicated on the article page online and as indicated below:

LastName, A.A.; LastName, B.B.; LastName, C.C. Article Title. *Journal Name* **Year**, *Volume Number*, Page Range.

ISBN 978-3-0365-3125-0 (Hbk)
ISBN 978-3-0365-3124-3 (PDF)

Contents

About the Editors

Moo-Yeon Lee

My name is Moo-Yeon Lee. Currently, I am professor in department of mechanical engineering and head of thermal energy management (TEM) lab. My general interests include, refrigeration and air-conditioning, green-energy cars, and heat transfer and thermal management systems. Currently, I am working on development of next generation cooling technology using nanofluids for high power density devices, cooling system for driving motor of electric vehicles and direct cooling system for battery thermal management. My scientific interests include electric vehicles, batteries, electric driving motor, HVAC system, renewable energy and cooling technology. My work is acknowledged by 2021 Seokdang Academic Award for excellent scientific contribution in Dong-A university, 2018 Marquis Who's Who in the World award, Yeonam International Cooperation Research Professor, LG Yonam Foundation, Award by Director of the Small and Medium Business Administration, Young Thermal Engineer Award, Contribution Award for Most Citations and Ministry of Trade, Industry & Energy minister Award.

Jae-Hyeong Seo

My name is Jae-Hyeong Seo. Currently, I am working as a senior researcher in thermal management R&D center of Korea Automotive Technology Institute (KATECH). My research interests include thermal management for xEVs, heat and mass transfer and nanofluid. My ongoing projects include development of HVAC system based on artificial intelligence technology.

Preface to "Heat Transfer in Engineering"

The advancements in research related to heat transfer have gathered much attention in recent decades following the quest for efficient thermal systems, interdisciplinary studies involving heat transfer, and energy research. Heat transfer, a fundamental transport phenomenon, has been considered one of the critical aspects for the development and advancement of many modern applications, including cooling, thermal systems which contain symmetry analysis, energy conservation and storage, and symmetry-preserving discretization of heat transfer in a complex turbulent flow. The purpose of this book is to present recent advances, as well as up-to-date progress in all areas of heat transfer in engineering and its influence on emerging technologies. The scope of the present book includes heat transfer and thermal phenomena at all scales; thermal systems and thermal management systems; an interdisciplinary study focusing on heat transfer; waste heat recovery and allied heat transfer applications; heat transfer in energy storage and energy conservation; experimental, numerical, and analytical studies focusing on heat transfer and thermal phenomena; and fundamental mechanisms and practical applications of heat transfer in a wide variety of processes. This book provides a reference platform for the emerging engineers and researchers for solving their heat transfer engineering problems and employing them in application technologies.

Moo-Yeon Lee, Jae-Hyeong Seo
Editors

symmetry

MDPI

Article

First and Second Law Thermodynamic Analyses of Hybrid Nanofluid with Different Particle Shapes in a Microplate Heat Exchanger

Kunal Sandip Garud [1], Seong-Guk Hwang [1], Taek-Kyu Lim [2], Namwon Kim [3] and Moo-Yeon Lee [1,*]

1. Department of Mechanical Engineering, Dong-A University, 37 Nakdong-Daero 550, Saha-gu, Busan 49315, Korea; 1876936@donga.ac.kr (K.S.G.); 2070392@donga.ac.kr (S.-G.H.)
2. Thermal Management R&D Center, KATECH, 303 Pungse-ro, Pungse-Myun, Cheonan 31214, Korea; tklim@katech.re.kr
3. Ingram School of Engineering, Texas State University, San Marcos, TX 78666, USA; namwonkim@txstate.edu

* Correspondence: mylee@dau.ac.kr; Tel.: +82-51-200-7642

Abstract: The improvement in the quantitative and qualitative heat transfer performances of working fluids is trending research in the present time for heat transfer applications. In the present work, the first and second law analyses of a microplate heat exchanger with single-particle and hybrid nanofluids are conducted. The microplate heat exchanger with single-particle and hybrid nanofluids is analyzed using the computational fluid dynamics approach with symmetrical heat transfer and fluid flow analyses. The single-particle Al_2O_3 nanofluid and the hybrid Al_2O_3/Cu nanofluid are investigated for different nanoparticles shapes of sphere (Sp), oblate spheroid (OS), prolate spheroid (PS), blade (BL), platelet (PL), cylinder (CY) and brick (BR). The first law characteristics of NTU, effectiveness and performance index and the second characteristics of thermal, friction and total entropy generation rates and Bejan number are compared for Al_2O_3 and Al_2O_3/Cu nanofluids with considered different-shaped nanoparticles. The OS- and PL-shaped nanoparticles show superior and worse first and second law characteristics, respectively, for Al_2O_3 and Al_2O_3/Cu nanofluids. The hybrid nanofluid presents better first and second law characteristics compared to single-particle nanofluid for all nanoparticle shapes. The Al_2O_3/Cu nanofluid with OS-shaped nanoparticles depicts maximum values of performance index and Bejan number as 4.07 and 0.913, respectively. The first and second law characteristics of the best combination of the Al_2O_3/Cu nanofluid with OS-shaped nanoparticles are investigated for various volume fractions, different temperature and mass flow rate conditions of hot and cold fluids. The first and second law characteristics are optimum at higher hot fluid temperature, lower cold fluid temperature, lower hot and cold fluid mass flow rates. In addition, the first and second law characteristics have improved with increase in volume fraction.

Keywords: Bejan number; hybrid nanofluid; microplate heat exchanger; particle shape; performance index; thermodynamic analysis

Citation: Garud, K.S.; Hwang, S.-G.; Lim, T.-K.; Kim, N.; Lee, M.-Y. First and Second Law Thermodynamic Analyses of Hybrid Nanofluid with Different Particle Shapes in a Microplate Heat Exchanger. *Symmetry* **2021**, *13*, 1466. https://doi.org/10.3390/sym13081466

Academic Editor: Toshio Tagawa

Received: 12 July 2021
Accepted: 4 August 2021
Published: 10 August 2021

1. Introduction

The nanofluids are novel working fluids which show significant improvement in thermophysical properties due to dispersion of nanoparticles into the base fluid. The nanofluids enable the superior thermal conductivity and convective heat transfer rate compared to conventional working fluids. Therefore, the research trend on applicability of nanofluids in thermal systems is a growing body of work since the last few years. The nanoparticle shapes have considerable effect on the thermal and hydraulic performance characteristics of thermal systems incorporated with nanofluids. Numerous research studies elaborate the influence of nanoparticle shapes on hydrothermal performance of single-particle and hybrid nanofluid flow in various thermal systems. The introduction is arranged as, the first paragraph summarizes various research studies on thermophysical properties of nanofluid with different-shaped nanoparticles, the second paragraph discusses various research studies on behavior of heat transfer characteristics of single-particle nanofluid with different-shaped nanoparticles, the third paragraph summarizes various research studies to present the effect of different-shaped nanoparticles on heat transfer characteristics of hybrid nanofluid and at the end, the key research gaps and objectives of the present study are highlighted in the last paragraph.

This paragraph presents the effect of nanoparticle shapes on the thermophysical properties of nanofluids. Kim et al. have presented improvement in the thermal conductivity by 16%, 23% and 28% for BL-, PL- and BR-shaped nanoparticles, respectively, compared to Sp-shaped nanoparticles for alumina nanofluid [1]. Xie et al. have shown enhancements of 22.9% and 18.5% in thermal conductivity of CY- and Sp-shaped silicon carbide nanoparticles, respectively [2]. Similarly, Murshed et al. have also presented the superiority in thermal conductivity of CY nanoparticles compared to Sp-shaped nanoparticles [3]. Timofeeva et al. and Maheshwary et al. have proved that the Al_2O_3 and TiO_2 nanofluids with CY-shaped nanoparticles present higher thermal conductivity and that those with Sp-shaped nanoparticles present lower viscosity [4,5]. Singh et al. have reported maximum enhancement in thermal conductivity for PL-shaped silicon carbide nanoparticles [6]. Jeong et al. have concluded that the ZnO nanofluid with rectangular-shaped nanoparticles present higher thermal conductivity and viscosity compared to that with Sp-shaped nanoparticles [7]. Zhang et al. have reported the thermal conductivities of 0.2619 W/m-K and 0.2843 W/m-K for silver nanofluid with Sp and nanowire nanoparticles, respectively [8]. Nithiyanantham et al. have presented enhancement in thermal conductivity by 16% and 12%, and in viscosity by 25% and 37% for Sp- and CY-shaped alumina nanoparticles [9].

This paragraph presents the summary of single-particle nanofluid with various nanoparticle shapes in heat transfer applications. Vanaki et al. have analyzed the heat transfer and flow characteristics of SiO_2 nanofluid flow in a wavy channel for various nanoparticle shapes and concentrations. The SiO_2 nanofluid with a PL shape presents the highest enhancement in heat transfer characteristics [10]. Mahian et al. have presented the first law analysis in terms of heat transfer coefficient and Nusselt number, and the second law analysis in terms of entropy generation for alumina nanofluid with nanoparticle shapes of BL, PL, CY and BR [11]. Akbar et al. have concluded that the PL nanoparticle shape presents maximum velocity, and the BR nanoparticle shape presents maximum enhancement in thermal conductivity for nanofluid flow in non-uniform channel [12]. Bahiraei et al. have proved that the Sp, BR, BL, CY and PL nanoparticle shapes present the descending order of entropy generation for alumina nanofluid flow in microchannel heat sink [13]. Sheikholeslami et al. have presented that the PL nanoparticle shape has the highest Nusselt number compared to Sp, CY and BR nanoparticle shapes for Fe_3O_4 nanofluid in a porous curved enclosure, as well as in a porous cavity [14,15]. Nguyen et al. have concluded that the PL-shaped nanoparticles show more than 55% enhancement in heat transfer rates compared to Sp-shaped nanoparticles for CuO nanofluid flow in a wavy channel with obstacles [16]. Hatami et al. have presented that the TiO_2 nanofluid with a PL nanoparticle shape results in superior engine cooling or heat recovery performance at the higher volume fraction [17]. Kim et al. have concluded that the acetone-based Al_2O_3 nanofluid with CY-, BR- and Sp-shaped nanoparticles present lower thermal resistance by 16%, 29% and 33%, respectively, compared with pure acetone [18]. Bahiraei et al. have investigated the thermal and hydraulic characteristics of alumina nanofluid in micro plate heat exchangers, considering nanoparticle shapes of CY, OS, BR, BL and PL [19]. Vo et al. have reported that the PL-shaped nanoparticles present the highest heat transfer rate and the best performance evaluation criteria, whereas BR-shaped nanoparticles show the lowest pressure drop [20]. Khan et al. have concluded that the nanoparticle shapes of CY, PL and BR have a significant effect on temperature distribution compared to velocity distribution for copper nanofluid flow in parallel channels [21]. Raza et al. have shown that the Sp-shaped nanoparticles have a higher heat transfer rate compared with CY- and lamina-shaped nanoparticles [22].

Gireesha et al. have concluded that the BL-shaped nanoparticles have a superior heat transfer rate and Sp-shaped nanoparticles have the highest entropy generation rate compared to BR-, PL- and CY-shaped nanoparticles [23]. Elias et al. have concluded that the CY-shaped nanoparticles show better heat transfer and entropy generation characteristics for shell and tube heat exchangers with and without baffles [24,25]. The PL-shaped nanoparticles show the maximum heat transfer rate and Sp-shaped nanoparticles show the minimum pumping power as concluded by Shahsavar et al. for laminar flow and that by Alsarraf et al. for turbulent flow in a mini channel heat exchanger [26,27]. Al-Rashed et al. have reported that the PL-shaped nanoparticles show the maximum entropy generation rate and Bejan number, and Sp-shaped nanoparticles show the minimum entropy generation rate for laminar flow, and reverse results are reported by Monfared et al. for turbulent flow in a mini channel heat exchanger [28,29]. Sadripour and Chamkha have presented the heat transfer and entropy generation comparison of various shapes of metallic and non-metallic nanoparticles for different nanofluids flow in a solar collector [30]. The heat flow path, heat transfer and entropy generation of CuO nanofluid with Sp-, CY-, BR- and PL-shaped nanoparticles are simulated by Liu et al. [31].

The open literature on nanofluid in heat transfer application presents that research is trending towards the hybrid nanofluids due to their improved thermophysical properties compared to single-particle nanofluids. Therefore, there are few studies which attempted to demonstrate the improvement in heat transfer performance of hybrid nanofluids under the consideration of different nanoparticle shapes. This paragraph presents the summary of various research studies on hybrid nanofluids with various nanoparticle shapes in heat transfer applications. Ghadikolaei et al. have compared the TiO_2/Cu nanofluid with CY-, BR- and PL-shaped nanoparticles and reported that the PL-shaped nanoparticles show the highest heat transfer rate [32]. Ghadikolaei et al. have also proved that the PL-shaped nanoparticles of Fe_3O_4/Ag nanofluid show the maximum heat transfer rate due to an increase in the shape factor [33]. Dinarvand et al. have investigated the heat transfer and fluid flow characteristics of TiO_2/CuO nanofluid with Sp-, CY-, PL- and BR-shaped nanoparticles and the shown maximum Nusselt number for PL-shaped nanoparticles [34]. Bhattad and Sarkar have proved that the BR- and PL-shaped nanoparticles present the best and worst hydrothermal performances, respectively, for the hybrid nanofluid with combinations of alumina, titania and copper oxide or copper with silica nanoparticles [35]. Benkhedda et al. have reported the highest heat transfer rate for BL-shaped nanoparticles and the highest friction factor for PL-shaped nanoparticles when the TiO_2/Ag nanofluid flows through a tube [36]. Ghobadi et al. have compared the magnetohydrodynamic heat transfer of hexahedron- and lamina-shaped nanoparticles for the Al_2O_3/TiO_2 nanofluid and reported that the Nusselt number is affected mostly by lamina-shaped nanoparticles [37]. Aziz et al. have proposed an inverse relation between the shape factor of the nanoparticle and heat transfer for Cu and Fe_3O_4/Cu nanofluids [38]. Ghadikolaei et al. have reported the highest heat transfer rate and Nusselt number for TiO_2/CuO and MoS_2/Ag nanofluids with BL-shaped nanoparticles because of an increase in the shape factor at a higher temperature [39,40]. Similar results are deducted for the GO/MoS_2 nanofluid by Ghadikolaei and Gholinia [41]. Maraj et al. have investigated the shape effect of nanoparticles on magnetohydrodynamic heat transfer and flow characteristics [42]. Sahu et al. have presented the energy and exergy analyses of various hybrid nanofluids with Sp-, CY- and PL-shaped nanoparticles [43].

The comprehensive literature review reveals that there is no concrete comparative study on the first and second law analyses of single-particle and hybrid nanofluids with different particle shapes in heat transfer applications. The objective of this study is to investigate the first and second law characteristics of the microplate heat exchanger incorporated with single-particle and hybrid nanofluids with different nanoparticle shapes under various volume fractions, temperatures and mass flow rates. The computational fluid dynamics approach with symmetrical heat transfer and fluid flow concept is adopted to evaluate the first and second law characteristics of the microplate heat exchanger. The

NTU, effectiveness, performance index, thermal entropy generation rate, friction entropy generation rate, total entropy generation rate and Bejan number are compared for Al_2O_3 and Al_2O_3/Cu nanofluids with Sp-, OS-, PS1-, PS2-, PS3-, PS4-, BL-, PL-, CY- and BR-shaped nanoparticles. In addition, the best combination of nanofluid with nanoparticle shape to archive an optimum heat transfer performance in the microplate heat exchanger is suggested based on the first and second law characteristics.

2. Numerical Method

2.1. Computational Geometry and Boundary Conditions

The numerical investigation based on the computational fluid dynamics (CFD) approach is conducted to evaluate the first and second law characteristics of the microplate heat exchanger with single-particle and hybrid nanofluids. The three-dimensional model of the microplate heat exchanger is depicted in Figure 1. The microplate heat exchanger comprises of two flow lines in the counter flow direction, the nanofluid as hot fluid and water as cold fluid. There are three flow passages for the nanofluid and three flow passages for water. Each flow passage is divided into 17 microchannels with a cross section of 0.25 mm $\times$ 0.32 mm for the nanofluid and that of 0.25 mm $\times$ 0.42 mm for water. Hence, there are total 51 microchannels with a length of 12.5 mm and wall thickness of 0.52 mm for the nanofluid and water. The microplate heat exchanger is made up of copper material. The microplate heat exchanger is symmetrical, which results in symmetrical heat transfer and fluid flow through the microchannel, hence one pair of microplates for hot and cold fluids is considered as the computational geometry to reduce the computational time. The single-particle nanofluid of Al_2O_3 and hybrid nanofluid of Al_2O_3/Cu with volume fractions of 0.5%, 1.0% and 2.0% are considered as hot fluid. The nanofluids are flowing with inlet temperatures of 90 °C, 80 °C and 70 °C and inlet mass flow rates of 10 kg/h, 20 kg/h and 30 kg/h. The cold fluid water is flowing with inlet temperatures of 10 °C, 20 °C and 30 °C and inlet mass flow rates of 10 kg/h, 20 kg/h and 30 kg/h. The pressure outlet boundary condition is applied at the outlet of the heat exchanger. The nanofluids and water enter with uniform velocity and uniform temperature. The external surfaces of the heat exchanger are assumed to be insulated and a no-slip condition is assumed at all walls of the heat exchanger. The conjugate heat transfer is considered in the present study in that the solid domain is subjected to a conduction mechanism and fluid domains are subjected to conduction and convection mechanisms. The computational geometry of the microplate heat exchanger presents a similar pattern and symmetry boundary and forms the mirror pattern and thermal and flow characteristics. Therefore, the symmetry boundary conditions are applied on the computational geometry of the microplate heat exchanger.

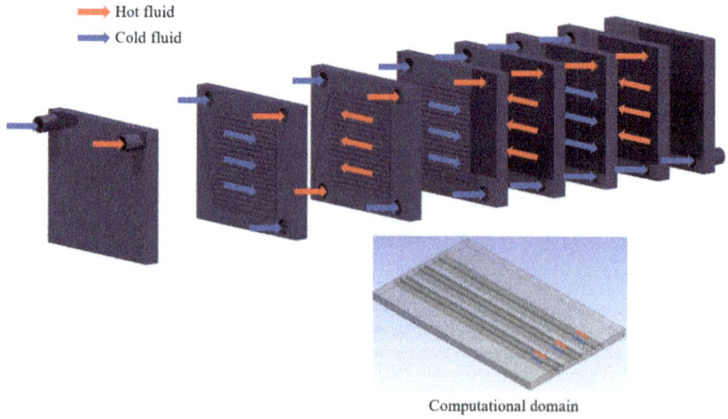

Figure 1. Three-dimensional model of the microplate heat exchanger.

2.2. Governing Equation and Meshing

The continuity, momentum and energy equations are solved using the computational fluid dynamics approach to analyze the thermodynamic attributes of the microplate heat exchanger with single-particle and hybrid nanofluids [44–46]. The continuity Equation (1) is expressed per unit of surface area (m^2). The governing equations are considered for the single pair microplates in the heat exchanger based on symmetrical heat transfer. While solving the equations, it is assumed that the flow is three-dimensional, steady, turbulent and incompressible. Additionally, the working fluids are assumed to be Newtonian [47].

Continuity equation

$$\nabla\cdot(\rho U) = 0 \tag{1}$$

Momentum equation

$$\nabla\cdot(\rho UU) = -\nabla p + \nabla\cdot(\mu\nabla \mathrm{U}) \tag{2}$$

Energy equation for fluid domains

$$\nabla\cdot(\rho Uh) = \nabla\cdot(\lambda\nabla T) \tag{3}$$

Energy equation for solid domains

$$\nabla^2 T = 0 \tag{4}$$

The working fluids in the microplate heat exchanger are exposed to heat exchange which results in entropy generation. The volumetric total entropy generation rate is the sum of the volumetric thermal entropy generation rate and the volumetric friction entropy generation rate as presented by Equation (5) [48].

$$\dot{S}_T = \dot{S}_{Th} + \dot{S}_{Fr} \tag{5}$$

The volumetric thermal entropy generation rate is calculated using Equation (6) as the summation of volumetric thermal entropy generations due to average and fluctuating temperature gradients [48].

$$\dot{S}_{Th} = \frac{\lambda}{\overline{T}^2}[(\frac{\partial \overline{T}}{\partial x})^2 + (\frac{\partial \overline{T}}{\partial y})^2 + (\frac{\partial \overline{T}}{\partial z})^2] + \frac{\lambda}{\overline{T}^2}[\overline{(\frac{\partial T'}{\partial x})^2} + \overline{(\frac{\partial T'}{\partial y})^2} + \overline{(\frac{\partial T'}{\partial z})^2}] \tag{6}$$

The first term at the right side of Equation (6) presents the volumetric thermal entropy generation rate due to time–mean temperature gradients. Whereas, the second term at right side of Equation (6) presents the volumetric thermal entropy generation rate due to fluctuating temperature gradients, which could also be expressed as Equation (7) [48].

$$\dot{S}_{Th'} = \frac{\lambda_t}{\overline{T}^2}[(\frac{\partial \overline{T}}{\partial x})^2 + (\frac{\partial \overline{T}}{\partial y})^2 + (\frac{\partial \overline{T}}{\partial z})^2] \tag{7}$$

Therefore, the volumetric thermal entropy generation rate could be reduced to Equation (8) after manipulating $\lambda_{eff} = \lambda + \lambda_t$.

$$\dot{S}_{Th} = \frac{\lambda_{eff}}{\overline{T}^2}[(\frac{\partial \overline{T}}{\partial x})^2 + (\frac{\partial \overline{T}}{\partial y})^2 + (\frac{\partial \overline{T}}{\partial z})^2] \tag{8}$$

The volumetric friction entropy generation rate is evaluated using Equation (9) as the summation of direct dissipation due to the average velocity gradient and indirect dissipation due to the fluctuating velocity gradient [48].

$$\dot{S}_{Fr} = \frac{\mu}{T}\{2[(\frac{\partial\overline{v}}{\partial x})^2 + (\frac{\partial\overline{v}}{\partial y})^2 + (\frac{\partial\overline{v}}{\partial z})^2] + (\frac{\partial\overline{v}_x}{\partial y} + \frac{\partial\overline{v}_y}{\partial x})^2 + (\frac{\partial\overline{v}_x}{\partial z} + \frac{\partial\overline{v}_z}{\partial x})^2 + (\frac{\partial\overline{v}_y}{\partial z} + \frac{\partial\overline{v}_z}{\partial y})^2\}$$
$$+\frac{\mu}{T}2[\overline{(\frac{\partial v'_x}{\partial x})^2} + \overline{(\frac{\partial v'_y}{\partial y})^2} + \overline{(\frac{\partial v'_z}{\partial z})^2}] + \overline{(\frac{\partial v'_x}{\partial y} + \frac{\partial v'_y}{\partial x})^2}$$
$$+\overline{(\frac{\partial v'_x}{\partial z} + \frac{\partial v'_z}{\partial x})^2} + \overline{(\frac{\partial v'_y}{\partial z} + \frac{\partial v'_z}{\partial y})^2} \quad (9)$$

The first term on right side of Equation (9) indicates the direct entropy generation due to dissipation in the mean flow field, which is commonly denoted as direct dissipation. Whereas, the second term on the right side of Equation (9) indicates turbulent or indirect dissipation due to fluctuating velocity gradients, which is also expressed as Equation (10) [48].

$$\dot{S}_{Fr'} = \frac{\rho\beta^* kO}{\overline{T}} \quad (10)$$

Therefore, the volumetric friction entropy generation rate could be reduced to Equation (11) [48].

$$\dot{S}_{Fr} = \frac{\mu}{T}\{2[(\frac{\partial\overline{v}}{\partial x})^2 + (\frac{\partial\overline{v}}{\partial y})^2 + (\frac{\partial\overline{v}}{\partial y})^2] + (\frac{\partial\overline{v}_x}{\partial y} + \frac{\partial\overline{v}_y}{\partial x})^2 + (\frac{\partial\overline{v}_x}{\partial z} + \frac{\partial\overline{v}_z}{\partial x})^2 + (\frac{\partial\overline{v}_y}{\partial z} + \frac{\partial\overline{v}_z}{\partial y})^2\} + \frac{\rho\beta^* kO}{\overline{T}} \quad (11)$$

Here, β^* is the model constant with value of 0.09.

The Bejan number is evaluated to quantify the contribution of the volumetric thermal entropy generation rate in the volumetric total entropy generation rate. The Bejan number (Be) is defined as the ratio of the volumetric thermal entropy generation rate to the volumetric total entropy generation rate, as presented by Equation (12) [48].

$$Be = \frac{\dot{S}_{Th}}{\dot{S}_T} \quad (12)$$

The governing equations are solved using the finite volume method and second order approach. The tetrahedron mesh elements are considered for the computational geometry of the heat exchanger and fluid domains. The inflation layers are provided on the fluid domains to consider the effect of boundary layers at the walls of the heat exchanger. The mesh independency test is carried out by generating five different mesh element numbers on the computational geometry. The results of performance index are evaluated for different mesh element numbers by considering the flow of water as hot and cold fluids. The inlet temperature and mass flow rate on the hot side are 90 °C and 20 kg/h, and those on the cold side are 20 °C and 20 kg/h. The mesh independency results for five different mesh elements are presented in Table 1. The temperature and pressure drop results for hot and cold fluids are significantly varying when the mesh element numbers are ranging from 157,649 to 732,993. However, beyond the mesh element number of 732,993, the simulated results of outlet temperatures and pressure drops of hot and cold fluids vary within ±1% [49]. Therefore, the computational geometry with a mesh element number of 732,993 is considered for the further numerical investigations. The mesh configuration of computational geometry with selected final mesh elements is depicted in Figure 2. The SIMPLE scheme with velocity-pressure coupling and the convergence criteria of 10^{-8} are considered for solving the governing equations.

Table 1. Mesh independency test results.

Mesh Elements	Hot Fluid-Temperature (°C)	Cold Fluid-Temperature (°C)	Hot Fluid-Pressure Drop(bar)	Cold Fluid-Pressure Drop (bar)
157,649	88.669	29.841	2.413	1.321
489,478	86.295	26.877	2.637	1.444
732,993	85.180	24.888	2.738	1.502
1,142,485	85.182	24.086	2.742	1.505
1,588,899	85.181	23.865	2.751	1.507

Figure 2. Mesh configuration for computational geometry.

3. Thermophysical Properties of Nanofluids with Nanoparticle Shapes

The thermophysical properties of single-particle and hybrid nanofluids with different nanoparticle shapes are evaluated using the models presented in Sections 3.1 and 3.2, respectively. The different nanoparticle shapes considered are Sp, OS, PS1, PS2, PS3, PS4, BL, PL, CY and BR, as depicted in Figure 3.

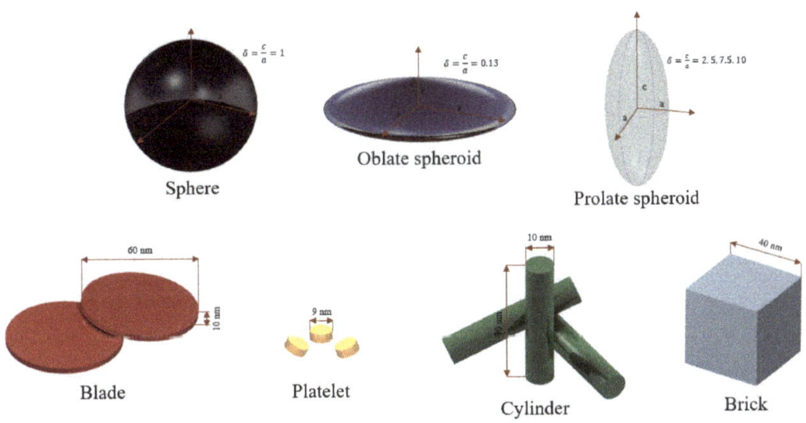

Figure 3. Different nanoparticle shapes considered in the present study.

3.1. *Single-Particle Nanofluid Properties*

The volume fraction of nanoparticles in nanofluid is calculated using Equation (13) [50].

$$O = \frac{V_{np}}{V_{bf} + V_{np}} \tag{13}$$

The density and specific heat of single-particle nanofluids are not affected by nanoparticle shapes and are calculated using Equations (14) and (15), respectively [51,52].

$$\rho_{nf} = (1 - O)\rho_{bf} + O\rho_{np} \tag{14}$$

$$C_{p,nf} = \frac{(1 - O)\rho_{bf}C_{p,bf} + O\rho_{np}C_{p,np}}{\rho_{nf}} \tag{15}$$

The nanoparticle shape affects the thermal conductivity and viscosity of single-particle nanofluids. The thermal conductivity of single-particle nanofluids with Sp-, OS-, PS1-, PS2-,

PS3- and PS4-shaped nanoparticles are calculated using the model proposed by Hamilton–Crosser, as presented by Equation (16). The model proposed by Timofeeva et al. [4], as presented by Equation (17), is used to calculate the thermal conductivity of single-particle nanofluids with BL-, PL-, CY-, and BR-shaped nanoparticles [53].

$$\frac{k_{nf}}{k_{bf}} = \frac{k_{np} + (n-1)k_{bf} - (n-1)\left(k_{bf} - k_{np}\right)O}{k_{np} + (n-1)k_{bf} + O\left(k_{bf} - k_{np}\right)} \tag{16}$$

$$\frac{k_{nf}}{k_{bf}} = 1 + \left(C_k^{shape} + C_k^{surface}\right)O = 1 + C_k O \tag{17}$$

Here, n is shape factor = $\frac{3}{\Psi}$. The values n for OS, PS1, PS2, PS3 and PS4 are calculated using the sphericity parameter Ψ, whose values for given nanoparticle shapes are reflected in Table 2. C_k^{shape} is the nanoparticle shape contribution to thermal conductivity, $C_k^{surface}$ is surface resistance that affects the thermal conductivity of nanofluid and influences by the solid/liquid interface.

Table 2. Parameters for calculating thermal conductivity and viscosity of spherical, oblate spheroid and prolate spheroid nanoparticle shape-based nanofluids.

Particle Shape	δ	Ψ	$\varnothing_m$
Sphere	1	1	0.599
Oblate spheroid	0.13	0.4904	0.575
Prolate spheroid 1	2	0.9287	0.546
Prolate spheroid 2	5	0.7321	0.432
Prolate spheroid 3	7.5	0.6453	0.368
Prolate spheroid 4	10	0.5883	0.321

The viscosity of single-particle nanofluids with Sp-, OS-, PS1-, PS2-, PS3- and PS4-shaped nanoparticles is evaluated using Equation (18). This equation is proposed by Muller et al., based on the mathematical model presented by Maron and Pierce. For calculating the viscosity of single-particle nanofluids with BL-, PL-, CY-, and BR-shaped nanoparticles, Timofeeva et al.'s [4] model, as presented by Equation (19), is used [53].

$$\frac{\mu_{nf}}{\mu_{bf}} = (1 - \frac{O}{O_m})^{-2} \tag{18}$$

$$\mu_{nf} = \mu_{bf}\left(1 + A_1 O + A_2 O^2\right) \tag{19}$$

Here, O is the volume fraction of nanoparticles in nanofluid and O_m is the packing fraction, which is calculated using Equation (20) [53]. A_1, A_2 are coefficients proposed by Timofeeva et al., and values of these coefficients are extracted from references [4,53].

$$O_m = \frac{2}{(0.321\delta + 3.02)} \tag{20}$$

The aspect ratio of spheroids is expressed as Equation (21). The parameters a and c are denoted as lengths of spheroid semi-axes, which could be seen in Figure 3. For OS- and PS-shaped nanoparticles, these parameters are presented by Equations (22) and (23), respectively [53].

$$\delta = \frac{c}{a} \tag{21}$$

$$OS = 1 - (\frac{c}{a})^2,\ c < a \tag{22}$$

$$PS = 1 - (\frac{a}{c})^2,\ c > a \tag{23}$$

3.2. Hybrid Nanofluid Properties

The open literature lacks presentation of the concrete models for evaluating the thermophysical properties of hybrid nanofluids with different nanoparticle shapes. In the present study, the updated equations for calculating the thermophysical properties of hybrid nanofluids with different nanoparticle shapes are proposed based on equations presented in Section 3.1 and references [54,55]. The density, specific heat, thermal conductivity and viscosity of hybrid nanofluids with different nanoparticle shapes are evaluated using Equations (24)–(30).

Volume fraction of hybrid nanofluid

$$O_{hnf} = O_{np1} + O_{np2} \tag{24}$$

Density of hybrid nanofluid

$$\rho_{hnf} = O_{np1}\rho_{np1} + O_{np2}\rho_{np2} + \left(1 - O_{hnf}\right)\rho_{bf} \tag{25}$$

Specific heat of hybrid nanofluid

$$C_{p,hnf} = \frac{O_{np1}\rho_{np1}C_{p,np1} + O_{np2}\rho_{np2}C_{p,np2} + \left(1 - O_{hnf}\right)\rho_{bf}C_{p,bf}}{\rho_{hnf}} \tag{26}$$

Thermal conductivity of hybrid nanofluid

For Sp-, OS-, PS1-, PS2-, PS3- and PS4-shaped nanoparticles

$$\frac{k_{hnf}}{k_{bf}} = \frac{\frac{O_{np1}k_{np1}+O_{np2}k_{np2}}{O_{hnf}} + (n-1)k_{bf} + (n-1)\left(O_{np1}k_{np1} + O_{np2}k_{np2}\right) - (n-1)O_{hnf}k_{bf}}{\frac{O_{np1}k_{np1}+O_{np2}k_{np2}}{O_{hnf}} + (n-1)k_{bf} - (n-1)\left(O_{np1}k_{np1} + O_{np2}k_{np2}\right) + O_{hnf}k_{bf}} \tag{27}$$

For BL-, PL-, CY-, and BR-shaped nanoparticles

$$\frac{k_{hnf}}{k_{bf}} = 1 + \left(C_k^{shape} + C_k^{surface}\right)O_{hnf} = 1 + C_kO_{hnf} \tag{28}$$

Viscosity of hybrid nanofluid

For Sp-, OS-, PS1-, PS2-, PS3- and PS4-shaped nanoparticles

$$\frac{\mu_{nf}}{\mu_{bf}} = (1 - \frac{O_{hnf}}{O_m})^{-2} \tag{29}$$

For BL-, PL-, CY-, and BR-shaped nanoparticles

$$\mu_{nf} = \mu_{bf}\left(1 + A_1O_{hnf} + A_2{O_{hnf}}^2\right) \tag{30}$$

The parameters for calculating thermal conductivity and viscosity of nanofluids with Sp-, OS- and PS-shaped nanoparticles are shown in Table 2 [53]. The parameters for calculating thermal conductivity and viscosity of nanofluids with BL-, PL-, CY- and BR-shaped nanoparticles are shown in Tables 3 and 4, respectively [53].

Table 5 depicts the properties of the considered nanoparticles of Boehmite alumina (Al_2O_3) and copper (Cu) and the base fluid of water [19,50]. These properties are computed in equations of Sections 3.1 and 3.2 to evaluate the density, specific heat, thermal conductivity and viscosity for Al_2O_3 and Al_2O_3/Cu nanofluids with different nanoparticle shapes.

Table 3. Parameters for calculating thermal conductivity of blade, platelet, cylinder and brick nanoparticle shape-based nanofluids.

Particle Shape	Aspect Ratio	C_k	C_k^{shape}	$C_k^{surface}$
Blade	1:6:1/12	2.74	8.26	−5.52
Platelet	1:1/8	2.61	5.72	−3.11
Cylinder	1:8	3.95	4.82	−0.87
Brick	1:1:1	3.37	3.72	−0.35

Table 4. Parameters for calculating viscosity of blade, platelet, cylinder and brick nanoparticle shape-based nanofluids.

Particle Shape	Coefficients	
	A_1	A_2
Blade	14.6	123.3
Platelet	37.1	612.6
Cylinder	13.5	904.4
Brick	1.90	471.4

Table 5. Properties of base fluid and nanoparticles.

Property	Water	Alumina	Copper
Density (kg/m^3)	997.1	3050	8933
Specific heat (J/kg·K)	4179	618.3	385
Thermal conductivity (W/m·K)	0.613	30	400
Viscosity (Pa·s)	0.001003	-	-

4. Data Reduction

The amount of heat released from the hot fluid (Nanofluid) and amount of heat gained by the cold fluid (Water) are calculated using Equations (31) and (32), respectively [56,57]. The heat absorbed by the working fluid is calculated using Equation (18) [36].

$$\dot{Q}_{hf} = \dot{m}_{hf} c_{p,hf} \left(T_{hf,i} - T_{hf,o} \right) \tag{31}$$

$$\dot{Q}_{cf} = \dot{m}_{cf} c_{p,cf} \left(T_{cf,o} - T_{cf,i} \right) \tag{32}$$

The average heat exchange between hot and cold fluids is calculated using Equation (33) [58].

$$\dot{Q} = \frac{\dot{Q}_{hf} + \dot{Q}_{cf}}{2} \tag{33}$$

The overall heat transfer coefficient for the microplate heat exchanger is evaluated using Equation (34) [58].

$$U = \frac{\dot{Q}}{A \Delta T_{LMTD}} \tag{34}$$

The hot and cold fluid flows are considered as a counter flow, hence the log mean temperature difference is calculated using Equation (35).

$$\Delta T_{LMTD} = \frac{\Delta T_2 - \Delta T_1}{\ln\left(\frac{\Delta T_2}{\Delta T_1}\right)} \tag{35}$$

$$\Delta T_1 = T_{hf,i} - T_{cf,o} \tag{36}$$

$$\Delta T_2 = T_{hf,o} - T_{cf,i} \tag{37}$$

The effectiveness of the microplate heat exchanger is evaluated using Equation (38) as the ratio of actual (average) heat transfer to maximum heat transfer between hot and cold fluids [59].

$$\varepsilon = \frac{\dot{Q}}{Q_{max}} \tag{38}$$

$$\dot{Q}_{max} = C_{min}\left(T_{hf,i} - T_{cf,i}\right) \tag{39}$$

$$C_{min} = \min\left(C_{hf}, C_{cf}\right) \tag{40}$$

$$C_{hf} = \dot{m}_{hf}c_{p,hf} \tag{41}$$

$$C_{cf} = \dot{m}_{cf}c_{p,cf} \tag{42}$$

The number of transfer units (NTU) of the microplate heat exchanger is calculated using Equation (43) [59,60].

$$NTU = \frac{UA}{C_{min}} \tag{43}$$

The performance index of the microplate heat exchanger is evaluated using Equation (44) as the ratio of average heat transferred between hot and cold fluids to total pumping power [61].

$$\eta = \frac{\dot{Q}}{P_{pump}} \tag{44}$$

The total pumping power is calculated by adding the pumping powers of hot and cold fluids, as expressed in Equation (45). The pump efficiency is assumed at 80%.

$$P_{pump} = \frac{\dot{m}_{hf}}{0.80}\frac{\Delta P_{hf}}{\rho_{hf}} + \frac{\dot{m}_{cf}}{0.80}\frac{\Delta P_{cf}}{\rho_{cf}} \tag{45}$$

The total entropy generation rate (W/K) for the microplate heat exchanger is defined as the sum of the entropy generation rate due to heat transfer (W/K) and the entropy generation rate due to pressure drop (W/K). The total entropy generation rate, entropy generation rate due to heat transfer and entropy generation rate due to pressure drop are calculated using Equations (46)–(48), respectively [62,63].

$$\dot{S}_{gen,total} = \dot{S}_{gen,heat\ transfer} + \dot{S}_{gen,pressure\ drop} \tag{46}$$

Entropy generation rate due to heat transfer

$$\dot{S}_{gen,heat\ transfer} = \dot{m}_{hf}C_{p,hf}\ln\left(\frac{T_{hf,o}}{T_{hf,i}}\right) + \dot{m}_{cf}C_{p,cf}\ln\left(\frac{T_{cf,o}}{T_{cf,i}}\right) \tag{47}$$

Entropy generation rate due to pressure drop

$$\dot{S}_{gen,f} = \frac{\dot{m}_{hf}\Delta P_{hf}}{\rho_{hf}T_{avg,hf}} + \frac{\dot{m}_{cf}\Delta P_{cf}}{\rho_{cf}T_{avg,cf}} \tag{48}$$

5. Results and Discussion

5.1. Validation

The experimental conditions applied by Alm et al. [64] for the same geometry and structure of the microplate heat exchanger are reflected in the present work to validate the accuracy of the numerical model. The experimental and numerical results are compared for the hot fluid inlet temperature and mass flow rate of 90 °C and 21 kg/h, respectively. The cold fluid mass flow rate ranges from 20 kg/h to 120 kg/h and the cold fluid inlet temperature is fixed at 12.5 °C for the comparison. The warm water as the hot fluid and cold water as the cold fluid are considered for the comparison. The outlet temperatures and pressure drops of hot and cold fluids are compared for the experimental and numerical methods, as presented in Figure 4. The trends for the experimental and numerical results are same for outlet temperatures and pressure drops of hot and cold fluids. The cold fluid outlet temperature decreases with increase in the cold fluid inlet mass flow rate, which results in an increase in the hot fluid outlet temperature. Whereas, the pressure drops for hot and cold fluids have increased with an increase in the cold fluid inlet mass flow rate. Over the variation range of the cold fluid inlet mass flow rate, the maximum deviation between the experimental and numerical results of the hot fluid outlet temperature is 4.64%, that of cold fluid outlet temperature is 4.93%, that of hot fluid pressure drop is 5.64% and that of the cold fluid pressure drop is 6.49%. The numerical results are in closer agreement with the corresponding experimental results, with a maximum deviation within ±10% for both thermal and flow characteristics of the microplate heat exchanger. Therefore, the numerical model is valid and reliable for the detailed thermodynamic investigations on the microplate heat exchanger.

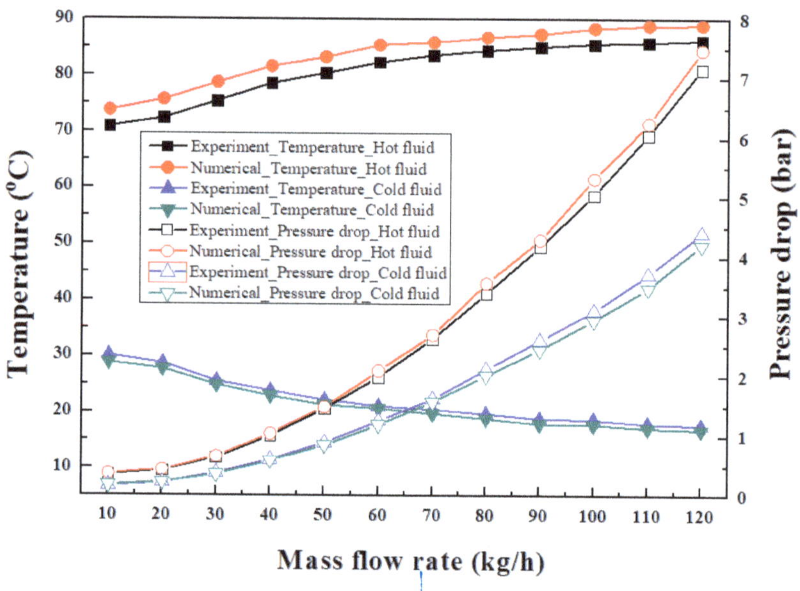

Figure 4. Comparison experimental and numerical results of temperature and pressure drop for hot and cold fluids.

5.2. *Evaluation of Nanofluid Thermophysical Properties for Different Nanoparticle Shapes*

The density and specific heat are not affected by changes in nanoparticle shape, unlike thermal conductivity and viscosity. Timofeeva et al. [4] have proved that the whole area of a solid-liquid interface greatly affects the thermal conductivity and viscosity of nanofluids. Therefore, the behavior of thermal conductivity and viscosity of single-particle and hybrid nanofluids with different nanoparticle shapes are depicted in Figure 5a,b, respectively. The behavior comparison is presented for 1.0% volume fraction of the nanoparticle in both single-particle and hybrid nanofluids. The nanoparticles of Al_2O_3 and Cu are mixed in the proposition of 50–50% in the hybrid nanofluid. The stability and agglomeration of different-shaped nanoparticles significantly affect the thermal conductivity of nanofluids. The thermal conductivity of both single-particle and hybrid nanofluids with all nanoparticle shapes are superior compared to water because of dispersion of high thermal conductivity nanoparticles into the base fluid. For nanoparticle shapes of Sp, OS, PS1, PS2, PS3 and PS3, the thermal conductivity of the hybrid nanofluid are better than the single-particle nanofluid. Whereas, in cases of nanoparticle shapes of BL, PL, CY and BR, the thermal conductivity values are the same for single-particle and hybrid nanofluids. Among all nanoparticle shapes, OS presents highest, and PL presents lowest thermal conductivity values in the respective cases of single-particle and hybrid nanofluids. The order of thermal conductivity is obtained based on the enhancement in the aspect ratio due to the fact that a rise in the contact area causes significant heat transfer when nanoparticles collide with each other [65]. Kim et al. have stated that BR-shaped nanoparticles present better thermal conductivity compared to BL-shaped nanoparticles due to rapid agglomeration [1]. The thermal conductivity values of Al_2O_3 and Al_2O_3/Cu nanofluids with OS-shaped nanoparticles are higher by 5.48% and 10.61%, respectively, compared to water. The thermal conductivity of Al_2O_3 and Al_2O_3/Cu nanofluids with PL-shaped nanoparticles is higher by 2.61% compared to water. The viscosity of single-particle and hybrid nanofluids with different nanoparticle shapes are higher than water because of the dispersion of nanoparticles into the base fluid. However, the viscosity values are same for single-particle and hybrid nanofluids with the same nanoparticle shape. The Al_2O_3 and Al_2O_3/Cu nanofluids with PL-shaped nanoparticles present the highest viscosity among all nanoparticle shapes, which is 43.23% superior to the viscosity of water. The Al_2O_3 and Al_2O_3/Cu nanofluids with Sp- and OS-shaped nanoparticles show the lowest and the second lowest values of viscosity, which are higher by 3.42% and 3.57%, respectively, compared to water. The PL-, CY- and BL-shaped nanoparticles present larger viscosity values compared to other-shaped nanoparticles due to limitation of rotational and Brownian motions. In addition, the PL- and CY-shaped nanoparticles stay in contact with one another for longer periods and interact between themselves significantly compared to other-shaped nanoparticles, which results in a higher viscosity in PL- and CY-shaped nanoparticles. Mahian et al. have presented the highest viscosity for PL-shaped nanoparticles, which increases with volume fraction [11]. The density and specific heat of the Al_2O_3 nanofluid are 1017.63 kg/m^3 and 4072.28 J/kg·K, respectively, and those of the Al_2O_3/Cu nanofluid are 1047.04 kg/m^3 and 3965.29 J/kg·K, respectively, for all nanoparticle shapes. From this comparison, it could be concluded that the single-particle and hybrid nanofluids with OS-shaped nanoparticles have excellent thermophysical properties compared to other nanoparticle shapes.

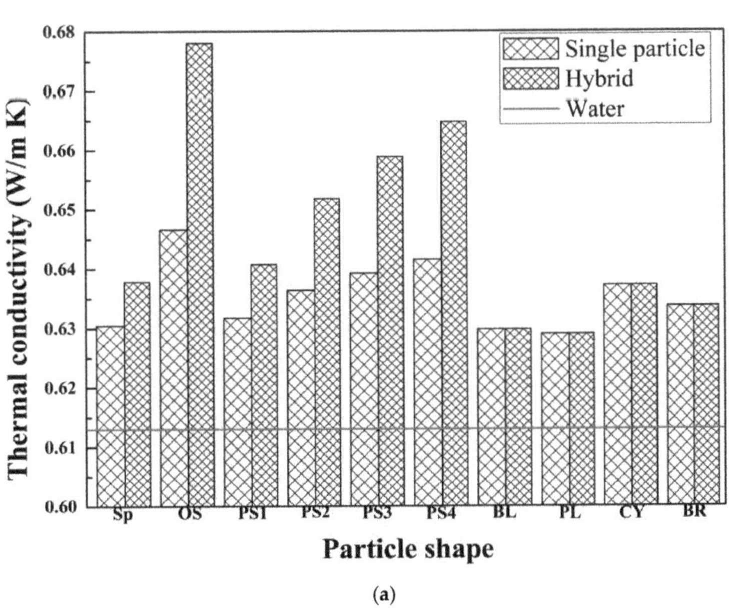

(a)

Figure 5. *Cont.*

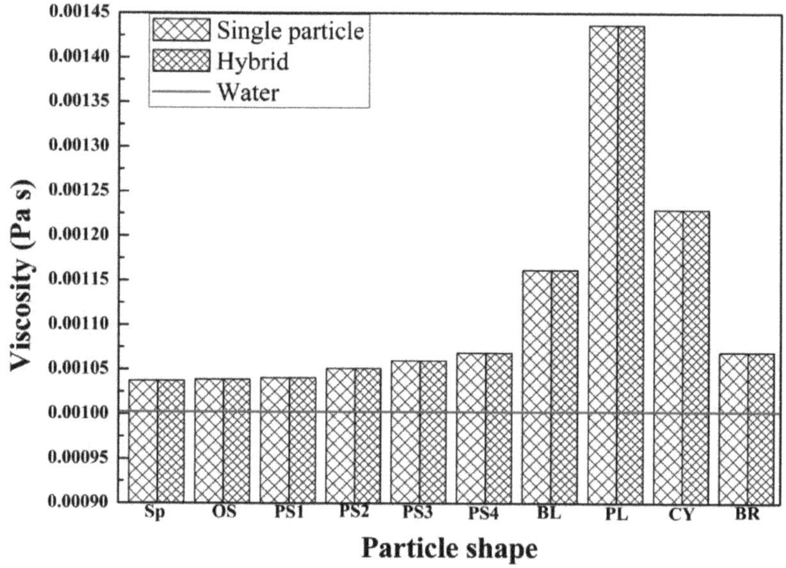

(b)

Figure 5. Behavior of (**a**) thermal conductivity and (**b**) viscosity of single-particle and hybrid nanofluids with different nanoparticle shapes.

5.3. Evaluation of First Law Characteristics for Different Nanoparticle Shapes

The comparison of NTU for single-particle and hybrid nanofluids with different nanoparticle shapes is shown in Figure 6. The NTU of single-particle and hybrid nanofluids are improved for nanoparticle shapes of Sp, OS, PS1, PS2, PS3, PS4, BL and BR compared to water because of improvement in the thermophysical properties of single-particle and hybrid nanofluids. In the case of PL-shaped nanoparticles, the NTU values are lower for both single-particle and hybrid nanofluids compared to water, due to a higher velocity of PL-shaped nanoparticles. The OS- and PL-shaped nanoparticles present the lowest and highest velocities, respectively, which results correspondingly into the lower and higher values of heat transfer coefficients. The lower velocity of OS-shaped nanoparticles results in a lower heat capacity, which dominates the lower heat transfer coefficients. Hence, as per Equation (43), it results in the highest NTU value. The higher velocity of PL-shaped nanoparticles results in a higher heat capacity and higher heat transfer coefficient; therefore, based on Equation (43), the combined effect of a higher heat transfer coefficient and a higher heat capacity presents a lower NTU for PL-shaped nanoparticles. The higher heat capacity dominates the higher heat transfer coefficient for PL-shaped nanoparticles. In the case of CY-shaped nanoparticles, the hybrid nanofluid shows superior NTU values and single-particle nanofluid shows poorer NTU values than water. In addition, for the same nanoparticle shape, the NTU of the Al_2O_3/Cu nanofluid is superior to the NTU of the Al_2O_3 nanofluid due to the addition of high thermal conductivity Cu nanoparticles to the Al_2O_3 nanofluid, which results in the thermal conductivity improvement of the Al_2O_3/Cu nanofluid. Bahiraei et al. have shown similar results of OS- and PL-shaped nanoparticles with the highest and lowest NTU, respectively, for single-particle nanofluids [19]. The NTU values of the Al_2O_3 and Al_2O_3/Cu nanofluids with OS-shaped nanoparticles are higher by 2.86% and 6.38%, respectively, compared to the NTU of water. The Al_2O_3 and Al_2O_3/Cu nanofluids with PL-shaped nanoparticles present the NTU values as lower by 3.99% and 1.82%, respectively, compared to the NTU of water.

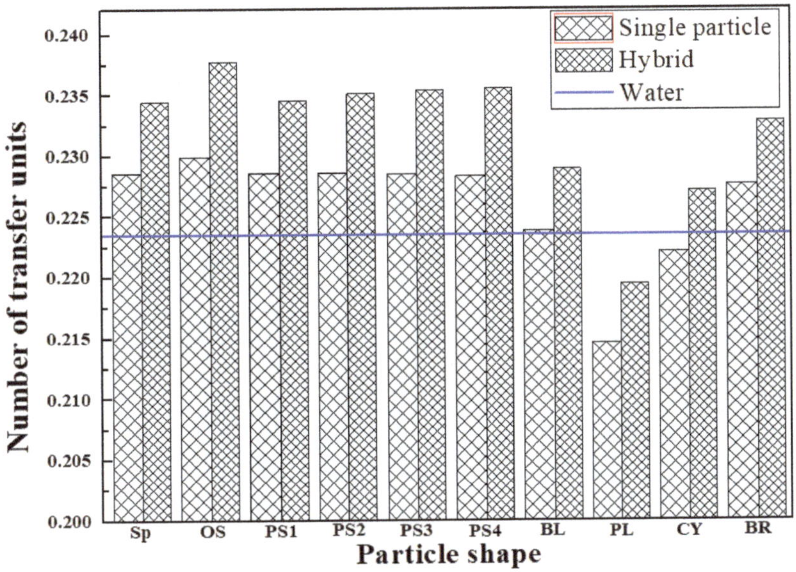

Figure 6. Comparison of NTU for single-particle and hybrid nanofluids with different nanoparticle shapes.

The effectiveness of single-particle and hybrid nanofluids with different nanoparticle shapes are compared in Figure 7. For the same nanoparticle shapes, the effectiveness of the Al_2O_3/Cu nanofluid are better than the Al_2O_3 nanofluid because of an increase in the thermal conductivity of the hybrid nanofluid by the dispersion of high thermal conductivity nanoparticles. The behavior of effectiveness is same as of the NTU for single-particle and hybrid nanofluids with all nanoparticle shapes. For the single-particle nanofluids, OS-shaped nanoparticles present the highest effectiveness, followed by Sp, PS1 = PS2, PS3, PS4, BR, BL, CY and PL in decreasing order of effectiveness, in which the effectiveness of CY- and PL-shaped nanoparticles are lower than water. In the case of the hybrid nanofluid, the decreasing order of effectiveness is OS, PS4, PS3, PS2, PS1, Sp, BR, BL, CY and PL, respectively, in which the effectiveness value of PL-shaped nanoparticles is lower than water. Shahsavar et al. have also presented a lower effectiveness for alumina nanofluids with PL-shaped nanoparticles [26]. The velocity of different-shaped nanoparticles is in the inverse relation with the temperature gradient. Therefore, the lower velocity of OS-shaped nanoparticles raises the temperature variation of nanofluids, which results in higher effectiveness. The effectiveness values of Al_2O_3 and Al_2O_3/Cu nanofluids with OS-shaped nanoparticles are higher by 2.75% and 6.10%, respectively, and those with PL-shaped nanoparticles are lower by 3.65% and 1.54%, respectively, compared to the effectiveness of water.

The performance index presents the combined effect of heat transfer and pressure drop characteristics. The comparison of the performance index of single-particle and hybrid nanofluids with different particle shapes is presented in Figure 8. The particle shape with the superior combination of thermal conductivity and viscosity shows the higher value of the performance index. Therefore, single-particle and hybrid nanofluids with OS-shaped nanoparticles show the highest values of the performance index among all nanoparticle shapes and water. The performance index values of single-particle and hybrid nanofluids with PL-shaped nanoparticles are lowest among all nanoparticle shapes, as well as water, due to poor thermal conductivity and viscosity. Despite the lower heat transfer rate in OS-shaped nanoparticles, the lowest pressure drop results in the highest performance index. Whereas, the higher pressure drop for PL-shaped nanoparticles results in the lowest performance index. Vo et al. have also illustrated that the pressure drop of PL-shaped nanoparticles is superior to other nanoparticle shapes, which increases as the volume fraction increases [20]. The OS- and PL-shaped nanoparticles show the maximum and minimum performance indexes for the alumina nanofluid, as proven by Bahiraei et al. and Arani et al. [19,53]. The single-particle and hybrid nanofluids with CY-shaped nanoparticles and the single-particle nanofluids with BL-shaped nanoparticles show a lower performance index than water despite better thermal conductivity because of higher viscosity and density. Apart from these combinations, other nanoparticle shapes present better performance index than water, in which hybrid nanofluids show a superior performance index than the single-particle nanofluid. The Al_2O_3 and Al_2O_3/Cu nanofluids with OS-shaped nanoparticles present the performance index as higher by 2.24% and 6.58%, respectively, and those with PL-shaped nanoparticles present the performance index as lower by 8.78% and 5.80%, respectively, than water. The single-particle and hybrid nanofluids with other nanoparticle shapes show the performance index values in a range between the highest and lowest values.

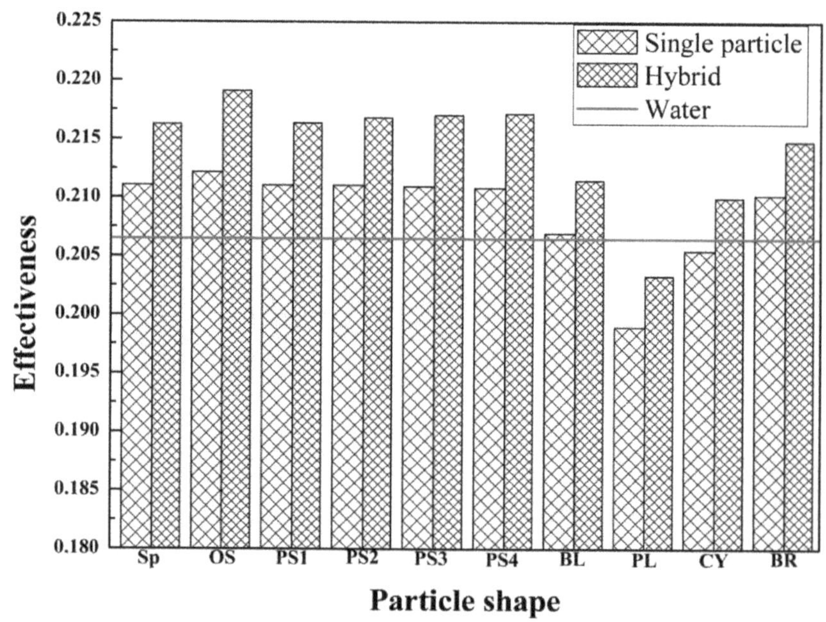

Figure 7. Effectiveness of single-particle and hybrid nanofluids with different nanoparticle shapes.

5.4. Evaluation of Second Law Characteristics for Different Nanoparticle Shapes

The thermal entropy generation rate for single-particle and hybrid nanofluids with different nanoparticle shapes is depicted in Figure 9. The thermal entropy generation is due to heat transfer, which depends on the temperature gradient. The thermal entropy generation rates of the hybrid nanofluid are better than that of single-particle nanofluids for OS-, PS2-, PS3- and PS4-shaped nanoparticles due to superior heat transfer properties and temperature gradients. However, in the case of Sp-, PS1-, BL-, PL-, CY- and BR-shaped nanoparticles, the thermal entropy generation rates of hybrid nanofluids are lower than single-particle nanofluids. The thermal entropy generation rates of all nanoparticle shapes except OS-shaped nanoparticles are lower than the water for both single-particle and hybrid nanofluids. The thermal entropy generation rates of single-particle and hybrid nanofluids with PL-shaped nanoparticles are lowest among all nanoparticle shapes. The PL-shaped nanoparticles have a higher velocity, which creates significant mixing and disturbance in the boundary layer. Hence, the temperature gradient decreases, which results in a lower heat transfer and thermal entropy generation. The opposite discussion could be applied for OS-shaped nanoparticles with lower velocity. Bahiraei et al. have also presented that the thermal entropy generation rates are highest and lowest for OS- and PL-shaped nanoparticles, respectively [13,48]. The Al_2O_3 and Al_2O_3/Cu nanofluids with OS-shaped nanoparticles show thermal entropy generation rates as higher by 0.14% and 0.70%, respectively, compared to water; however, the percentage increase is not significantly higher. The thermal entropy generation rates of Al_2O_3 and Al_2O_3/Cu nanofluids with PL-shaped nanoparticles are lower by 6.08% and 6.53%, respectively, compared to water.

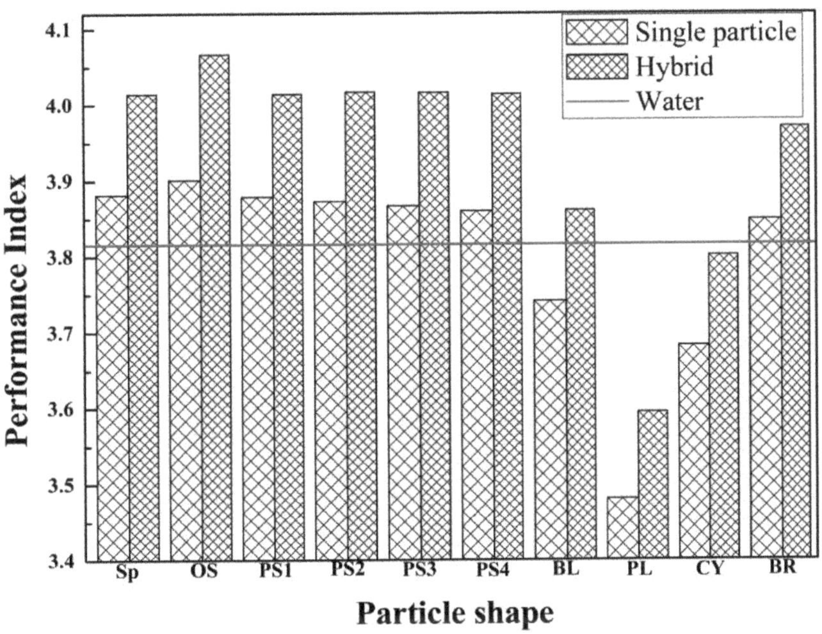

Figure 8. Comparison of performance index of single-particle and hybrid nanofluids with different particle shapes.

The comparison of the friction entropy generation rate for single-particle and hybrid nanofluids with different nanoparticle shapes is shown in Figure 10. The friction entropy generation is due to a pressure drop, which depends on the density and viscosity of nanofluids with various nanoparticle shapes. The friction entropy generation rates are lower for hybrid nanofluids compared to single-particle nanofluids for all nanoparticle shapes. Except for single-particle nanofluids with PL- and CY-shaped nanoparticles, the friction entropy generation rates of other combinations are lower than water. The single-particle and hybrid nanofluids with Sp- and OS-shaped nanoparticles present the lowest and second lowest values of friction entropy generation rates because of the same order of viscosity behavior for both nanoparticle shapes. The single-particle and hybrid nanofluids with PL-shaped nanoparticles show the highest values of friction entropy generation rates due to superior values of viscosity among all nanoparticles. The higher and lower values of friction entropy generation rates correspond to higher and lower velocity gradients of different nanoparticle shapes. Mahian et al. have also shown a trend in similar results, in that the friction entropy generation rate for PL-shaped nanoparticles is superior, followed by CY-, BL- and BR-shaped nanoparticles in the decreasing order [11]. The friction entropy generation rates of Al_2O_3 and Al_2O_3/Cu nanofluids with Sp-shaped nanoparticles are lower by 1.93% and 5.13%, respectively, and those with OS-shaped nanoparticles are lower by 1.91% and 5.11%, respectively, than water. The friction entropy generation rate of the

Al_2O_3 nanofluid with PL-shaped nanoparticles is higher by 2.73% and that of the Al_2O_3/Cu nanofluid with PL-shaped nanoparticles are lower by 0.74% compared to water.

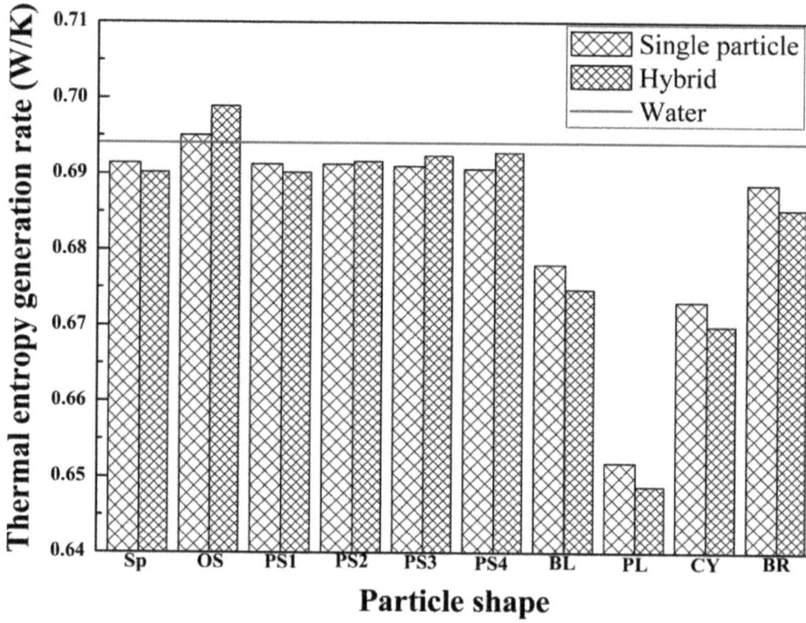

Figure 9. Thermal entropy generation rate for single-particle and hybrid nanofluids with different nanoparticle shapes.

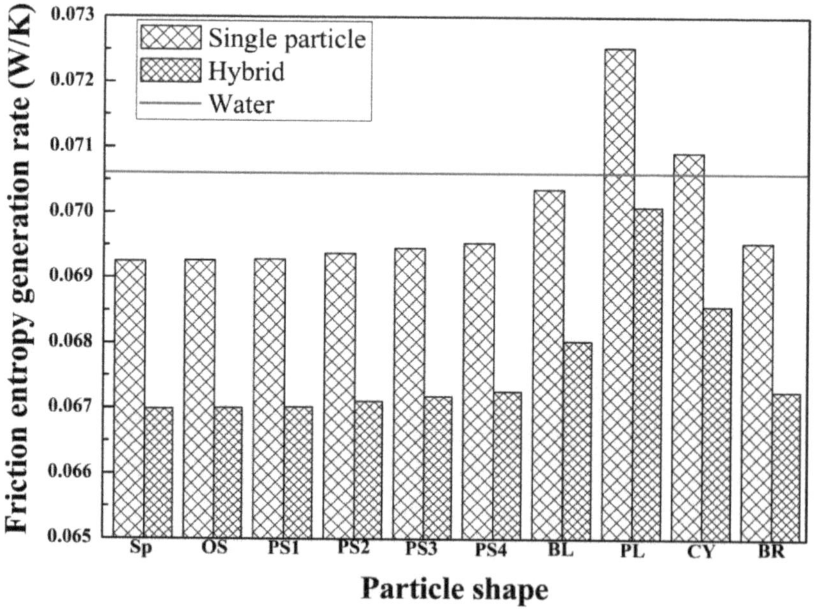

Figure 10. Comparison of friction entropy generation rate for single-particle and hybrid nanofluids with different nanoparticle shapes.

The behavior of the Bejan number for single-particle and hybrid nanofluids with different nanoparticle shapes is depicted in Figure 11. The Bejan number presents the contribution of thermal entropy generation in total entropy generation. The Bejan numbers of hybrid nanofluids are superior to single-particle nanofluids for all nanoparticle shapes. In addition, except the single-particle and hybrid nanofluids with PL- and CY-shaped nanoparticles and single-particle nanofluids with BL-shaped nanoparticles, all other combinations present higher values of Bejan numbers compared to water. The OS-shaped nanoparticles show the highest values of Bejan numbers, followed by Sp, PS1, PS2, PS3, PS4, BR, BL, CY and PL, respectively, in the decreasing order of the Bejan number for single-particle nanofluids. In the case of hybrid nanofluids, the decreasing order of Bejan numbers are OS-, PS2 = PS3-, Sp-, PS4-, PS1-, BR-, BL-, CY- and PL-shaped nanoparticles, respectively. The OS-shaped nanoparticles show a higher contribution of the thermal entropy generation rate and a lower contribution of the friction entropy generation rate, which results in the highest value of the Bejan number. The opposite discussion could be applied for PL-shaped nanoparticles for the lowest value of the Bejan number. Al-Rashed et al. and Monfared et al. have also shown the lower Bejan number for PL-shaped nanoparticles [28,29]. The Bejan number is maximum for OS-shaped nanoparticles and minimum for PL-shaped nanoparticles, as presented by Bahiraei et al. [48]. Compared to water, the Al_2O_3 and Al_2O_3/Cu nanofluids with OS-shaped nanoparticles depict the Bejan number as higher by 0.19% and 0.54%, respectively, and those with PL-shaped nanoparticles depict the Bejan number as lower by 0.86% and 0.57%, respectively.

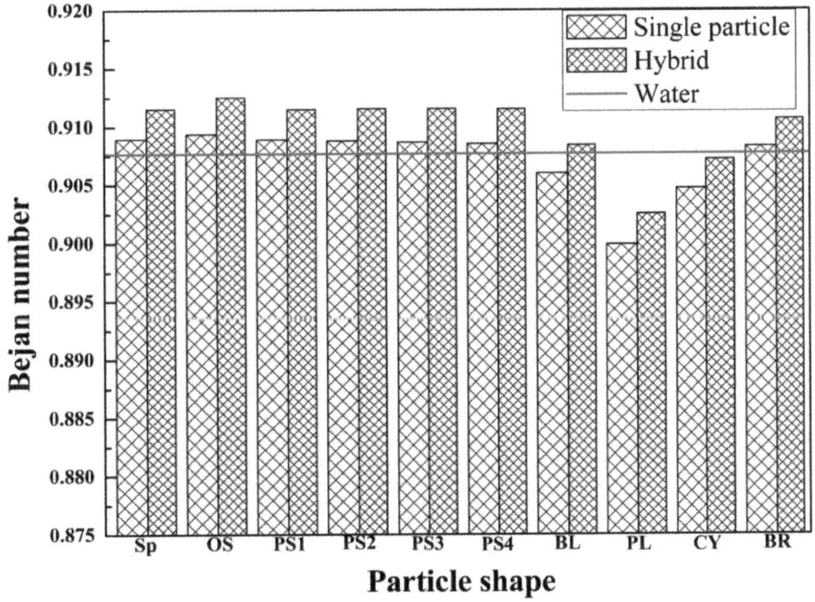

Figure 11. The behavior of Bejan number for single-particle and hybrid nanofluids with different nanoparticle shapes.

The combination of hybrid nanofluid with OS shaped nanoparticles show the excellent first law and second law characteristics compared to other combinations as well as water. Therefore, the first law characteristic namely, performance index and the second law characteristic namely, Bejan number of hybrid nanofluid (Al_2O_3/Cu) with OS shaped nanoparticles are further investigated for different temperatures and mass flow rates of hot and cold fluids. In addition, the influence of volume fraction is also integrated while investigating the effect of temperature and mass flow rate on different characteristics. The performance index presents the combined effect of heat transfer and pressure drop whereas, the Bejan number presents the combined effect of thermal and friction entropy generations hence, these two parameters are considered as the first and second law characteristics under the influence of various boundary parameters.

5.5. Effect of Hot Fluid Temperature on First and Second Law Characteristics

The behavior of first and second law characteristics of the Al_2O_3/Cu nanofluid with OS-shaped nanoparticles for various volume fractions and hot fluid temperatures is presented in Figure 12. The hot fluid temperature varies at 90 °C, 80 °C, and 70 °C and the volume fraction varies at 0.5%, 1.0% and 2.0%. The performance index increases with an increase in the volume fraction as well as hot fluid temperature. The heat transfer and pressure drop both increase with an increase in volume fraction, but the increase in the heat transfer dominates compared to the increase in the pressure drop, hence, as a result, the performance index increases as the volume fraction increases for all hot fluid temperatures. The hot fluid at the higher temperature could transfer more heat compared to hot fluid at a lower temperature. The pressure drop remains almost the same for various hot fluid temperatures, whereas the heat transfer increases with a rise in the temperature, which shows an enhancement in the performance index with an increase in the hot fluid temperature for all volume fractions. The performance index of Al_2O_3/Cu with OS-shaped nanoparticles increases by 20% and 40% as the hot fluid temperature increases from 70 °C to 80 °C and 70 °C to 90 °C, respectively, for each volume fraction. With the increase in volume fraction from 0.5% to 2.0%, the performance index of Al_2O_3/Cu with OS-shaped nanoparticles enhances by 9.42% for each hot fluid temperature. As elaborated, the heat transfer enhances with an increase in the volume fraction and hot fluid temperature due to an increase in the temperature gradient. Hence, the thermal entropy generation rate increases with an increase in the volume fraction and hot fluid temperature. The increase in the thermal entropy generation rate with the volume fraction is not significantly high, but could not be neglected. The friction entropy generation depends on the pressure drop, but as per the formula, the friction entropy generation rate is evaluated based on the ratio of pressure drop and average temperature. The pressure drop and average temperature both increase as the volume fraction increases, but as explained before, the dominance of the heat transfer is more than pressure drop with an increase in the volume fraction, which leads to a higher increase rate of average temperature than the pressure drop. Therefore, the friction entropy generation rate decreases as the volume fraction increases for all hot fluid temperatures. The pressure drop shows negligible change and heat transfer shows significant enhancement with a rise in the hot fluid temperature. Therefore, the dominance of average temperature rise is higher than the pressure drop as the hot fluid temperature increases, which results in a decrease in the friction entropy generation rate with an increase in the hot fluid temperature for all volume fractions. The Bejan number increases with an increase in the volume fraction and an increase in the hot fluid temperature, because the thermal entropy generation rate increases and the friction entropy generation rate decreases as the volume fraction and hot fluid temperature have increased. The Bejan number is at maximum at the higher hot fluid temperature and higher volume fraction. The Bejan number of Al_2O_3/Cu with OS-shaped nanoparticles increases by 2.12%, 2.06% and 1.95% as the hot fluid temperature increases from 70 °C to 80 °C, and that increases by 3.69%, 3.58% and 3.40% when the hot fluid temperature increases from 70 °C to 90 °C for volume fractions of 0.5%, 1.0% and 2.0%, respectively. As the volume fraction increases from 0.5% to 2.0%, the Bejan number of Al_2O_3/Cu with OS-shaped nanoparticles increases by 1.01%, 0.85% and 0.73% for hot fluid temperatures of 70 °C, 80 °C, and 90 °C, respectively. Singh and Sarkar have presented the improvement in Nusselt number and reduction in friction factor with an increase in hot fluid temperature [62]. The contributions of thermal and friction entropy generation rates for Al2O3/Cu with OS-shaped nanoparticles at various hot fluid temperatures are depicted in Figure 13.

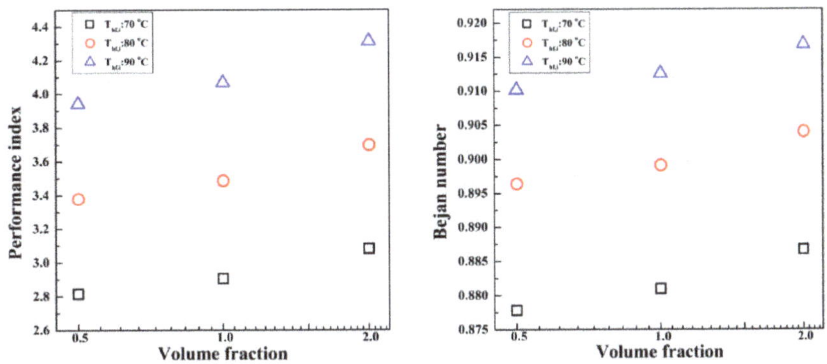

Figure 12. Behavior of first and second law characteristics of Al_2O_3/Cu nanofluid with OS-shaped nanoparticles for various volume fractions and hot fluid temperatures.

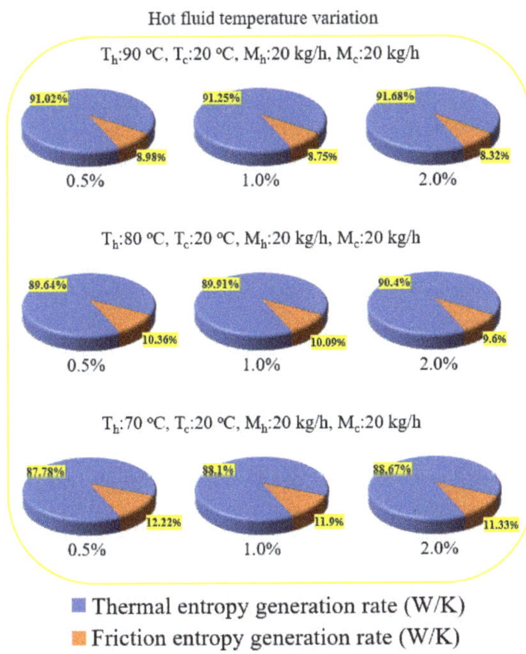

Figure 13. Contributions of thermal and friction entropy generation rates for Al_2O_3/Cu with OS-shaped nanoparticles at various hot fluid temperatures.

5.6. Effect of Hot Fluid Mass Flow Rate on First and Second Law Characteristics

The effect of the hot fluid mass flow rate and volume fraction on first and second law characteristics of hybrid nanofluids with OS-shaped nanoparticles is presented in Figure 14. The hot fluid mass flow rate varies at 10 kg/h, 20 kg/h and 30 kg/h. The performance index increases with the volume fraction for the same hot fluid mass flow rate because of higher dominance of the heat transfer increase compared to the pumping power increase as the volume fraction increases. Moreover, the steepness of the increasing trend of the performance index with the volume fraction becomes sharp at the higher hot fluid mass flow rate. For the same volume fraction, the performance index decreases with an increase in the hot fluid mass flow rate because the dominance of the increase in the pumping power is superior compared to the increase in heat transfer with an increase in the hot fluid mass flow rate. The performance index of Al_2O_3/Cu with OS-shaped nanoparticles increases by 3.99%, 9.43% and 12.27% for hot fluid mass flow rates of 10 kg/h, 20 kg/h and 30 kg/h, respectively, as the volume fraction increases from 0.5% to 2.0%. The performance index of Al_2O_3/Cu with OS-shaped nanoparticles decreases by 39.37%, 38.29% and 36.20% as the hot fluid mass flow rate increases from 10 kg/h to 20 kg/h, and that decreases by 70.31%, 69.51% and 67.94% as the hot fluid mass flow rate increases from 10 kg/h to 30 kg/h for volume fractions of 0.5%, 1.0% and 2.0%, respectively. The thermal entropy generation rate increases with an increase in the hot fluid mass flow rate for all volume fractions, due to increase in the temperature gradient and heat transfer. Similarly, the heat transfer enhances with the volume fraction, which results in an increase in the thermal entropy generation rate with an increase in the volume fraction for all hot fluid mass flow rates. However, the increasing trends of thermal entropy generation rates with volume fractions are not steep for higher hot fluid mass flow rates. The pressure drop and average temperature have both increased with an increase in volume fraction and hot fluid mass flow rate. With an increase in volume fraction, the increase in average temperature is more dominant than the increase in pressure drop, hence the friction entropy generation rate decreases as the volume fraction increases for all hot fluid mass flow rates. This decreasing trend becomes steeper with the volume fraction at the higher mass flow rates. On the other side, with the increase in the hot fluid mass flow rate, the increase in the pressure drop is dominating compared to the increase in average temperature; therefore, the friction entropy generation rate is high at a higher mass flow rate and vice versa for all volume fractions. Based on the trends for the thermal and friction entropy generation rates with volume fractions and hot fluid mass flow rate, the Bejan number is evaluated. The Bejan number increases with the increase in volume fraction and decrease in the hot fluid mass flow rate. The thermal entropy generation increases, and the friction entropy generation decreases with the increase in volume fraction, which results in an increase in the Bejan number with an increase in the volume fraction. The thermal and friction entropy generation rates have both increased with the increase in the hot fluid mass flow rate but the increasing rate of the friction entropy generation rate is significantly higher than the thermal entropy generation rate. Therefore, the Bejan number decreases with the increase in the hot fluid mass flow rate for each volume fraction. The Bejan number of the Al_2O_3/Cu nanofluid with OS-shaped nanoparticles increases by 0.21%, 0.73% and 1.74% for hot fluid mass flow rates of 10 kg/h, 20 kg/h and 30 kg/h, respectively, as the volume fraction increases from 0.5% to 2.0%. The Bejan number of Al_2O_3/Cu with OS-shaped nanoparticles decreases by 3.06%, 2.89% and 2.57% as the hot fluid mass flow rate increases from 10 kg/h to 20 kg/h, and that decreases by 10.72%, 10.24% and 9.37% as the hot fluid mass flow rate increases from 10 kg/h to 30 kg/h for volume fractions of 0.5%, 1.0% and 2.0%, respectively. Soroush and Chamkha have also shown that the first and second law characteristics of single-particle nanofluids enhances as the volume fraction increases for all nanoparticle shapes [30]. The contributions of thermal and friction entropy generation rates for Al2O3/Cu with OS-shaped nanoparticles at various hot fluid mass flow rates are depicted in Figure 15.

5.7. Effect of Cold Fluid Temperature on First and Second Law Characteristics

The variation in first and second law characteristics of hybrid nanofluid with OS-shaped nanoparticles for various volume fractions and cold fluid temperatures is depicted in Figure 16. The cold fluid temperature is varied as 10 °C, 20 °C and 30 °C. The cold fluid at the lower temperature absorbs more heat and presents the higher temperature gradient and heat transfer rate. The pressure drop is not significantly affected by change in the cold fluid temperature. Therefore, the performance index increases as the cold fluid temperature decreases. With the increase in volume fraction, the dominance of the increase in heat transfer is superior to the increase in the pressure drop; therefore, the performance index increases with the increase in the volume fraction. The performance index of the Al_2O_3/Cu nanofluid with OS-shaped nanoparticles decreases by 12.5% and 24.99% when the cold fluid temperature increases from 10 °C to 20 °C and 10 °C to 30 °C, respectively, for all volume fractions. With an increase in the volume fraction from 0.5% to 2.0%, the performance index increases by 9.43% for all cold fluid temperatures. The thermal entropy generation increases with the increase in volume fraction and decrease in the cold fluid temperature because the heat transfer rate increases with the increase in volume fraction and decrease in the cold fluid temperature. The friction entropy generation rate decreases with the increase in volume fraction despite an increase in pressure drop, because the increase in the average temperature with an increase in the volume fraction is dominant compared to an increase in pressure drop. Similar to the hot fluid temperature, the lower cold fluid temperature presents higher values of the friction entropy generation rate, and vice versa. The higher Bejan number is obtained at the lower cold fluid temperature because the dominance of the thermal entropy generation increase is higher than the friction entropy generation increase with a decrease in the cold fluid temperature. The thermal entropy generation increases and the friction entropy generation decreases with an increase in the volume fraction, which results in an increase in the Bejan number. The Bejan number of the Al_2O_3/Cu nanofluid with OS-shaped nanoparticles decreases by 1.16%, 1.12% and 1.07% as the cold fluid temperature increases from 10 °C to 20 °C, and that decreases by 2.66%, 2.59% and 2.45% as the cold fluid temperature increases from 10 °C to 30 °C for volume fractions of 0.5%, 1.0% and 2.0%, respectively. With the increase in volume fraction from 0.5% to 2.0%, the Bejan number of Al_2O_3/Cu nanofluid with OS-shaped nanoparticles increases by 0.63%, 0.73% and 0.85% for cold fluid temperatures of 10 °C, 20 °C and 30 °C, respectively. Garud et al. have proved that the performance of the heat exchanger is optimum for the lower mass flow rate of cold fluid [66]. The contributions of thermal and friction entropy generation rates for Al2O3/Cu with OS-shaped nanoparticles at various cold fluid temperatures are depicted in Figure 17.

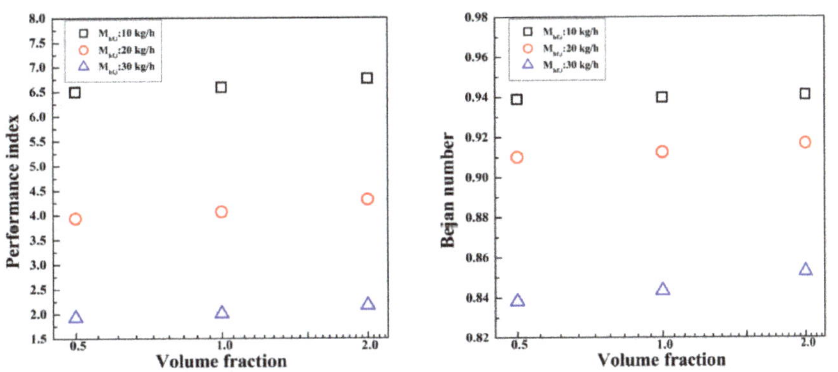

Figure 14. Effect of hot fluid mass flow rate and volume fraction on first and second law characteristics of hybrid nanofluid with OS-shaped nanoparticles.

5.8. Effect of Cold Fluid Mass Flow Rate on First and Second Law Characteristics

The behavior of first and second law characteristics of the hybrid nanofluid with OS-shaped nanoparticles for various volume fractions and cold fluid mass flow rates is presented in Figure 18. The cold fluid mass flow rate is varied at 10 kg/h, 20 kg/h and 30 kg/h. The ratio of heat transfer to pumping power is dominating at a higher volume fraction and lower cold fluid mass flow rate. Therefore, the performance index increases with the increase in volume fraction and decrease in cold fluid mass flow rate. The performance index of the Al_2O_3/Cu nanofluid with OS-shaped nanoparticles increases by 13.41%, 9.43% and 5.79% for cold fluid mass flow rates of 10 kg/h, 20 kg/h and 30 kg/h, respectively, as the volume fraction increases from 0.5% to 2.0%. The performance index of the Al_2O_3/Cu nanofluid with OS-shaped nanoparticles decreases by 6.41%, 7.52% and 9.70% as the cold fluid mass flow rate increases from 10 kg/h to 20 kg/h, and that decreases by 38.71%, 40.13% and 42.82% as the cold fluid mass flow rate increases from 10 kg/h to 30 kg/h for volume fractions of 0.5%, 1.0% and 2.0%, respectively. The thermal entropy generation rate increases with the volume fraction and cold fluid mass flow rate due to an increase in the heat transfer at a higher volume fraction and higher cold fluid mass rate. For the same cold fluid mass flow rate, the ratio of pressure drop to average temperature is less dominant at a higher volume fraction; therefore, the friction entropy generation rate decreases with an increase in volume fraction. The friction entropy generation increases with an increase in the cold fluid mass flow rate for all volume fractions because the ratio of pressure drop to average temperature is highly dominant at higher cold fluid mass flow rates. The ratio of thermal entropy generation rate to total entropy generation rate presents an increasing trend of Bejan numbers with an increase in volume fraction and a decrease in the cold fluid mass flow rate. With the increase in volume fraction from 0.5% to 2.0%, the Bejan number of the Al_2O_3/Cu nanofluid with OS-shaped nanoparticles increases by 0.94%, 0.73% and 0.69% for cold fluid mass flow rates of 10 kg/h, 20 kg/h and 30 kg/h, respectively. The Bejan number of the Al_2O_3/Cu nanofluid with OS-shaped nanoparticles decreases by 1.06%, 1.13% and 1.26% as the cold fluid mass flow rate increases from 10 kg/h to 20 kg/h, and that decreases by 6.21%, 6.30% and 6.45% as the cold fluid mass flow rate increases from 10 kg/h to 30 kg/h for volume fractions of 0.5%, 1.0% and 2.0%, respectively. Tiwari et al. have proved that the lower cold fluid mass flow rate shows the better first law characteristics compared to the higher cold fluid mass flow rate [59,61]. The contributions of thermal and friction entropy generation rates for Al_2O_3/Cu with OS-shaped nanoparticles at various cold fluid mass flow rates are depicted in Figure 19.

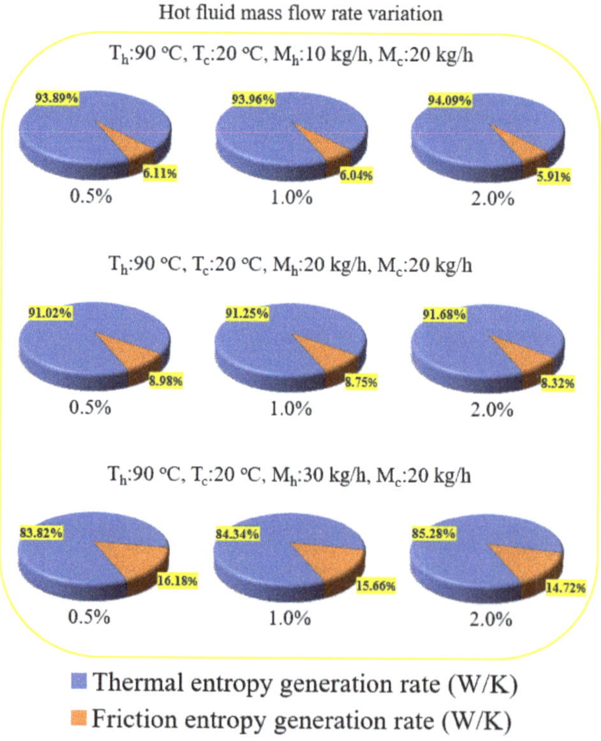

Figure 15. Contributions of thermal and friction entropy generation rates for Al_2O_3/Cu with OS-shaped nanoparticles at various hot fluid mass flow rates.

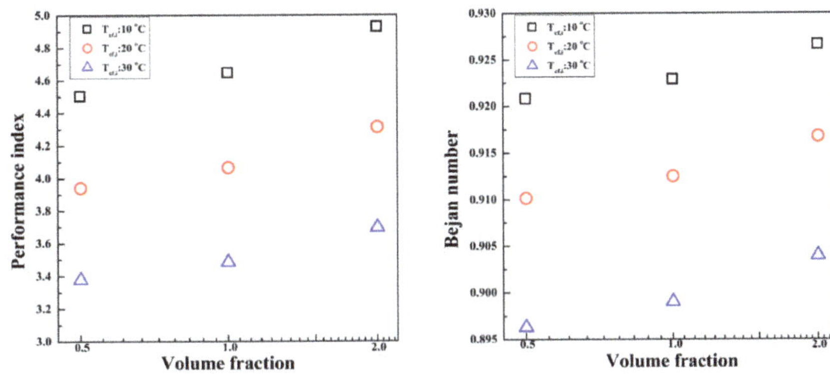

Figure 16. Variation in first and second law characteristics of hybrid nanofluid with OS-shaped nanoparticles with various volume fractions and cold fluid temperatures.

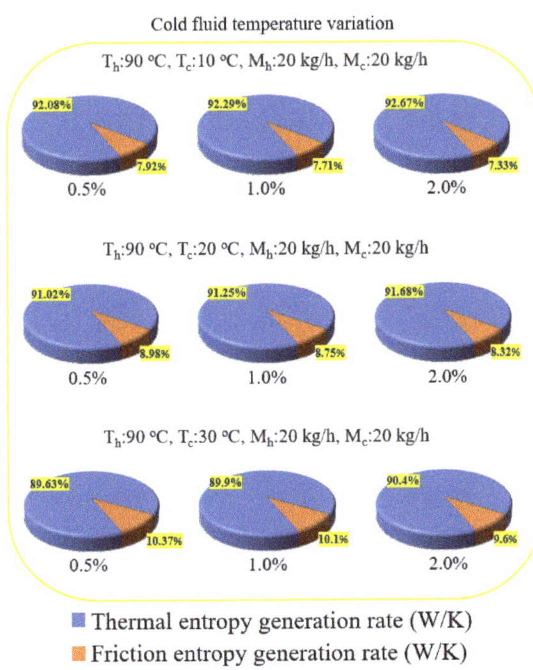

Figure 17. Contributions of thermal and friction entropy generation rates for Al_2O_3/Cu with OS-shaped nanoparticles at various cold fluid temperatures.

The above-mentioned results in Sections 5.3–5.8 are summarized as the comparison of single-particle and hybrid nanofluids with different nanoparticle shapes, based on numerous first and second law characteristics of the microplate heat exchanger. The Al_2O_3/Cu nanofluid with OS-shaped nanoparticles presents the excellent first and second law characteristics among all combinations of single-particle and hybrid nanofluids with nanoparticle shapes. The first and second law characteristics in terms of performance index and Bejan number are investigated under various conditions of volume fraction, temperature and mass flow rate for the best combination of Al_2O_3/Cu nanofluid with OS-shaped nanoparticles. The performance index and Bejan number of the Al_2O_3/Cu

nanofluid with OS-shaped nanoparticles are maximum at a higher hot fluid temperature, lower cold fluid temperature and lower mass flow rates of hot and cold fluids.

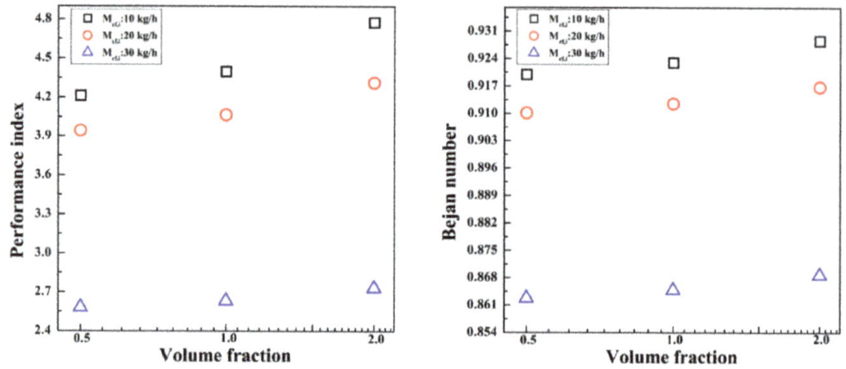

Figure 18. Behavior of first and second law characteristics of hybrid nanofluid with OS-shaped nanoparticles for various volume fractions and cold fluid mass flow rates.

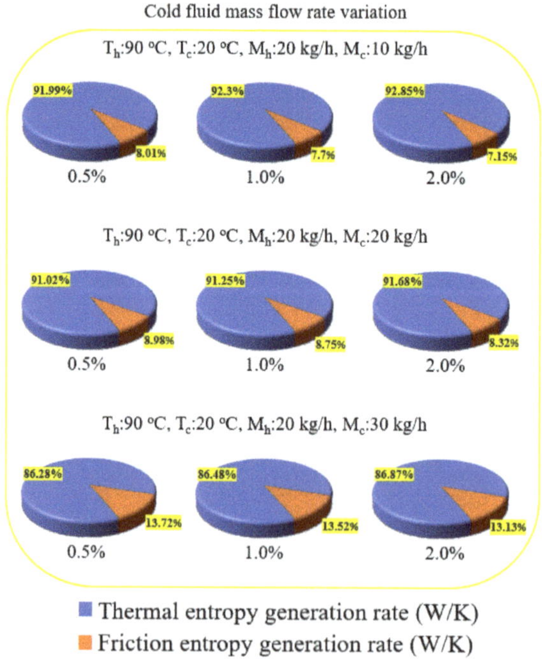

Figure 19. Contributions of thermal and friction entropy generation rates for Al_2O_3/Cu with OS-shaped nanoparticles at various cold fluid mass flow rates.

6. Conclusions

The first and second law analyses have been conducted on the microplate heat exchanger, comprising of single-particle and hybrid nanofluids with different-shaped nanoparticles. Firstly, the first and second law characteristics are compared for different-shaped nanoparticles, and then the effect of various volume fractions, temperatures and mass flow rates are investigated on the first and second law characteristics of the optimum-shaped nanoparticles. The following key findings are highlighted from the present study.

(a) The decreasing order of first law characteristics is evaluated as hybrid nanofluid, single-particle nanofluid and water, respectively, for all nanoparticle shapes. The Al_2O_3/Cu nanofluid with OS-shaped nanoparticles shows maximum values of NTU, effectiveness and performance index, which are higher by 6.38%, 6.10% and 6.58%, respectively, compared to water. The Al_2O_3 nanofluid with PL-shaped nanoparticles shows minimum values of NTU, effectiveness and performance index, which are lower by 3.99%, 3.65% and 8.78%, respectively, compared to water. The Al_2O_3/Cu nanofluid with OS-shaped nanoparticles shows the optimum values of first law characteristics.

(b) The thermal entropy generation rates of OS-shaped nanoparticles are at a maximum, which are 0.14% and 0.70% higher for Al_2O_3 and Al_2O_3/Cu nanofluids, respectively, compared to water. The friction entropy generation rates are maximum for PL-shaped nanoparticles which are 2.73% higher and 0.74% lower for Al_2O_3 and Al_2O_3/Cu nanofluids, respectively, compared to water. The increasing order of Bejan numbers are water, single-particle nanofluid and hybrid nanofluid, respectively, for all nanoparticle shapes. The Bejan number of the Al_2O_3/Cu nanofluid with OS-shaped nanoparticles is the maximum, and that of the Al_2O_3 nanofluid with PL-shaped nanoparticles is the minimum, which are 0.54% higher and 0.86% lower compared to water. The Al_2O_3/Cu nanofluid with OS-shaped nanoparticles shows the optimum values of second law characteristics.

(c) The first law characteristic performance index of the Al_2O_3/Cu nanofluid with OS-shaped nanoparticles has increased with an increase in volume fraction for various temperature and mass flow rate conditions of hot and cold fluids. The performance index increases with the increase in the hot fluid temperature and decrease in the cold fluid temperature for all volume fractions. The performance index has decreased with the increase in the hot and cold fluid mass flow rates.

(d) The second law characteristic Bejan number of the Al_2O_3/Cu nanofluid with OS-shaped nanoparticles has increased with the increase in the volume fraction for all temperature and mass flow rate conditions of hot and cold fluids. The Bejan number has increased with the increase in hot fluid temperature, whereas with the increase in the cold fluid temperature, the Bejan number has decreased for all volume fractions. The Bejan number has decreased with the increase in hot and cold fluid mass flow rates for all volume fractions.

Author Contributions: Conceptualization, K.S.G.; and M.-Y.L.; methodology, K.S.G.; S.-G.H.; and M.-Y.L.; formal analysis, K.S.G.; S.-G.H.; and M.-Y.L.; investigation, K.S.G.; S.-G.H.; T.-K.L.; N.K.; and M.-Y.L.; resources, K.S.G.; and M.-Y.L.; data curation, K.S.G.; and T.-K.L.; validation, K.S.G.; software, K.S.G.; writing—original draft preparation, K.S.G.; and M.-Y.L.; writing—review and editing, K.S.G.; T.-K.L.; N.K.; and M.-Y.L.; visualization, K.S.G.; and M.-Y.L.; supervision, M.-Y.L.; project administration, M.-Y.L.; funding acquisition, M.-Y.L. All authors have read and agreed to the published version of the manuscript.

Funding: This research received no external funding.

Institutional Review Board Statement: Not applicable.

Informed Consent Statement: Not applicable.

Data Availability Statement: The data presented in this study will be available on request to the corresponding author.

Acknowledgments: This work was supported by the Dong-A University research fund.

Conflicts of Interest: The authors declare no conflict of interest.

Nomenclature

A	total heat transfer area (m^2)
C_{cf}	cold fluid heat capacity (W/K)
C_{hf}	hot fluid heat capacity (W/K)
C_{min}	minimum heat capacity (W/K)
$C_{p,bf}$	specific heat of base fluid (J/kg·K)
$C_{p,cf}$	cold fluid specific heat (J/kg·K)
$C_{p,hf}$	hot fluid specific heat (J/kg·K)
$C_{p,hnf}$	specific heat of hybrid nanofluid (J/kg·K)
$C_{p,nf}$	specific heat of nanofluid (J/kg·K)
$C_{p,np}$	specific heat of nanoparticles (J/kg·K)
$C_{p,np1}$	specific heat of nanoparticle1 (J/kg·K)
$C_{p,np2}$	specific heat of nanoparticle2 (J/kg·K)
h	enthalpy (J/kg)
k	turbulent kinetic energy (J/kg)
k_{bf}	thermal conductivity of base fluid (W/m·K)
k_{hnf}	thermal conductivity of hybrid nanofluid (W/m·K)
k_{nf}	thermal conductivity of nanofluid (W/m·K)
k_{np}	thermal conductivity of nanoparticle (W/m·K)
k_{np1}	thermal conductivity of nanoparticle1 (W/m·K)
k_{np2}	thermal conductivity of nanoparticle2 (W/m·K)
$\dot{m}_{cf}$	cold fluid mass flow rate (kg/s)
$\dot{m}_{hf}$	hot fluid mass flow rate (kg/s)
m_{np}	mass of nanoparticles (kg)
p	static pressure (Pa)
Q_{max}	maximum possible heat transfer rate (W)
$\dot{S}_{Fr}$	Volumetric friction entropy generation rate (W/m^3 K)
$\dot{S}_{Th}$	Volumetric thermal entropy generation rate (W/m^3 K)
$\dot{S}_{T}$	Volumetric total entropy generation rate (W/m^3 K)
$T_{cf,i}$	cold fluid inlet temperature (K)
$T_{cf,o}$	cold fluid outlet temperature (K)
$T_{hf,i}$	hot fluid inlet temperature (K)
$T_{hf,o}$	hot fluid output temperature (K)
$\overline{T}$	average temperature (K)
T'	temperature fluctuation (K)
U	average velocity (m/s)
V_{bf}	volume of base fluid (L)
V_{np}	volume of nanoparticles = $\frac{m_{np}}{\rho_{np}}$ (L)
$\overline{v}$	average velocity (m/s)
v'	fluctuating velocity (m/s)
ρ	density (kg/m^3)
ρ_{bf}	density of base fluid (kg/m^3)
ρ_{hnf}	density of hybrid nanofluid (kg/m^3)
ρ_{nf}	density of nanofluid (kg/m^3)
ρ_{np}	density of nanoparticles (kg/m^3)
ρ_{np1}	density of nanoparticle1 (kg/m^3)
ρ_{np2}	density of nanoparticle2 (kg/m^3)
μ	dynamic viscosity (Pa·s)
μ_{bf}	viscosity of base fluid (Pa·s)
μ_{nf}	viscosity of nanofluid (Pa·s)
O_{hnf}	volume fraction of hybrid nanofluid (%)
O_{np1}	volume fraction of nanoparticle1 (%)
O_{np2}	volume fraction of nanoparticle2 (%)
λ	thermal conductivity (W/m·K)
λ_{eff}	effective thermal conductivity (W/m·K)

λ_t turbulent thermal conductivity (W/m·K)
∇ gradient operator
O specific dissipation rate (s^{-1})

References

1. Kim, H.J.; Lee, S.H.; Lee, J.H.; Jang, S.P. Effect of particle shape on suspension stability and thermal conductivities of water-based bohemite alumina nanofluids. *Energy* **2015**, *90*, 1290–1297. [CrossRef]
2. Xie, H.Q.; Wang, J.C.; Xi, T.G.; Liu, Y. Thermal conductivity of suspensions containing nanosized SiC particles. *Int. J. Thermophys.* **2002**, *23*, 571–580. [CrossRef]
3. Murshed, S.M.S.; Leong, K.C.; Yang, C. Enhanced thermal conductivity of TiO_2–Water based nanofluids. *Int. J. Therm. Sci.* **2005**, *44*, 367–373. [CrossRef]
4. Timofeeva, E.V.; Routbort, J.L.; Singh, D. Particle shape effects on thermophysical properties of alumina nanofluids. *J. Appl. Phys.* **2009**, *106*, 014304. [CrossRef]
5. Maheshwary, P.B.; Handa, C.C.; Nemade, K.R. A comprehensive study of effect of concentration, particle size and particle shape on thermal conductivity of titania/water based nanofluid. *Appl. Therm. Eng.* **2017**, *119*, 79–88. [CrossRef]
6. Singh, D.; Timofeeva, E.; Yu, W.; Routbort, J.; France, D.; Smith, D.; Lopez-Cepero, J.M. An investigation of silicon carbide-water nanofluid for heat transfer applications. *J. Appl. Phys.* **2009**, *105*, 064306. [CrossRef]
7. Jeong, J.; Li, C.; Kwon, Y.; Lee, J.; Kim, S.H.; Yun, R. Particle shape effect on the viscosity and thermal conductivity of ZnO nanofluids. *Int. J. Refrig.* **2013**, *36*, 2233–2241. [CrossRef]
8. Zhang, L.; Yu, W.; Zhu, D.; Xie, H.; Huang, G. Enhanced thermal conductivity for nanofluids containing silver nanowires with different shapes. *J. Nanomater.* **2017**, *2017*, 5802016. [CrossRef]
9. Nithiyanantham, U.; González-Fernández, L.; Grosu, Y.; Zaki, A.; Igartua, J.M.; Faik, A. Shape effect of Al_2O_3 nanoparticles on the thermophysical properties and viscosity of molten salt nanofluids for TES application at CSP plants. *Appl. Therm. Eng.* **2020**, *169*, 114942. [CrossRef]
10. Vanaki, S.M.; Mohammed, H.A.; Abdollahi, A.; Wahid, M.A. Effect of nanoparticle shapes on the heat transfer enhancement in a wavy channel with different phase shifts. *J. Mol. Liq.* **2014**, *196*, 32–42. [CrossRef]
11. Mahian, O.; Kianifar, A.; Heris, S.Z.; Wongwises, S. First and second laws analysis of a minichannel-based solar collector using boehmite alumina nanofluids: Effects of nanoparticle shape and tube materials. *Int. J. Heat Mass Transf.* **2014**, *78*, 1166–1176. [CrossRef]
12. Akbar, N.S.; Butt, A.W.; Tripathi, D. Nanoparticle shapes effects on unsteady physiological transport of nanofluids through a finite length non-uniform channel. *Results Phys.* **2017**, *7*, 2477–2484. [CrossRef]
13. Bahiraei, M.; Monavari, A.; Naseri, M.; Moayedi, H. Irreversibility characteristics of a modified microchannel heat sink operated with nanofluid considering different shapes of nanoparticles. *Int. J. Heat Mass Transf.* **2020**, *151*, 119359. [CrossRef]
14. Sheikholeslami, M.; Shamlooei, M.; Moradi, R.J.C.E. Numerical simulation for heat transfer intensification of nanofluid in a porous curved enclosure considering shape effect of Fe_3O_4 nanoparticles. *Chem. Eng. Process.* **2018**, *124*, 71–82. [CrossRef]
15. Sheikholeslami, M.; Shehzad, S.A. Numerical analysis of Fe_3O_4–H_2O nanofluid flow in permeable media under the effect of external magnetic source. *Int. J. Heat Mass Transf.* **2018**, *118*, 182–192. [CrossRef]
16. Nguyen, T.K.; Saidizad, A.; Jafaryar, M.; Sheikholeslami, M.; Gerdroodbary, M.B.; Moradi, R.; Li, Z. Influence of various shapes of CuO nanomaterial on nanofluid forced convection within a sinusoidal channel with obstacles. *Chem. Eng. Res. Des.* **2019**, *146*, 478–485. [CrossRef]
17. Hatami, M.; Jafaryar, M.; Zhou, J.; Jing, D. Investigation of engines radiator heat recovery using different shapes of nanoparticles in $H_2O/(CH_2OH)_2$ based nanofluids. *Int. J. Hydrogen Energy* **2017**, *42*, 10891–10900. [CrossRef]
18. Kim, H.J.; Lee, S.H.; Kim, S.B.; Jang, S.P. The effect of nanoparticle shape on the thermal resistance of a flat-plate heat pipe using acetone-based Al_2O_3 nanofluids. *Int. J. Heat Mass Transf.* **2016**, *92*, 572–577. [CrossRef]
19. Bahiraei, M.; Monavari, A. Thermohydraulic characteristics of a micro plate heat exchanger operated with nanofluid considering different nanoparticle shapes. *Appl. Therm. Eng.* **2020**, *179*, 115621. [CrossRef]
20. Vo, D.D.; Alsarraf, J.; Moradikazerouni, A.; Afrand, M.; Salehipour, H.; Qi, C. Numerical investigation of γ-AlOOH nano-fluid convection performance in a wavy channel considering various shapes of nanoadditives. *Powder Technol.* **2019**, *345*, 649–657. [CrossRef]
21. Khan, U.; Ahmed, N.; Mohyud-Din, S.T. Analysis of magnetohydrodynamic flow and heat transfer of Cu–water nanofluid between parallel plates for different shapes of nanoparticles. *Neural Comput. Appl.* **2018**, *29*, 695–703. [CrossRef]
22. Raza, J.; Mebarek-Oudina, F.; Chamkha, A.J. Magnetohydrodynamic flow of molybdenum disulfide nanofluid in a channel with shape effects. *Multidiscip. Model. Mater. Struct.* **2019**, *15*, 737–757. [CrossRef]
23. Gireesha, B.J.; Sindhu, S. Entropy generation analysis of nanoliquid flow through microchannel considering heat source and different shapes of nanoparticle. *Int. J. Numer. Methods Heat Fluid Flow* **2019**, *30*, 1457–1477. [CrossRef]
24. Elias, M.M.; Miqdad, M.; Mahbubul, I.M.; Saidur, R.; Kamalisarvestani, M.; Sohel, M.R.; Amalina, M.A. Effect of nanoparticle shape on the heat transfer and thermodynamic performance of a shell and tube heat exchanger. *Int. Commun. Heat Mass Transf.* **2013**, *44*, 93–99. [CrossRef]

25. Elias, M.M.; Shahrul, I.M.; Mahbubul, I.M.; Saidur, R.; Rahim, N.A. Effect of different nanoparticle shapes on shell and tube heat exchanger using different baffle angles and operated with nanofluid. *Int. J. Heat Mass Transf.* **2014**, *70*, 289–297. [CrossRef]
26. Shahsavar, A.; Rahimi, Z.; Salehipour, H. Nanoparticle shape effects on thermal-hydraulic performance of boehmite alumina nanofluid in a horizontal double-pipe minichannel heat exchanger. *Heat Mass Transf.* **2019**, *55*, 1741–1751. [CrossRef]
27. Alsarraf, J.; Moradikazerouni, A.; Shahsavar, A.; Afrand, M.; Salehipour, H.; Tran, M.D. Hydrothermal analysis of turbulent boehmite alumina nanofluid flow with different nanoparticle shapes in a minichannel heat exchanger using two-phase mixture model. *Phys. A Stat. Mech. Appl.* **2019**, *520*, 275–288. [CrossRef]
28. Al-Rashed, A.A.; Ranjbarzadeh, R.; Aghakhani, S.; Soltanimehr, M.; Afrand, M.; Nguyen, T.K. Entropy generation of boehmite alumina nanofluid flow through a minichannel heat exchanger considering nanoparticle shape effect. *Phys. A Stat. Mech. Appl.* **2019**, *521*, 724–736. [CrossRef]
29. Monfared, M.; Shahsavar, A.; Bahrebar, M.R. Second law analysis of turbulent convection flow of boehmite alumina nanofluid inside a double-pipe heat exchanger considering various shapes for nanoparticle. *J. Therm. Anal. Calorim.* **2019**, *135*, 1521–1532. [CrossRef]
30. Sadripour, S.; Chamkha, A.J. The effect of nanoparticle morphology on heat transfer and entropy generation of supported nanofluids in a heat sink solar collector. *Therm. Sci. Eng. Prog.* **2019**, *9*, 266–280. [CrossRef]
31. Liu, J.B.; Bayati, M.; Abbas, M.; Rahimi, A.; Naderi, M. Mesoscopic approach for simulating nanofluid flow and heat transfer in a finned multi-pipe heat exchanger. *Int. J. Numer. Methods Heat Fluid Flow* **2019**, *29*, 2822–2839. [CrossRef]
32. Ghadikolaei, S.S.; Yassari, M.; Sadeghi, H.; Hosseinzadeh, K.; Ganji, D.D. Investigation on thermophysical properties of TiO_2–Cu/H_2O hybrid nanofluid transport dependent on shape factor in MHD stagnation point flow. *Powder Technol.* **2017**, *322*, 428–438. [CrossRef]
33. Ghadikolaei, S.S.; Hosseinzadeh, K.; Ganji, D.D. Investigation on three dimensional squeezing flow of mixture base fluid (ethylene glycol-water) suspended by hybrid nanoparticle (Fe_3O_4-Ag) dependent on shape factor. *J. Mol. Liq.* **2018**, *262*, 376–388. [CrossRef]
34. Dinarvand, S.; Rostami, M.N.; Pop, I. A novel hybridity model for TiO_2-CuO/water hybrid nanofluid flow over a static/moving wedge or corner. *Sci. Rep.* **2019**, *9*, 1–11. [CrossRef]
35. Bhattad, A.; Sarkar, J. Effects of nanoparticle shape and size on the thermohydraulic performance of plate evaporator using hybrid nanofluids. *J. Therm. Anal. Calorim.* **2019**, *143*, 767–779. [CrossRef]
36. Benkhedda, M.; Boufendi, T.; Tayebi, T.; Chamkha, A.J. Convective heat transfer performance of hybrid nanofluid in a horizontal pipe considering nanoparticles shapes effect. *J. Therm. Anal. Calorim.* **2020**, *140*, 411–425. [CrossRef]
37. Ghobadi, A.H.; Hassankolaei, M.G. A numerical approach for MHD Al_2O_3–TiO_2/H_2O hybrid nanofluids over a stretching cylinder under the impact of shape factor. *Heat Transf. Asian Res.* **2019**, *48*, 4262–4282. [CrossRef]
38. Aziz, A.; Jamshed, W.; Ali, Y.; Shams, M. Heat transfer and entropy analysis of Maxwell hybrid nanofluid including effects of inclined magnetic field, Joule heating and thermal radiation. *Discret. Contin. Dyn. Syst. S* **2020**, *13*, 2667. [CrossRef]
39. Ghadikolaei, S.S.; Hosseinzadeh, K.; Ganji, D.D. Investigation on ethylene glycol-water mixture fluid suspend by hybrid nanoparticles (TiO_2-CuO) over rotating cone with considering nanoparticles shape factor. *J. Mol. Liq.* **2018**, *272*, 226–236. [CrossRef]
40. Ghadikolaei, S.S.; Gholinia, M.; Hoseini, M.E.; Ganji, D.D. Natural convection MHD flow due to MoS_2–Ag nanoparticles suspended in $C_2H_6O_2H_2O$ hybrid base fluid with thermal radiation. *J. Taiwan Inst. Chem. Eng.* **2019**, *97*, 12–23. [CrossRef]
41. Ghadikolaei, S.S.; Gholinia, M. 3D mixed convection MHD flow of GO-MoS_2 hybrid nanoparticles in H_2O–$(CH_2OH)_2$ hybrid base fluid under the effect of H2 bond. *Int. Commun. Heat Mass Transf.* **2020**, *110*, 104371. [CrossRef]
42. Maraj, E.N.; Iqbal, Z.; Azhar, E.; Mehmood, Z. A comprehensive shape factor analysis using transportation of MoS_2-SiO_2/H_2O inside an isothermal semi vertical inverted cone with porous boundary. *Results Phys.* **2018**, *8*, 633–641. [CrossRef]
43. Sahu, M.; Sarkar, J. Steady-state energetic and exergetic performances of single-phase natural circulation loop with hybrid nanofluids. *J. Heat Transf.* **2019**, *141*, 082401. [CrossRef]
44. Garud, K.S.; Seo, J.H.; Patil, M.S.; Bang, Y.M.; Pyo, Y.D.; Cho, C.P.; Lee, M.Y. Thermal–electrical–structural performances of hot heat exchanger with different internal fins of thermoelectric generator for low power generation application. *J. Therm. Anal. Calorim.* **2021**, *143*, 387–419. [CrossRef]
45. Garud, K.S.; Seo, J.H.; Cho, C.P.; Lee, M.Y. Artificial neural network and adaptive neuro-fuzzy interface system modelling to predict thermal performances of thermoelectric generator for waste heat recovery. *Symmetry* **2020**, *12*, 259. [CrossRef]
46. Dadsetani, R.; Sheikhzadeh, G.A.; Safaei, M.R.; Leon, A.S.; Goodarzi, M. Cooling enhancement and stress reduction opti-mization of disk-shaped electronic components using nanofluids. *Symmetry* **2020**, *12*, 931. [CrossRef]
47. Seo, J.H.; Garud, K.S.; Lee, M.Y. Grey relational based Taguchi analysis on thermal and electrical performances of thermoelectric generator system with inclined fins hot heat exchanger. *Appl. Therm. Eng.* **2021**, *184*, 116279. [CrossRef]
48. Bahiraei, M.; Monavari, A.; Moayedi, H. Second law assessment of nanofluid flow in a channel fitted with conical ribs for utilization in solar thermal applications: Effect of nanoparticle shape. *Int. J. Heat Mass Transf.* **2020**, *151*, 119387. [CrossRef]
49. Lee, M.Y.; Seo, J.H.; Lee, H.S.; Garud, K.S. Power generation, efficiency and thermal stress of thermoelectric module with leg geometry, material, segmentation and two-stage arrangement. *Symmetry* **2020**, *12*, 786. [CrossRef]
50. Ooi, E.H.; Popov, V. Numerical study of influence of nanoparticle shape on the natural convection in Cu-water nanofluid. *Int. J. Therm. Sci.* **2013**, *65*, 178–188. [CrossRef]

51. Garud, K.S.; Lee, M.Y. Numerical Investigations on Heat Transfer Characteristics of Single Particle and Hybrid Nanofluids in Uniformly Heated Tube. *Symmetry* **2021**, *13*, 876. [CrossRef]
52. Sanches, M.; Marseglia, G.; Ribeiro, A.P.; Moreira, A.L.; Moita, A.S. Nanofluids Characterization for Spray Cooling Applications. *Symmetry* **2021**, *13*, 788. [CrossRef]
53. Arani, A.A.A.; Sadripour, S.; Kermani, S. Nanoparticle shape effects on thermal-hydraulic performance of boehmite alumina nanofluids in a sinusoidal–wavy mini-channel with phase shift and variable wavelength. *Int. J. Mech. Sci.* **2017**, *128*, 550–563. [CrossRef]
54. Ghachem, K.; Aich, W.; Kolsi, L. Computational analysis of hybrid nanofluid enhanced heat transfer in cross flow micro heat exchanger with rectangular wavy channels. *Case Stud. Therm. Eng.* **2021**, *24*, 100822. [CrossRef]
55. Bhattad, A.; Sarkar, J.; Ghosh, P. Energy-economic analysis of plate evaporator using brine-based hybrid nanofluids as secondary refrigerant. *Int. J. Air-Cond. Refrig.* **2018**, *26*, 1850003. [CrossRef]
56. Arun, K.R.; Kunal, G.; Srinivas, M.; Kumar, C.S.; Mohanraj, M.; Jayaraj, S. Drying of untreated Musa nendra and Momordica charantia in a forced convection solar cabinet dryer with thermal storage. *Energy* **2020**, *192*, 116697. [CrossRef]
57. Kwon, J.T.; Nahm, T.H.; Kim, T.W.; Kwon, Y.C. An Experimental study on pressure drop and heat transfer coefficient of laminar Ag nanofluid flow in mini-tubes. *J. Korea Acad. Ind. Coop. Soc.* **2009**, *10*, 3525–3532.
58. Zheng, D.; Wang, J.; Chen, Z.; Baleta, J.; Sundén, B. Performance analysis of a plate heat exchanger using various nanofluids. *Int. J. Heat Mass Transf.* **2020**, *158*, 119993. [CrossRef]
59. Tiwari, A.K.; Ghosh, P.; Sarkar, J. Heat transfer and pressure drop characteristics of CeO_2/water nanofluid in plate heat exchanger. *Appl. Therm. Eng.* **2013**, *57*, 24–32. [CrossRef]
60. Kim, N.H. An experimental investigation on the air-side performance of fin-and-tube heat exchangers having radial slit fins. *Int. J. Air-Cond. Refrig.* **2015**, *23*, 1550021. [CrossRef]
61. Tiwari, A.K.; Ghosh, P.; Sarkar, J. Performance comparison of the plate heat exchanger using different nanofluids. *Exp. Therm. Fluid Sci.* **2013**, *49*, 141–151. [CrossRef]
62. Singh, S.K.; Sarkar, J. Improving hydrothermal performance of hybrid nanofluid in double tube heat exchanger using tapered wire coil turbulator. *Adv. Powder Technol.* **2020**, *31*, 2092–2100. [CrossRef]
63. Raj, A.K.; Kunal, G.; Srinivas, M.; Jayaraj, S. Performance analysis of a double-pass solar air heater system with asymmetric channel flow passages. *J. Therm. Anal. Calorim.* **2019**, *136*, 21–38. [CrossRef]
64. Alm, B.; Imke, U.; Knitter, R.; Schygulla, U.; Zimmermann, S. Testing and simulation of ceramic micro heat exchangers. *Chem. Eng. J.* **2008**, *135*, S179–S184. [CrossRef]
65. Ghosh, M.M.; Ghosh, S.; Pabi, S.K. Effects of particle shape and fluid temperature on heat-transfer characteristics of nanofluids. *J. Mater. Eng. Perform.* **2013**, *22*, 1525–1529. [CrossRef]
66. Garud, K.S.; Seo, J.H.; Lee, M.Y. Effect of Guide Fin Structures and Boundary Parameters on Thermal Performances of Heat Exchanger for Waste Heat Recovery Thermoelectric Generator. *J. Korea Acad. Ind. Coop. Soc.* **2021**, *22*, 30–35.

symmetry

MDPI

Article

Numerical Investigations on Heat Transfer Characteristics of Single Particle and Hybrid Nanofluids in Uniformly Heated Tube

Kunal Sandip Garud and **Moo-Yeon Lee** *

Department of Mechanical Engineering, Dong-A University, 37 Nakdong-Daero 550, Saha-gu, Busan 49315, Korea; 1876936@donga.ac.kr
* Correspondence: mylee@dau.ac.kr; Tel.: +82-51-200-7642

Abstract: In the present study, the heat transfer characteristics, namely, heat transfer coefficient, Nusselt number, pressure drop, friction factor and performance evaluation criteria are evaluated for water, Al_2O_3 and Al_2O_3/Cu nanofluids. The effects of Reynolds number, volume fraction and composition of nanoparticles in hybrid nanofluid are analyzed for all heat transfer characteristics. The single particle and hybrid nanofluids are flowing through a plain straight tube which is symmetrically heated under uniform heat flux condition. The numerical model is validated for Nusselt number within 7.66% error and friction factor within 8.83% error with corresponding experimental results from the previous literature study. The thermophysical properties of hybrid nanofluid are superior to the single particle nanofluid and water. The heat transfer coefficient, Nusselt number and pressure drop show increasing trend with increase in the Reynolds number and volume fraction. The friction factor shows the parabolic trend, and the performance evaluation criteria shows small variations with change in Reynolds number. However, both friction factor and performance evaluation criteria have increased with increase in the volume fraction. The 2.0% Al_2O_3/Cu with equal composition of both nanoparticles (50/50%) have presented superior heat transfer characteristics among all working fluids. Further, the heat transfer characteristics of 2.0% Al_2O_3/Cu hybrid nanofluid are enhanced by changing the nanoparticle compositions. The performance evaluation criteria for 2.0% Al_2O_3, 2.0% Al_2O_3/Cu (50/50%), 2.0% Al_2O_3/Cu (75/25%) and 2.0% Al_2O_3/Cu (25/75%) are evaluated as 1.08, 1.11, 1.10 and 1.12, respectively.

Keywords: heat transfer characteristics; hybrid nanofluid; performance evaluation criteria; single particle nanofluid; uniformly heated tube

Citation: Garud, K.S.; Lee, M.-Y. Numerical Investigations on Heat Transfer Characteristics of Single Particle and Hybrid Nanofluids in Uniformly Heated Tube. *Symmetry* **2021**, *13*, 876. https://doi.org/10.3390/sym13050876

Academic Editor: Mikhail Sheremet

Received: 21 April 2021
Accepted: 12 May 2021
Published: 14 May 2021

1. Introduction

Currently, energy saving in thermal systems has grabbed considerable attention to reduce the heat losses and improve the heat transfer performance [1]. Due to restrictions on the further improvement in the thermophysical properties of conventional working fluids, the research trend is drastically shifting towards nanofluids applications in heat transfer systems [2]. Numerous studies were reported on heat transfer applications of nanofluids in the thermal systems such as tubes.

Firoozi et al. have conducted a numerical study to investigate the heat transfer and flow characteristics of Al_2O_3/water nanofluid flowing through tubes incorporated with various dimple configurations [3]. Ledari et al. have investigated the heat transfer and friction factor characteristics of oil based CuO and Fe_3O_4 nanofluids flowing through the U-tube under the influence of various mass concentrations, flow rates and inclination angles [4]. Chaurasia and Sarviya have conducted experimental and numerical studies to analyze the thermal hydraulic and entropy generation performances of nanofluid flowing through helical screw insert tube with single and double strips [5]. Ying et al. have concluded that Al_2O_3/water nanofluid with 0.063% mass concentration shows improvement in heat transfer coefficient and Nusselt number by 7.29% and 6.90%, respectively under the cosine heat flux condition and that by 7.25% and 6.85%, respectively under the Gaussian-cosine heat flux condition [6]. Saedodin et al. have compared the heat transfer performances of SiO_2, Al_2O_3, CuO and TiO_2 nanofluids flowing through the straight tube with twisted turbulators [7]. Kristiawan et al. have investigated the thermal and friction factor characteristics of helical microfin tube comprises of TiO_2/water nanofluid with various volume concentrations [8]. Tiwari et al. have compared thermal characteristics of triple tube heat exchanger with porous plate, twisted tape and rib type inserts using WO_3/water nanofluid [9]. Shahsavar et al. have proposed PVT system with rifled serpentine tube comprises of three ribs and six ribs as the replacement of plain serpentine tube [10]. Mukherjee et al. have presented 94% enhancement in forced convective flow boiling heat transfer of Al_2O_3/water nanofluid flowing through a horizontal tube compared to water [11]. Heyhat et al. have analyzed the heat transfer and pressure drop characteristics of SiO_2/water nanofluid flowing through a conically coiled tube considering the effects of volume concentrations, cone angle and cone pitch [12]. Ho et al. have experimentally and numerically explored the cooling performance characteristics of Al_2O_3/water nanofluid flowing through a copper tube under constant heat flux condition [13]. Sun et al. have compared the convective heat transfer and flow characteristics of smooth and inner grooved copper tubes considering flow of Fe_2O_3/water nanofluid [14]. Behzadnia et al. have proposed an optimum geometry of rectangular corrugated tube to maximize the heat transfer efficiency using Al_2O_3/water nanofluid [15]. Kaood et al. have investigated the thermal, hydraulic and energy performances of Al_2O_3/water and SiO_2/water nanofluids flowing through a corrugated tube with curved ribs, rectangular, triangular and trapezoidal geometries [16]. Safaei et al. have numerically investigated the influences of volume fraction, particle size and velocity on pressure and friction factor of Cu/water nanofluid flow inside a pipe bend [17]. Qureshi et al. have investigated the heat transfer and entropy characteristics of Williamson nanofluid based on magnetohydrodynamics [18].

Yildirim et al. have proved that SiO_2 and Cu based mono and hybrid nanofluids show 15% better thermal performance compared with water for U-tube incorporated in an evacuated tube solar collector [19]. Saleh and Sundar have presented improvement in the thermal performance factor, Nusselt number, frictional entropy generation, friction factor and exergy efficiency by 14.19%, 19.67%, 210.6%, 15.11% and 17.54%, respectively and 22.93% decrease in the thermal entropy generation for Fe_3O_4/nanodiamond nanofluid compared to base fluid for circular tube [20]. Ramadhan et al. have shown the behavior of Nusselt number and friction factor of TiO_2/SiO_2 nanofluid flowing through a plain tube for various Reynolds number and volume concentrations of nanofluid [21]. Han et al. have investigated the heat transfer performances of n-decane-ZnO nanofluid which is pressurized below supercritical pressure and flowing through a horizontal tube. The heat transfer and Nusselt number of n-decane-ZnO nanofluid have increased 20% under a supercritical pressure [22]. Akbar et al. have concluded that the hybrid nanofluid of alumina and titanium with low volume concentrations presents 30% enhanced heat transfer compared to water for a horizontal heated tube [23]. Azmi et al. have studied the effect of various composition ratios of TiO_2/SiO_2 hybrid nanofluid on thermal-hydraulic performance of flow in a tube with wire coils [24]. Moldoveanu et al. have experimentally investigated the thermal conductivity of Al_2O_3/SiO_2 hybrid nanofluid under the influence of temperature and volume fraction [25]. Yang et al. have concluded that Al_2O_3/TiO_2 hybrid nanofluid exhibits superior Nusselt number, friction factor and performance evaluation criteria compared to Al_2O_3/ZrO_2 hybrid nanofluid for a parallel channel flow under the same pumping power [26]. Adriana has presented the enhancement of 12% in thermal conductivity of hybrid nanofluid and develop the Nusselt number correlation in terms of Reynolds number, Prandtl number and volume fraction [27].

There are numerous studies in the open literature which show the suitability of single particle nanofluids in heat transfer applications. The research work in heat transfer enhancement using hybrid nanofluids is comparatively less. Specifically, in the field of circular tubes used for solar receiver/collectors and multitube heat exchanger, the investigations on the trade-off comparison between heat transfer characteristics considering different compositions of hybrid nanoparticles is not explored detailly and compared with single particle nanofluid and conventional working fluid results. There are very few studies which present the comprehensive comparison of conventional working fluid, single particle, and hybrid nanofluids based on heat transfer characteristics considering the effect of Reynolds number, volume fraction and composition of hybrid nanoparticles. The objective of the present study is to compare heat transfer coefficient, Nusselt number, pressure drop, friction factor and performance evaluation criteria as heat transfer characteristics for water, Al_2O_3 and Al_2O_3/Cu nanofluids in a plain straight tube under symmetrical and uniform heat flux condition. The comparison is carried out for various Reynolds number, volume fractions and compositions of Al_2O_3/Cu nanoparticles. The novelty of the present work is to summarize the effects on heat transfer characteristics of hybrid nanofluid by mixing two nanoparticles, one with higher stability and lower thermal conductivity (Al_2O_3) and other one with lower stability and higher thermal conductivity (Cu) for different compositions of both nanoparticles. In addition, these results could give a guideline on how much propositions of both nanoparticles could be mixed in hybrid nanofluid to achieve the effective balance between heat transfer and pressure drop.

2. Numerical Method

The numerical model based on computational fluid dynamics (CFD) is employed to analyze the heat transfer characteristics including thermal and flow attributes of single particle and hybrid nanofluids flow in uniformly heater tube. The tube with water, single particle and hybrid nanofluids as working fluids is considered under the constant heat flux condition. The constant heat flux applied on a tube is symmetrically distributed on the surface of tube to analyze the heat transfer characteristics of single particle and hybrid nanofluids. The 3D computational domain of tube is shown in Figure 1. The 3D computational domain is considered to account the effect of uniform heat flux boundary condition [28]. The tube has length of 1500 mm, inner diameter of 16 mm and outer diameter of 19 mm. The tube is made up of copper. The physical properties of copper considered for the numerical analysis as, density of 8940 kg/m^3, specific heat of 376.8 $J/kg{\cdot}K$ and thermal conductivity of 401 $W/m{\cdot}K$ [29]. The constant heat flux of 7957 W/m^2 is applied uniformly on the tube outer surface [30]. To analyze the heat transfer characteristics of water and nanofluids, the tube under constant heat flux condition is simulated in ANSYS commercial software. The meshing with tetrahedron mesh elements and five different sizes is generated for the considered computational domain to show the mesh independency of the simulated results [31]. The Nusselt number and friction factor are simulated for five different mesh element numbers ranging from 100,000 to 700,000. The variation of Nusselt number (Nu) and friction factor (f) for various mesh element numbers is presented in Figure 2. The variation of the simulated results of Nusselt number and friction factor are within $\pm 1\%$ beyond the mesh element number of 425,691. After this mesh element number, the simulated results are independent of number of mesh elements. Hence, the mesh element number of 425,691 corresponding to the sizing of 2 mm is selected as the final meshing configuration for the numerical analysis on the considered computational domain. The inflation layers are employed on the fluid domain to account the effect of boundary layer.

The continuity, momentum and energy equations are solved for the considered tube with flow of various working fluids to simulate the thermal and flow characteristics [32,33].

Continuity equation

$$\nabla \cdot (\rho U) = 0 \tag{1}$$

Momentum equation

$$\nabla \cdot (\rho U U) = -\nabla P + \nabla \tau \tag{2}$$

Stress tensor τ is expressed in terms of strain rate as follows

$$\tau = \mu(\nabla U + (\nabla U)^T - \frac{2}{3}\delta\nabla\cdot U) \tag{3}$$

Energy equation

$$\nabla\cdot(\rho U h) = \nabla\cdot(k\nabla T) + \tau : \nabla U \tag{4}$$

Here, ρ (kg/m^3) is density, U (m/s) is velocity, P (Pa) is static pressure, μ (Pa·s) is viscosity, h (J) is enthalpy, k (W/m·K) is thermal conductivity and ∇T is temperature gradient.

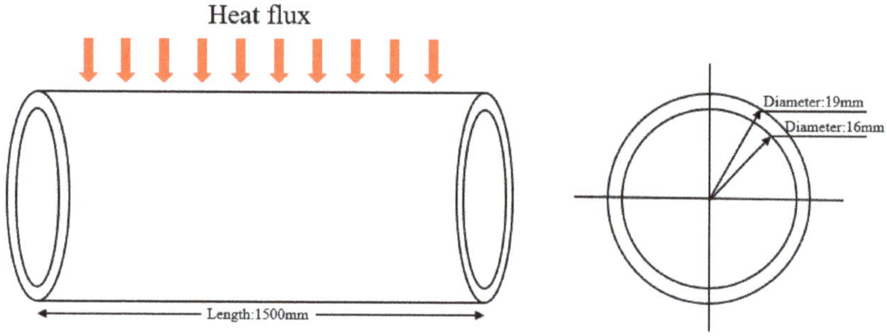

Figure 1. 3D computational domain of tube.

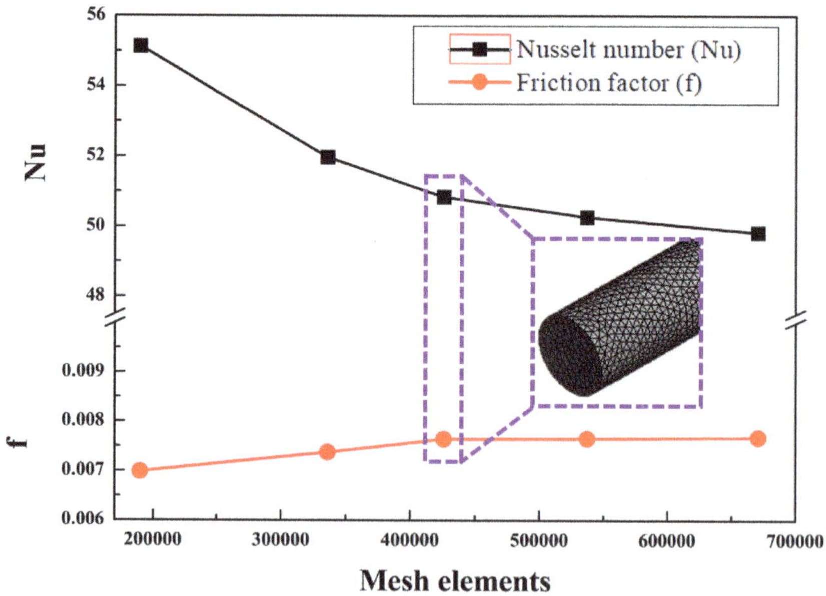

Figure 2. Variation of Nusselt number and friction factor for various mesh element numbers.

The k-ε turbulence model could be expressed by Equations (5) and (6), respectively [34].

$$\frac{\partial(\rho U_i k)}{\partial x_i} = \frac{\partial}{\partial x_i}\left[\left(v + \frac{v_t}{\sigma_k}\right)\frac{\partial k}{\partial x_i}\right] + P_k - \rho\varepsilon \tag{5}$$

$$\frac{\partial(\rho U_i \varepsilon)}{\partial x_i} = \frac{\partial}{\partial x_i}\left[\left(v + \frac{v_t}{\sigma_\varepsilon}\right)\frac{\partial \varepsilon}{\partial x_i}\right] + \rho C_1 S\varepsilon - \rho C_2\frac{\varepsilon^2}{k + \sqrt{v\varepsilon}} \tag{6}$$

Here, k presents the turbulence kinetic energy, v_t presents the turbulent eddy viscosity, ε presents the dissipation rate of turbulence energy, S presents the modulus of the mean rate of strain tensor, P_k stands for the shear production of turbulent kinetic energy σ_k and σ_ε are 1 and 1.2, respectively.

The boundary conditions considered in the numerical analysis are constant heat flux of 7957 W/m^2 at tube wall with symmetrical distribution, velocity inlet with various Reynolds number of working fluid and pressure outlet. The working fluids considered in the numerical analysis are conventional fluid water, single particle nanofluid namely, Al_2O_3 and hybrid nanofluid namely, Al_2O_3/Cu. The Reynolds number is ranging from 2000 to 12000 and inlet fluid temperature is considered as 298.15 K in the numerical analysis. The thermophysical properties of single particle and hybrid nanofluids are evaluated using the thermal properties of base fluid and respective nanoparticles. Three volume fractions of 0.5%, 1.0% and 2.0% are considered in the present numerical analysis. In case of hybrid nanofluids, the composition of both nanoparticles is mixed in the proportions of 75/25%, 50/50% and 25/75%. While solving equations using the considered boundary conditions, it is assumed that the flow is uniform, steady and incompressible [35]. The laminar and standard k-ε turbulence models are considered for the numerical analysis. The convergence criteria are selected as 10^{-8} for all equations.

3. Thermophysical Properties of Nanofluids

The properties of base fluid, water and considered nanoparticles, Al_2O_3 and Cu are depicted in Table 1 [36,37]. The considered Al_2O_3 and Cu nanoparticles are of spherical shape with 100 nm size. The thermophysical properties of single particle nanofluid and hybrid nanofluid are calculated using equations presented in Subsections (a) and (b) based on the depicted properties of basefluid and nanoparticles in Table 1. The widely used models in the open literature are employed to evaluate the thermophysical properties of single particle and hybrid nanofluids. All thermophysical properties are calculated by assuming the constant temperature hence, the temperature effect is neglected [5].

Table 1. Properties of base fluid and nanoparticles.

Property	Water	Al_2O_3	Cu
Density (kg/m^3)	998.2	3970	8300
Specific heat (J/kg·K)	4182	765	420
Thermal conductivity (W/m·K)	0.6	40	401
Viscosity (Pa·s)	0.001	–	–

(a) Single particle nanofluid properties [38,39];

Volume fraction of nanoparticles in nanofluid

$$\varnothing = \frac{V_{np}}{V_{bf} + V_{np}} \tag{7}$$

Density of nanofluid

$$\rho_{nf} = (1 - \varnothing)\rho_{bf} + \varnothing\rho_{np} \tag{8}$$

Specific heat of nanofluid

$$C_{p,nf} = \frac{(1 - \varnothing)\rho_{bf}C_{p,bf} + \varnothing\rho_{np}C_{p,np}}{\rho_{nf}} \tag{9}$$

Thermal conductivity of nanofluid

$$\frac{k_{nf}}{k_{bf}} = \frac{\left(k_{np} + 2k_{bf}\right) - 2\varnothing\left(k_{bf} - k_{np}\right)}{\left(k_{np} + 2k_{bf}\right) + \varnothing\left(k_{bf} - k_{np}\right)} \tag{10}$$

Viscosity of nanofluid

$$\mu_{nf} = \mu_{bf}\frac{1}{(1-\varnothing)^{2.5}} \quad (11)$$

Here, $\varnothing$ is volume fraction of nanoparticles in nanofluid, V_{bf} (L) is volume of basefluid, V_{np} (L) is volume of nanoparticles = $\frac{m_{np}}{\rho_{np}}$, m_{np} (kg) is mass of nanoparticles, ρ_{np} (kg/m^3) is density of nanoparticles, ρ_{nf} (kg/m^3) is density of nanofluid, ρ_{bf} (kg/m^3) is density of base fluid, $C_{p,nf}$ (J/kg·K) is specific heat of nanofluid, $C_{p,bf}$ (J/kg·K) is specific heat of basefluid, $C_{p,np}$ (J/kg·K) is specific heat of nanoparticles, k_{nf} (W/m·K) is thermal conductivity of nanofluid, k_{bf} (W/m·K) is thermal conductivity of basefluid, k_{np} (W/m·K) is thermal conductivity of nanoparticle, μ_{nf} (Pa·s) is viscosity of nanofluid and μ_{bf} (Pa·s) is viscosity of basefluid.

(b) Hybrid nanofluid properties [40];

Volume fraction of nanoparticles in nanofluid

$$\varnothing_{hnf} = \varnothing_{np1} + \varnothing_{np2} \quad (12)$$

Density of hybrid nanofluid

$$\rho_{hnf} = \varnothing_{np1}\rho_{np1} + \varnothing_{np2}\rho_{np2} + \left(1-\varnothing_{hnf}\right)\rho_{bf} \quad (13)$$

Specific heat of hybrid nanofluid

$$C_{p,hnf} = \frac{\varnothing_{np1}\rho_{np1}C_{p,np1} + \varnothing_{np2}\rho_{np2}C_{p,np2} + \left(1-\varnothing_{hnf}\right)\rho_{bf}C_{p,bf}}{\rho_{hnf}} \quad (14)$$

Thermal conductivity of hybrid nanofluid

$$k_{hnf} = \frac{\frac{\varnothing_{np1}k_{np1}+\varnothing_{np2}k_{np2}}{\varnothing_{hnf}} + 2k_{bf} + 2(\varnothing_{np1}k_{np1} + \varnothing_{np2}k_{np2}) - 2\varnothing_{hnf}k_{bf}}{\frac{\varnothing_{np1}k_{np1}+\varnothing_{np2}k_{np2}}{\varnothing_{hnf}} + 2k_{bf} - 2(\varnothing_{np1}k_{np1} + \varnothing_{np2}k_{np2}) + \varnothing_{hnf}k_{bf}} \quad (15)$$

Viscosity of hybrid nanofluid

$$\mu_{hnf} = \mu_{bf}\frac{1}{(1-\varnothing_{np1}-\varnothing_{np2})^{2.5}} \quad (16)$$

Here, $\varnothing_{hnf}$ is volume fraction of hybrid nanofluid, $\varnothing_{np1}$ is volume fraction of nanoparticle1, $\varnothing_{np2}$ is volume fraction of nanoparticle2, ρ_{hnf} (kg/m^3) is density of hybrid nanofluid, ρ_{np1} (kg/m^3) is density of nanoparticle1, ρ_{np2} (kg/m^3) is density of nanoparticle2, $C_{p,hnf}$ (J/kg·K) is specific heat of hybrid nanofluid, $C_{p,np1}$ (J/kg·K) is specific heat of nanoparticle1, $C_{p,np2}$ (J/kg·K) is specific heat of nanoparticle2, k_{hnf} (W/m·K) is thermal conductivity of hybrid nanofluid, k_{np1} (W/m·K) is thermal conductivity of nanoparticle1, k_{np2} (W/m·K) is thermal conductivity of nanoparticle2 and μ_{hnf} (Pa·s) is viscosity of hybrid nanofluid.

The volume fractions of 0.5%, 1.0% and 1.5% as well as density, specific heat, thermal conductivity and viscosity of water, Al_2O_3 and Cu nanoparticles as shown in Table 1 are imported in Equations (7)–(16) to calculate the density, specific heat, thermal conductivity and viscosity of single particle Al_2O_3 and hybrid Al_2O_3/Cu nanofluids. The calculated properties of Al_2O_3 and Al_2O_3/Cu nanofluids are elaborated in Section 5.2.

4. Data Reduction

The average heat transfer coefficient for the considered heated tube with various working fluids is calculated using Equation (17) [12].

$$h = \frac{Q}{A\left(T_s - T_{f,ave}\right)} \tag{17}$$

The heat absorbed by working fluid is calculated using Equation (18) [41,42].

$$Q = m_f c_p \left(T_{f,o} - T_{f,i}\right) \tag{18}$$

The average Nusselt number considering calculated average heat transfer coefficient, hydraulic diameter of tube and thermal conductivity of working fluid is evaluated using Equation (19) [23].

$$Nu = \frac{hD}{k} \tag{19}$$

Here, T_s (K) presents average temperature of wall surface, $T_{f,o}$ (K) presents outlet fluid temperature, $T_{f,i}$ (K) presents inlet fluid temperature, $T_{f,ave}$ (K) presents average fluid temperature, D (mm) presents hydraulic diameter, k (W/m·K) presents thermal conductivity of working fluid, m_f (kg/s) presents mass flow rate of fluid and c_p (J/kg·K) presents specific heat of fluid.

The pressure drop of working fluid across the tube is calculated using Equation (20) [43].

$$\Delta P = P_{f,i} - P_{f,o} \tag{20}$$

The friction factor for a tube with various working fluids is evaluated using pressure drop, length of tube, hydraulic diameter, density and average velocity as presented by Equation (21) [21,23].

$$f = \frac{\Delta P}{\frac{L}{D} \times \frac{\rho U^2}{2}} \tag{21}$$

Here, ΔP (Pa) is pressure drop, $P_{f,i}$ (Pa) is inlet fluid pressure, $P_{f,o}$ (Pa) is outlet fluid pressure, f is friction factor, L (mm) is length of tube, ρ (kg/m^3) is density of fluid and U (m/s) is the average velocity.

Considering the Nusselt numbers and friction factors of base fluid and nanofluids, the performance evaluation criteria is defined as presented by Equation (22) [44].

$$PEC = \frac{\left(\frac{Nu_{nf}}{Nu_{bf}}\right)}{\left(\frac{f_{nf}}{f_{bf}}\right)^{\frac{1}{3}}} \tag{22}$$

where Nu_{nf} is Nusselt number of nanofluid, Nu_{bf} is Nusselt number of basefluid, f_{nf} is friction factor of nanofluid and f_{nf} is friction factor of nanofluid.

5. Results and Discussion

5.1. Validation

The numerical model based on CFD is validated with the experimental results presented by Hamid et al. for the same geometry (dimensions and structure) and boundary conditions [30]. The experimental and numerical results of Nusselt number and friction factor are compared as the heat transfer characteristics for various Reynolds number. The comparison of Nusselt number and friction factor for previous experimental study and present numerical model for various Reynolds number (Re) is depicted in Figure 3. The comparison is presented for water/EG flow in the considered tube geometry under the constant heat flux of 7957 W/m^2. The trends of experimental and numerical results are similar, Nusselt number increases, and friction factor decreases with increase in the Reynolds number. The average deviation between the experimental and numerical results of Nusselt number is 7.66% and that of friction factor is 8.83% for all considered Reynolds number variation. The presented numerical model based on CFD approach is validated within 10% error with previous experimental results for heat transfer characteristics. Therefore, the validated numerical model could be used for the detail comparison of heat transfer characteristics of single particle and hybrid nanofluids flow in uniformly heated plain straight tube.

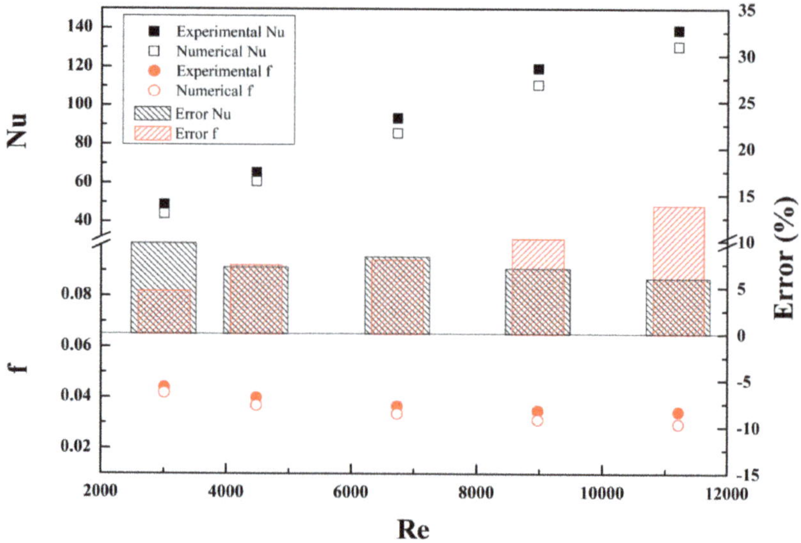

Figure 3. Comparison of Nusselt number and friction factor for previous experimental study and present numerical model for various Reynolds numbers.

5.2. Comparison of Nanofluid Properties

The thermophysical properties namely, density, specific heat, thermal conductivity and viscosity of single particle and hybrid nanofluids are calculated using equations presented in Section 3. The calculated thermophysical properties of single particle and hybrid nanofluids are depicted in Table 2. In case of hybrid nanofluid, the composition of Al_2O_3 and Cu are mixed in the proposition of 50/50%. The density and thermal conductivity of hybrid nanofluid are higher than those of single particle nanofluid for all volume fractions. This is because the Cu nanoparticles have higher density and thermal conductivity compared to Al_2O_3 nanoparticles. However, the specific heat of single particle nanofluid is superior to the specific heat of hybrid nanofluid for all volume fractions due to lower specific heat of Cu nanoparticles compared to specific heat of Al_2O_3 nanoparticles. There is no difference for the viscosity of single particle and hybrid nanofluids for each volume fraction because the viscosity of base fluid is same for both single particle and hybrid nanofluids. However, the viscosity of nanofluids increases with increase in the volume fraction. The density and thermal conductivity of single particle and hybrid nanofluids have increased and the specific heat has decreased with increase in the volume fraction. This is because the density and thermal conductivity of nanoparticles enhances, and the specific heat of nanoparticles reduces with increase in volume fraction of nanoparticle in both single particle and hybrid nanofluids. The density and thermal conductivity of nanoparticles are higher than water which shows superior density and thermal conductivity of nanofluids than base fluid which further increases with increase in the volume fraction. Whereas the specific heat of nanoparticles is lower than the water which results into lower specific heat of nanofluids than base fluid and further decreases with increase in the volume fraction. This is because the water has lower density, lower thermal conductivity and higher specific heat compared to Al_2O_3 and Cu nanoparticles. The density enhances by 4.4% and 7.5% for Al_2O_3 and Al_2O_3/Cu nanofluids, respectively with increase in volume fraction from 0.5% to 2.0%. The thermal conductivity improves by 4.3% and 6.1% for Al_2O_3 and Al_2O_3/Cu nanofluids, respectively with increase in volume fraction from 0.5% to 2.0%. The specific heat of Al_2O_3 and Al_2O_3/Cu nanofluids decease by 4.6% and 7.3%, respectively with increase in volume fraction from 0.5% to 2.0%. The viscosity increases by 3.9% for Al_2O_3 and Al_2O_3/Cu nanofluids with increase in volume fraction from 0.5% to 2.0%. Addition of Cu nanoparticle along with Al_2O_3 nanoparticle improves the thermophysical properties of hybrid nanofluid compared to single particle nanofluid.

Table 2. Properties of base fluid and nanoparticles.

Nanofluid	Density (kg/m^3)	Specific Heat (J/kg·K)	Thermal Conductivity (W/m·K)	Viscosity (Pa·s)
0.5% Al_2O_3	1013.059	4115.047	0.608648	0.001013
0.5% Al_2O_3/Cu	1023.884	4072.637	0.612030	0.001013
1.0% Al_2O_3	1027.918	4050.029	0.617380	0.001025
1.0% Al_2O_3/Cu	1049.568	3968.626	0.624304	0.001025
2.0% Al_2O_3	1057.636	3925.475	0.635098	0.001052
2.0% Al_2O_3/Cu	1100.936	3775.164	0.649614	0.001052

5.3. Comparison of Heat Transfer Characteristics

In this section, the heat transfer characteristics including thermal attributes of heat transfer coefficient and Nusselt number and flow attributes of pressure drop and friction factor are compared for single particle and hybrid nanofluids. In addition, the comparison of single particle and hybrid nanofluids is extended for the combined thermal and flow attributes by defining the parameter namely, performance evaluation criteria (PEC).

5.3.1. Heat Transfer Coefficient

The heat transfer coefficient (h) of water, single particle and hybrid nanofluids flow inside tube for various volume fractions and Reynolds number is presented in Figure 4. The heat transfer coefficients of all working fluids have increased with increase in the Reynolds number and the variation trends are linear. Saedodin et al. have also presented the increasing linear trend of heat transfer coefficient for various working fluids flow inside tube with Reynolds number [7]. The turbulence increases with increase in the Reynolds number for all working fluids. The heat transfer rate increases as the turbulence increases. Higher turbulence at higher Reynolds number of working fluids absorbs higher heat and presents higher heat transfer rate compared with lower turbulence at lower Reynolds number of working fluids. Therefore, the heat transfer coefficient of working fluids is higher at the higher Reynolds number and lower at lower Reynolds number. The heat transfer coefficient of single particle and hybrid nanofluids for all volume fractions are higher than water because of improved thermophysical properties of single particle and hybrid nanofluids compared to water. The heat transfer coefficient improves further with increase in the volume fractions. In addition, for the same volume fraction, the heat transfer coefficient of hybrid nanofluid is better than the single particle nanofluid due to addition of Cu nanoparticles with higher thermal conductivity in hybrid nanofluid. The heat transfer coefficients of Al_2O_3 nanofluid are higher by 2.5%, 4.9% and 10.2% and those of Al_2O_3/Cu nanofluids are higher by 3.6%, 7.2% and 14.8% for volume fractions of 0.5%, 1.0% and 2.0%, respectively compared with water. The Al_2O_3/Cu nanofluid with 2.0% volume fraction shows highest heat transfer coefficient among all working fluids. The heat transfer coefficient increases by 546.1% for Al_2O_3/Cu nanofluid with 2.0% volume fraction when the Reynolds number increases from 2000 to 12,000.

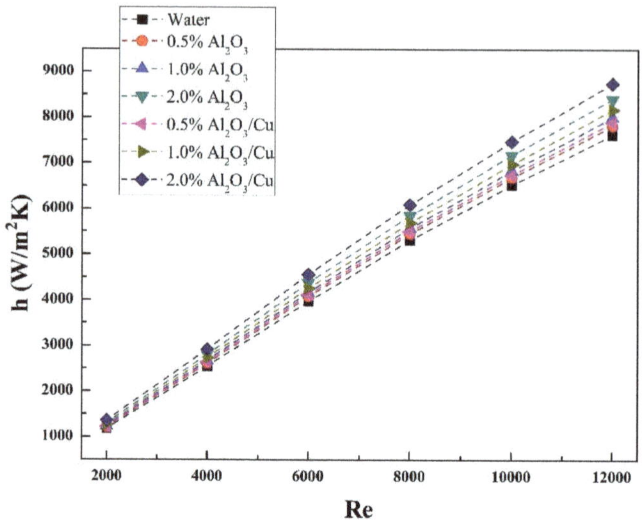

Figure 4. Heat transfer coefficient of water, single particle and hybrid nanofluids flow inside tube for various volume fractions and Reynolds number.

5.3.2. Nusselt Number

The comparison of Nusselt number of various working fluids with Reynolds number is presented in Figure 5. The Nusselt number depends on the heat transfer coefficient hence, the variation trend of Nusselt number is similar as of heat transfer coefficient for each working fluid. The Nusselt number increases linearly with Reynolds number for all working fluids because of increase in the turbulence with increase in the Reynolds number. The Nusselt number of hybrid nanofluid is better than the single particle nanofluid for the same volume fraction due to higher heat transfer coefficient of hybrid nanofluid compared to single particle nanofluid. In addition, due to the improved thermophysical properties of nanofluids by dispersion of nanoparticles in water, the Nusselt number of single particle and hybrid nanofluids are superior compared to water. Addition of solid nanoparticles into water has enhanced the thermal conductivity which improves continuously as larger amounts of nanoparticles are dispersed, which results in increase in heat transfer coefficient. Therefore, the Nusselt number increases with increase in the volume fraction for both single particle and hybrid nanofluids. The Nusselt numbers of 50.83, 52.13, 53.36, 56.02, 52.65, 54.49 and 58.29 are evaluated for water, 0.5% Al_2O_3, 1.0% Al_2O_3, 2.0% Al_2O_3, 0.5% Al_2O_3/Cu, 1.0% Al_2O_3/Cu and 2.0% Al_2O_3/Cu, respectively at Reynolds number of 12000. The 2.0% Al_2O_3/Cu nanofluid presents superior value of Nusselt number for all Reynolds number compared to other working fluids. The Nusselt number of 2.0% Al_2O_3/Cu nanofluid increases from 9.02 to 58.29 as the Reynolds number increases from 2000 to 12000. Firoozi et al. have shown the Nusselt number ranging from 10 to 40 for the variation of Reynolds number up to 5000 [3]. Torii and Hajime have reported the Nusselt number enhancement for graphene-oxide nanofluid as twice than water for horizontal circular tube under constant heat flux [45].

5.3.3. Pressure Drop

The variation of pressure drop for water, single particle and hybrid nanofluids with various volume fractions and Reynolds number is depicted in Figure 6. The pressure drop increases with increase in the Reynolds number because the turbulence increases as the Reynolds number increases for all working fluids. However, the variation trend of pressure drop with Reynolds number is not linear for all working fluids. The variation trend is exponential as confirmed by many research studies in the open literature. Kristiawan et al.

have presented the exponential (non-linear) variation of pressure drop with Reynolds number and pressure drop increases with Reynolds number for various working fluids flow in a tube [8]. Pressure drop of single particle and hybrid nanofluids are higher than water because the viscosity of single particle and hybrid nanofluids are higher compared to water. In addition, the viscosity of single particle and hybrid nanofluids increases as volume fraction increases due to dispersion of larger volume of nanoparticles into base fluid water. This results into increase of pressure drop with volume fraction for single particle and hybrid nanofluids. Despite of the same viscosity of single particle and hybrid nanofluids at the same volume fraction, the pressure drop for the hybrid nanofluid are lower than the single particle nanofluids for all volume fractions. The pressure drop of 2.0% Al_2O_3 nanofluid is highest among all working fluids. The pressure drops of 2.0% Al_2O_3 nanofluid and 2.0% Al_2O_3/Cu nanofluid are higher by 4.5% and 0.34%, respectively compared to water.

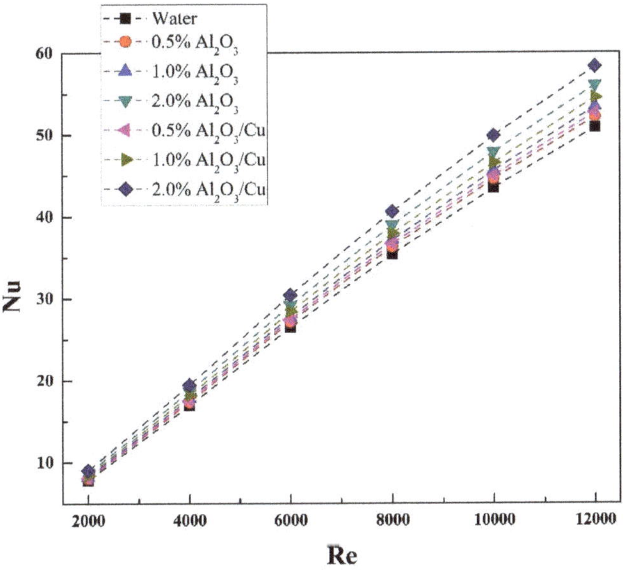

Figure 5. Comparison of Nusselt number of various working fluids with Reynolds number.

5.3.4. Friction Factor

The effect of Reynolds number and volume fraction on friction factor of water, single particle and hybrid nanofluids is shown in Figure 7. The friction factor is affected by the pressure drop and velocity of working fluid. Therefore, the variation trend of friction factor for various working fluids are not same as of pressure drop. The considered Reynolds number range shows the transition from laminar to turbulent regime which results into parabolic variation trend of friction factor with Reynolds number for all working fluids. The friction factor increases up to the Reynolds number of 6000 and decreases beyond this value for all working fluids as can be seen from Figure 7. There is a critical Reynolds number below which the friction factor increases with increase in the Reynolds number due to change of flow from laminar to transition regime and above critical Reynolds number, the friction factor decreases with increase in the Reynolds number because of change of flow from transition to turbulent regime. However, it is important to note that the critical point (Re:6000) for friction factor variation over the Reynolds number is not same as commonly suggested Reynolds number for transition of flow to turbulent (Re:4000). The reason for different critical point of friction factor than commonly suggested is dependency of friction factor on ratio of pressure drop to square of velocity as presented by Equation (21). Kristiawan et al. have presented the parabolic variation of friction factor with Reynolds number, the friction factor increases up to the critical Reynolds number of 5000 then decreases as the Reynolds number increases beyond the critical value [8]. Kaood et al. have also presented the non-linear trend of the friction factor with Reynolds number, the friction factor decreases as the Reynolds number increases in the turbulent regime [16]. The friction factors of single particle and hybrid nanofluids are higher than water which are increasing with increase in the volume fraction. This is because of higher viscosity of single particle and hybrid nanofluids compared to water which continuously increases with volume fraction. The density of the hybrid nanofluid is higher than the density of single particle nanofluid which results into lower velocity of hybrid nanofluid compared with single particle nanofluid. Therefore, the ratio of pressure drop to square of velocity the hybrid nanofluid shows higher friction factor than single particle nanofluid at the same volume fraction. Despite of lower pressure drop for hybrid nanofluid than single particle nanofluid, the higher density has caused higher friction factor. The friction factors of 0.5% Al_2O_3, 1.0% Al_2O_3, 2.0% Al_2O_3, 0.5% Al_2O_3/Cu, 1.0% Al_2O_3/Cu and 2.0% Al_2O_3/Cu nanofluids are higher by 1.5%, 2.9%, 5.9%, 2.6%, 5.2% and 10.3%, respectively compared to water. The 2.0% Al_2O_3/Cu nanofluid shows highest value of friction factor among all working fluids. The friction factor of 2.0% Al_2O_3/Cu nanofluid increases by 1.9% as the Reynolds number increases from 2000 to 12,000. Firoozi et al. have also presented the friction factor variation range from 0.02 to 0.14 for the maximum Reynolds number variation up to 5000 [3].

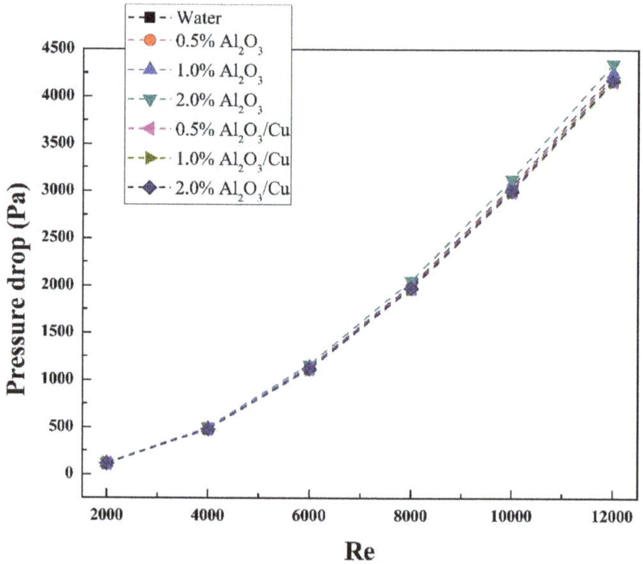

Figure 6. Variation of pressure drop for water, single particle and hybrid nanofluids with various volume fractions and Reynolds number.

5.3.5. Performance Evaluation Criteria

The performance evaluation criteria present the combined effect of thermal and flow characteristics because it is calculated based on the Nusselt numbers and friction factors of water and single particle and hybrid nanofluids. The comparison of performance evaluation criteria (PEC) for single particle and hybrid nanofluids with various volume fractions and Reynolds number is depicted in Figure 8. The effect of Reynolds number on the performance evaluation criteria of single particle and hybrid nanofluids is very small. This is because both Nusselt number and friction factor are affected significantly due to variation in Reynolds number. Hence, the overall effect of Reynolds number on performance evaluation criteria is nullified. The Nusselt number and friction factor have improved with increase in volume fraction of single particle and hybrid nanofluids as a result of improvement in the thermophysical properties. Therefore, the performance evaluation criteria enhance for both single particle and hybrid nanofluids as volume fraction increases. Azmi et al. have proved that the variation in the performance evaluation criteria is very small and non-linear with Reynolds number. In addition, the performance evaluation criteria increase with increase in volume fraction of nanoparticles [24]. The performance evaluation criteria for hybrid nanofluid are superior compared to single particle nanofluid for the same volume fraction because the increase in the Nusselt number is highly dominant compared to the increase in the friction factor for hybrid nanofluid. The 0.5% Al_2O_3 nanofluid presents the lowest value of performance evaluation criteria and 2.0% Al_2O_3/Cu nanofluid presents the highest value of performance evaluation criteria among all nanofluids. Compared to the performance evaluation criteria of 0.5% Al_2O_3 nanofluid, the performance evaluation criteria of 1.0% Al_2O_3, 2.0% Al_2O_3, 0.5% Al_2O_3/Cu, 1.0% Al_2O_3/Cu and 2.0% Al_2O_3/Cu nanofluids are higher by 1.9%, 5.9%, 0.7%, 3.3% and 8.9%, respectively. Overall, 2.0% Al_2O_3/Cu nanofluid with composition of 50/50% has presented the superior heat transfer characteristics among all working fluids. The 2.0% Al_2O_3 nanofluid in single particle case and 2.0% Al_2O_3/Cu nanofluids in hybrid case present the superior performances in the respective groups. Therefore, the simulated contours of temperature and velocity for water, 2.0% Al_2O_3 single particle nanofluid and 2.0% Al_2O_3/Cu hybrid nanofluid with composition of 50/50% are depicted in Figure 9. The most important section to observe the behavior of simulated results is the outlet of tube. Therefore, the temperature and velocity contours are presented for the working fluid domain at the tube outlet cross section considering the Reynolds numbers of 2000 and 12,000. As can be seen from Figure 9, the distribution of temperature and velocity contours at the outlet section of tube are symmetrical for water and nanofluids. The heat transfer characteristics of 2.0% Al_2O_3/Cu nanofluid is further investigated with additional two compositions of 75/25% and 25/75% and compared with composition of 50/50% as well as water and 2.0% Al_2O_3 nanofluid.

5.4. Hybrid Nanofluid with Different Compositions

The comparison of heat transfer characteristics of water, 2.0% Al_2O_3, 2.0% Al_2O_3/Cu (50/50%), 2.0% Al_2O_3/Cu (75/25%) and 2.0% Al_2O_3/Cu (25/75%) is presented in this section. The variation of heat transfer coefficient, Nusselt number, pressure drop, friction factor and performance evaluation criteria with various Reynolds number for above considered working fluids is presented in Figure 10. The heat transfer coefficient and Nusselt number of various working fluid have increased linearly with Reynolds number. This is because of increase in turbulence at higher Reynolds number has resulted into higher heat transfer rate. The hybrid nanofluids with all compositions show higher heat transfer coefficient and Nusselt number than single particle nanofluid and water. This is because of superior thermal conductivity and heat transfer performance of the single particle and hybrid nanofluids compared to water. The increase in the heat transfer coefficient and Nusselt number of hybrid nanofluid enhances with higher portion of Cu nanoparticle than Al_2O_3 nanoparticle because the thermal conductivity of Cu nanoparticle is superior to Al_2O_3 nanoparticle. This means 2.0% Al_2O_3/Cu (25/75%) nanofluid presents highest values of heat transfer coefficient and Nusselt number followed by 2.0% Al_2O_3/Cu (50/50%), 2.0% Al_2O_3/Cu (75/25%), 2.0% Al_2O_3 nanofluids and water, respectively. The heat transfer coefficient enhances by 546.4% and Nusselt number increases from 9.14 to 59.09 for 2.0% Al_2O_3/Cu (25/75%) nanofluid with increase in the Reynolds number from 2000 to 12,000. The heat transfer coefficients of 2.0% Al_2O_3, 2.0% Al_2O_3/Cu (50/50%), 2.0% Al_2O_3/Cu (75/25%) and 2.0% Al_2O_3/Cu (25/75%) nanofluids are higher by 10.2%, 14.8%, 13.3% and 16.3%, respectively compared with water. The water, 2.0% Al_2O_3, 2.0% Al_2O_3/Cu (50/50%), 2.0% Al_2O_3/Cu (75/25%) and 2.0% Al_2O_3/Cu (25/75%) nanofluids show the Nusselt number of 50.83, 56.02, 58.29, 57.48 and 59.09, respectively at the Reynolds number of 12,000. The variation of pressure drop for various working fluid are not linear but increasing with increase in the Reynolds number. This is because the turbulence has increased with increase in the Reynolds number. The decreasing order of pressure drop is 2.0% Al_2O_3, 2.0% Al_2O_3/Cu (75/25%) 2.0% Al_2O_3/Cu (50/50%), water and 2.0% Al_2O_3/Cu (25/75%), respectively. The increase in Cu nanoparticle portion in hybrid nanofluid composition reduces the pressure drop. The 2.0% Al_2O_3/Cu (25/75%) nanofluid shows the lowest pressure drop among all working fluids. Compared to water, the pressure drops of 2.0% Al_2O_3, 2.0% Al_2O_3/Cu (50/50%) and 2.0% Al_2O_3/Cu (75/25%) nanofluids are higher by 4.5%, 0.34% and 2.4%, respectively and pressure drop of 2.0% Al_2O_3/Cu (25/75%) nanofluid is lower by 1.6%. The variation in friction factor of various working fluids is parabolic with Reynolds number due to change of flow regime from laminar to transition and transition to turbulent with critical Reynolds number. The critical Reynolds number is 6000 where the friction factor is maximum for all working fluids. The 2.0% Al_2O_3/Cu (25/75%) nanofluid shows highest friction factor followed by 2.0% Al_2O_3/Cu (50/50%), 2.0% Al_2O_3/Cu (75/25%), 2.0% Al_2O_3 nanofluids and water respectively in the decreasing order. This is due to higher viscosity as well as higher density of hybrid nanofluids, the density increases as the portion of Cu nanoparticle increases which results into lower velocity. The friction factors of 2.0% Al_2O_3, 2.0% Al_2O_3/Cu (50/50%), 2.0% Al_2O_3/Cu (75/25%) and 2.0% Al_2O_3/Cu (25/75%) nanofluids are higher by 5.9%, 10.3%, 8.1% and 12.5%, respectively compared to water. The performance evaluation criteria of 2.0% Al_2O_3, 2.0% Al_2O_3/Cu (50/50%), 2.0% Al_2O_3/Cu (75/25%) and 2.0% Al_2O_3/Cu (25/75%) nanofluids are affected very less by the variation in the Reynolds number due to significant effect of Reynolds number on both Nusselt number and friction factor. The increase of Cu nanoparticle portion in composition of hybrid nanofluid has shown significant enhancement in the performance evaluation criteria due to overall improvement in the thermophysical properties of hybrid nanofluid. Compared to 2.0% Al_2O_3 nanofluid, the performance evaluation criteria of 2.0% Al_2O_3/Cu (50/50%), 2.0% Al_2O_3/Cu (75/25%) and 2.0% Al_2O_3/Cu (25/75%) nanofluids have improved by 2.8%, 2.1% and 3.5%, respectively. The 2.0% Al_2O_3/Cu (25/75%) nanofluid presents the highest value of performance evaluation criteria.

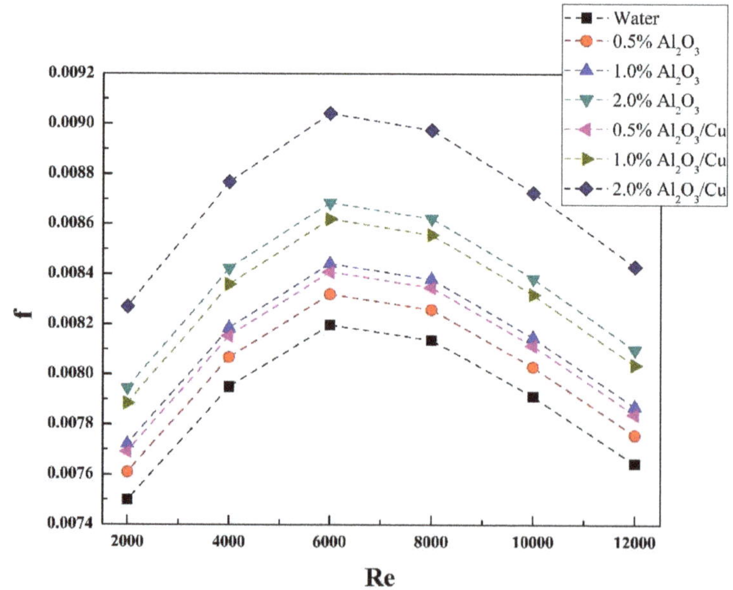

Figure 7. Effect of Reynolds number and volume fraction on friction factor of water, single particle and hybrid nanofluids.

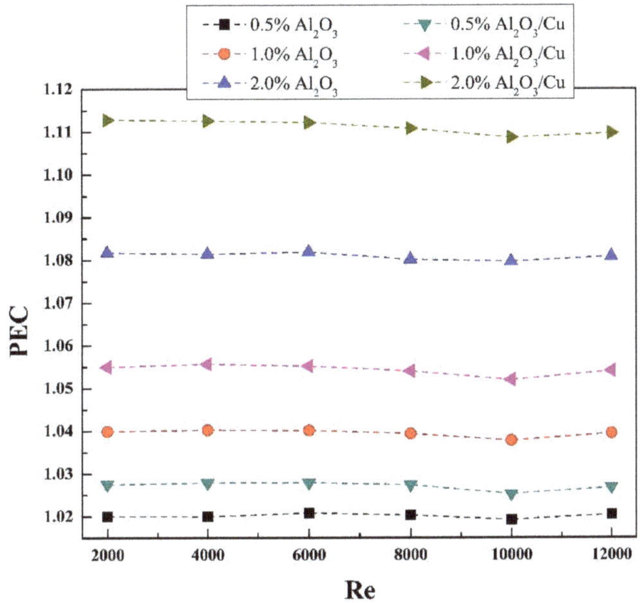

Figure 8. Comparison of performance evaluation criteria for single particle and hybrid nanofluids with various volume fractions and Reynolds number.

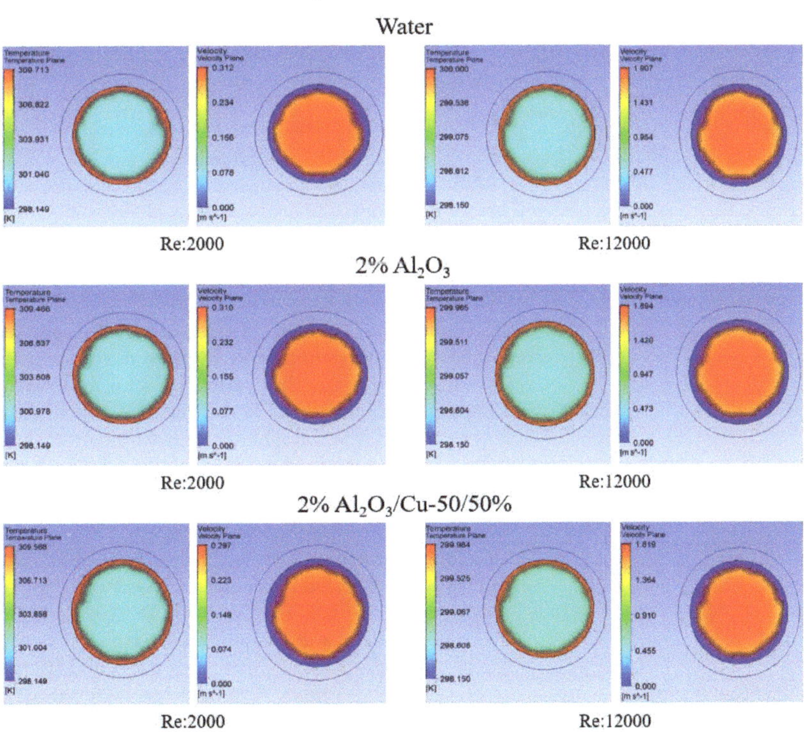

Figure 9. Temperature and velocity contours for water, 2% Al2O3 single particle nanofluid and 2% Al2O3/Cu hybrid nanofluid with 50/50% composition at Re:2000 and Re:12,000.

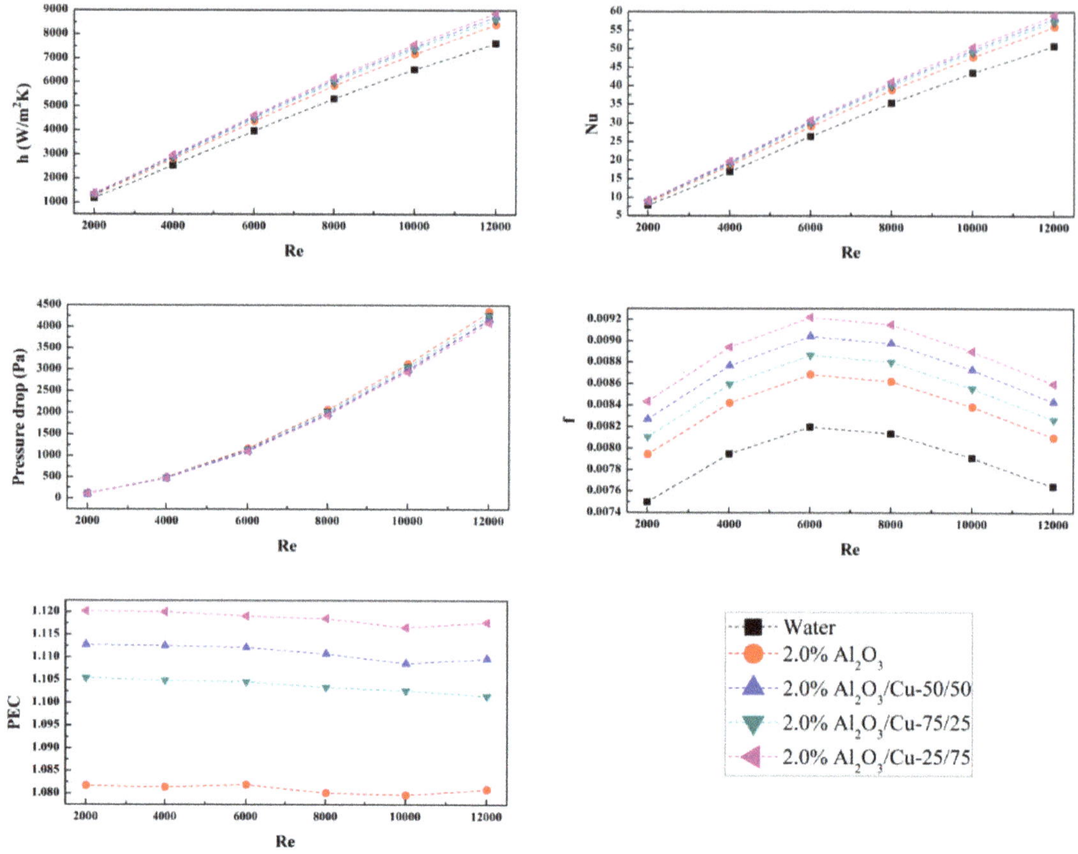

Figure 10. Comparison of heat transfer characteristics of water, 2.0% Al_2O_3, 2.0% Al_2O_3/Cu (50/50%), 2.0% Al_2O_3/Cu (75/25%) and 2.0% Al_2O_3/Cu (25/75%).

6. Conclusions

The heat transfer characteristics of water, single particle and hybrid nanofluids in uniformly tube under constant heat flux condition are compared. The heat transfer coefficient, Nusselt number, pressure drop, friction factor and performance evaluation criteria are investigated for various Reynolds number, volume fractions and nanoparticle compositions of hybrid nanofluid. The valuable findings of the present numerical study are summarized as follows.

(a) The heat transfer coefficient and Nusselt number have increased with Reynolds number and volume fraction for single particle and hybrid nanofluids. The heat transfer coefficients of 2.0% Al_2O_3 and 2.0% Al_2O_3/Cu nanofluids are higher by 10.2% and 14.8%, respectively compared to water. The 2.0% Al_2O_3 and 2.0% Al_2O_3/Cu nanofluids present Nusselt numbers of 56.02 and 58.29, respectively.

(b) The pressure drop and friction factor have increased with volume fraction. However, the pressure drop shows increasing trend and friction factor shows parabolic trend with Reynolds number. Compared to water, the 2.0% Al_2O_3 and 2.0% Al_2O_3/Cu nanofluids show pressure drops higher by 4.5% and 0.3%, respectively and those show friction factor higher by 5.9% and 10.3%, respectively.

(c) The 0.5% Al_2O_3, 1.0% Al_2O_3, 2.0% Al_2O_3, 0.5% Al_2O_3/Cu, 1.0% Al_2O_3/Cu and 2.0% Al_2O_3/Cu nanofluids present the performance evaluation criteria of 1.02, 1.04, 1.08, 1.03, 1.05 and 1.11, respectively.
(d) Addition of higher portion of Cu nanoparticle in the hybrid nanofluid increases the heat transfer coefficient, Nusselt number, friction factor and performance evaluation criteria and decreases the pressure drop. The 2.0% Al_2O_3, 2.0% Al_2O_3/Cu (50/50%), 2.0% Al_2O_3/Cu (75/25%) and 2.0% Al_2O_3/Cu (25/75%) present performance evaluation criteria of 1.08, 1.11, 1.10 and 1.12, respectively.
(e) Based on the numerical results, the best nanofluids could be prepared and used in the plain straight tube to extract the excellent experimental output results. The proposed plain straight tube incorporated with nanofluid flow could be used as an efficient solar receiver tubes to achieve the superior heat transfer performances.

Author Contributions: Conceptualization, K.S.G.; and M.-Y.L.; methodology, K.S.G.; and M.-Y.L.; formal analysis, K.S.G.; and M.-Y.L.; investigation, K.S.G.; and M.-Y.L.; resources, K.S.G.; and M.-Y.L.; data curation, K.S.G.; validation, K.S.G.; software, K.S.G.; writing—original draft preparation, K.S.G.; and M.-Y.L.; writing—review and editing, K.S.G.; and M.-Y.L.; visualization, K.S.G.; and M.-Y.L.; supervision, M.-Y.L.; project administration, M.-Y.L.; funding acquisition, M.-Y.L. All authors have read and agreed to the published version of the manuscript.

Funding: This research received no external funding.

Institutional Review Board Statement: Not applicable.

Informed Consent Statement: Not applicable.

Data Availability Statement: The data presented in this study will be available on request to the corresponding author.

Acknowledgments: This work was supported by the National Research Foundation of Korea (NRF) grant funded by the Korea government (MSIT) (No. 2020R1A2C1011555).

Conflicts of Interest: The authors declare no conflict of interest.

References

1. Dadhich, M.; Prajapati, O.S.; Sharma, V. Investigation of boiling heat transfer of titania nanofluid flowing through horizontal tube and optimization of results utilizing the desirability function approach. *Powder Technol.* **2021**, *378*, 104–123. [CrossRef]
2. Sheikh, N.A.; Ching, D.L.C.; Khan, I. A Comprehensive review on theoretical aspects of nanofluids: Exact solutions and analysis. *Symmetry* **2020**, *12*, 725. [CrossRef]
3. Firoozi, A.; Majidi, S.; Ameri, M. A numerical assessment on heat transfer and flow characteristics of nanofluid in tubes en-hanced with a variety of dimple configurations. *Therm. Sci. Eng. Prog.* **2020**, *19*, 100578. [CrossRef]
4. Ledari, B.H.; Sabzpooshani, M.; Khayat, M. An experimental investigation on the thermo-hydraulic properties of CuO and Fe_3O_4 oil-based nanofluids in inclined U-tubes: A comparative study. *Powder Technol.* **2021**, *379*, 191–202. [CrossRef]
5. Chaurasia, S.R.; Sarviya, R. Comparative thermal performance analysis with entropy generation on helical screw insert in tube with number of strips with nanofluid at laminar flow regime. *Int. Commun. Heat Mass Transf.* **2021**, *122*, 105138. [CrossRef]
6. Ying, Z.; He, B.; Su, L.; Kuang, Y.; He, D.; Lin, C. Convective heat transfer of molten salt-based nanofluid in a receiver tube with non-uniform heat flux. *Appl. Eng.* **2020**, *181*, 115922. [CrossRef]
7. Saedodin, S.; Zaboli, M.; Rostamian, S.H. Effect of twisted turbulator and various metal oxide nanofluids on the thermal performance of a straight tube: Numerical study based on experimental data. *Chem. Eng. Process. Process. Intensif.* **2020**, *158*, 108106. [CrossRef]
8. Kristiawan, B.; Rifa'I, A.I.; Enoki, K.; Wijayanta, A.T.; Miyazaki, T. Enhancing the thermal performance of TiO_2/water nanofluids flowing in a helical microfin tube. *Powder Technol.* **2020**, *376*, 254–262. [CrossRef]
9. Tiwari, A.K.; Javed, S.; Oztop, H.F.; Said, Z.; Pandya, N.S. Experimental and numerical investigation on the thermal per-formance of triple tube heat exchanger equipped with different inserts with WO_3/water nanofluid under turbulent condition. *Int. J. Therm. Sci.* **2021**, *164*, 106861. [CrossRef]
10. Shahsavar, A.; Jha, P.; Arıcı, M.; Estellé, P. Experimental investigation of the usability of the rifled serpentine tube to improve energy and exergy performances of a nanofluid-based photovoltaic/thermal system. *Renew. Energy* **2021**, *170*, 410–425. [CrossRef]
11. Mukherjee, S.; Jana, S.; Mishra, P.C.; Chaudhuri, P.; Chakrabarty, S. Experimental investigation on thermo-physical properties and subcooled flow boiling performance of Al_2O_3/water nanofluids in a horizontal tube. *Int. J. Sci.* **2021**, *159*, 106581. [CrossRef]

12. Heyhat, M.; Jafarzad, A.; Changizi, P.; Asgari, H.; Valizade, M. Experimental research on the performance of nanofluid flow through conically coiled tubes. *Powder Technol.* **2020**, *370*, 268–277. [CrossRef]
13. Ho, C.; Cheng, C.-Y.; Yang, T.-F.; Rashidi, S.; Yan, W.-M. Experimental study on cooling performance of nanofluid flow in a horizontal circular tube. *Int. J. Heat Mass Transf.* **2021**, *169*, 120961. [CrossRef]
14. Sun, B.; Lei, W.; Yang, D. Flow and convective heat transfer characteristics of Fe_2O_3–water nanofluids inside copper tubes. *Int. Commun. Heat Mass Transf.* **2015**, *64*, 21–28. [CrossRef]
15. Behzadnia, H.; Jin, H.; Najafian, M.; Hatami, M. Geometry optimization for a rectangular corrugated tube in supercritical water reactors (SCWRs) using alumina-water nanofluid as coolant. *Energy* **2021**, *221*, 119850. [CrossRef]
16. Kaood, A.; Hassan, M.A. Thermo-hydraulic performance of nanofluids flow in various internally corrugated tubes. *Chem. Eng. Process. Process. Intensif.* **2020**, *154*, 108043. [CrossRef]
17. Safaei, M.R.; Mahian, O.; Garoosi, F.; Hooman, K.; Karimipour, A.; Kazi, S.N.; Gharehkhani, S. Investigation of micro- and nanosized particle erosion in a 90° pipe bend using a two-phase discrete phase model. *Sci. World J.* **2014**, *2014*, 1–12. [CrossRef]
18. Amer Qureshi, M. Numerical simulation of heat transfer flow subject to mhd of williamson nanofluid with thermal ra-diation. *Symmetry* **2021**, *13*, 10. [CrossRef]
19. Yıldırım, E.; Yurddaş, A. Assessments of thermal performance of hybrid and mono nanofluid U-tube solar collector system. *Renew. Energy* **2021**, *171*, 1079–1096. [CrossRef]
20. Saleh, B.; Sundar, L.S. Entropy generation and exergy efficiency analysis of ethylene glycol-water based nanodiamond + Fe_3O_4 hybrid nanofluids in a circular tube. *Powder Technol.* **2021**, *380*, 430–442. [CrossRef]
21. Ramadhan, A.; Azmi, W.; Mamat, R.; Hamid, K. Experimental and numerical study of heat transfer and friction factor of plain tube with hybrid nanofluids. *Case Stud. Eng.* **2020**, *22*, 100782. [CrossRef]
22. Han, Z.; Zhou, W.; Yu, W.; Jia, Z. Experimental investigation on heat transfer of n-decane-ZnO nanofluids in a horizontal tube under supercritical pressure. *Int. Commun. Heat Mass Transf.* **2021**, *121*, 105108. [CrossRef]
23. Akbar, H.M.; Mohammed, A.S.; Ezzat, S.B. *Hybrid. Nanofluid to Improve Heat Transfer and Pressure Drop through Horizontal Tube*; Elsevier BV: Amsterdam, The Netherlands, 2021; Volume 42, pp. 1885–1888.
24. Azmi, W.; Hamid, K.A.; Ramadhan, A.; Shaiful, A. Thermal hydraulic performance for hybrid composition ratio of TiO_2–SiO_2 nanofluids in a tube with wire coil inserts. *Case Stud. Eng.* **2021**, *25*, 100899. [CrossRef]
25. Moldoveanu, G.M.; Huminic, G.; Minea, A.A.; Huminic, A. Experimental study on thermal conductivity of stabilized Al_2O_3 and SiO_2 nanofluids and their hybrid. *Int. J. Heat Mass Transf.* **2018**, *127*, 450–457. [CrossRef]
26. Yang, C.; Wu, X.; Zheng, Y.; Qiu, T. Heat transfer performance assessment of hybrid nanofluids in a parallel channel under identical pumping power. *Chem. Eng. Sci.* **2017**, *168*, 67–77. [CrossRef]
27. Minea, A.A. Hybrid nanofluids based on Al_2O_3, TiO_2 and SiO_2: Numerical evaluation of different approaches. *Int. J. Heat Mass Transf.* **2017**, *104*, 852–860. [CrossRef]
28. Patil, M.S.; Seo, J.-H.; Lee, M.-Y. Numerical study on geometric parameter effects of power generation performances for segmented thermoelectric generator. *Int. J. Air Cond. Refrig.* **2018**, *26*, 1850004. [CrossRef]
29. Seo, J.H.; Garud, K.S.; Lee, M.Y. Grey relational based Taguchi analysis on thermal and electrical performances of thermoe-lectric generator system with inclined fins hot heat exchanger. *Appl. Therm. Eng.* **2021**, *184*, 116279. [CrossRef]
30. Hamid, K.A.; Azmi, W.; Nabil, M.; Mamat, R. Experimental investigation of nanoparticle mixture ratios on TiO_2–SiO_2 nanofluids heat transfer performance under turbulent flow. *Int. J. Heat Mass Transf.* **2018**, *118*, 617–627. [CrossRef]
31. Lee, M.-Y.; Seo, J.-H.; Lee, H.-S.; Garud, K.S. Power generation, efficiency and thermal stress of thermoelectric module with leg geometry, material, segmentation and two-stage arrangement. *Symmetry* **2020**, *12*, 786. [CrossRef]
32. Garud, K.S.; Seo, J.-H.; Cho, C.-P.; Lee, M.-Y. Artificial neural network and adaptive neuro-fuzzy interface system modelling to predict thermal performances of thermoelectric generator for waste heat recovery. *Symmetry* **2020**, *12*, 259. [CrossRef]
33. Dadsetani, R.; Sheikhzadeh, G.A.; Safaei, M.R.; Leon, A.S.; Goodarzi, M. Cooling enhancement and stress reduction opti-mization of disk-shaped electronic components using nanofluids. *Symmetry* **2020**, *12*, 931. [CrossRef]
34. Minea, A.A. Challenges in hybrid nanofluids behavior in turbulent flow: Recent research and numerical comparison. *Renew. Sustain. Energy Rev.* **2017**, *71*, 426–434. [CrossRef]
35. Garud, K.S.; Seo, J.-H.; Patil, M.S.; Bang, Y.-M.; Pyo, Y.-D.; Cho, C.-P.; Lee, M.-Y. Thermal–electrical–structural performances of hot heat exchanger with different internal fins of thermoelectric generator for low power generation application. *J. Anal. Calorim.* **2021**, *143*, 387–419. [CrossRef]
36. Al-Baghdadi, M.A.S.; Noor, Z.M.; Zeiny, A.; Burns, A.; Wen, D. CFD analysis of a nanofluid-based microchannel heat sink. *Sci. Eng. Prog.* **2020**, *20*, 100685. [CrossRef]
37. Devakki, B.; Thomas, S. Experimental Investigation on absorption performance of nanofluids for CO_2 capture. *Int. J. Air Cond. Refrig.* **2020**, *28*, 2050017. [CrossRef]
38. Kim, M.B.; Park, H.G.; Park, C.Y. Change of thermal conductivity and cooling performance for water based Al_2O_3-surfactant nanofluid with time lapse. *Int. J. Air Cond. Refrig.* **2018**, *26*, 1850009. [CrossRef]
39. Sanches, M.; Marseglia, G.; Ribeiro, A.P.; Moreira, A.L.; Moita, A.S. Nanofluids characterization for spray cooling appli-cations. *Symmetry* **2021**, *13*, 788. [CrossRef]
40. Ghachem, K.; Aich, W.; Kolsi, L. Computational analysis of hybrid nanofluid enhanced heat transfer in cross flow micro heat exchanger with rectangular wavy channels. *Case Stud. Eng.* **2021**, *24*, 100822. [CrossRef]

41. Kwon, J.-T.; Nahm, T.-H.; Kim, T.-W.; Kwon, Y.-C. An Experimental study on pressure drop and heat transfer coefficient of laminar ag nanofluid flow in mini-tubes. *J. Korea Acad. Coop. Soc.* **2009**, *10*, 3525–3532. [CrossRef]
42. Bhattad, A.; Sarkar, J.; Ghosh, P. Energy-economic analysis of plate evaporator using brine-based hybrid nanofluids as secondary refrigerant. *Int. J. Air Cond. Refrig.* **2018**, *26*, 1850003. [CrossRef]
43. Garud, K.S.; Seo, J.H.; Lee, M.Y. Effect of guide fin structures and boundary parameters on thermal performances of heat exchanger for waste heat recovery thermoelectric generator. *J. Korea Acad. Ind.* **2021**, *22*, 30–35.
44. Singh, S.K.; Sarkar, J. Thermohydraulic behavior of concentric tube heat exchanger inserted with conical wire coil using mono/hybrid nanofluids. *Int. Commun. Heat Mass Transf.* **2021**, *122*, 105134. [CrossRef]
45. Torii, S.; Yoshino, H.; Hajime, Y. Effect of aqueous suspension of graphene-oxide nanoparticles on thermal fluid flow transport in circular tube. *Int. J. Air-Cond. Refrig.* **2015**, *23*, 1550005. [CrossRef]

symmetry

MDPI

Article

Energy Saving and Economic Evaluations of Exhaust Waste Heat Recovery Hot Water Supply System for Resort

Nghia-Huu Nguyen [1,†], Dong-Yeon Lee [2,†], Kunal Sandip Garud [3] and Moo-Yeon Lee [3,*]

1 Faculty of Mechanical Engineering, Nha Trang University, 02 Nguyen Dinh Chieu Street, Nha Trang City 650000, Vietnam; nghianh@ntu.edu.vn
2 Department of Robotics and Intelligent Machine, Yeungnam University, 280 Daehak-ro, Gyeongsan 38541, Korea; dylee@ynu.ac.kr
3 Department of Mechanical Engineering, Dong-A University, 37 Nakdong-daero 550, Saha-gu, Busan 49315, Korea; 1876936@donga.ac.kr
* Correspondence: mylee@dau.ac.kr; Tel.: +82-51-200-7642
† These authors contributed equally to this work.

Abstract: The objective of this study is to investigate the energy savings and economics of the hot water supply system for the luxury resort. The hot water was generated using the waste heat from the exhaust gas heat (EGH) of internal combustion engine (ICE) installed at the luxury resort. The capacity and characteristics of waste heat source, flow demand and supply system of hot water were surveyed, and data is collected from the real system. The new heat exchanger system which utilizes the EGH to produce the hot water is designed considering the dew point temperature and the back pressure of exhaust gas system. The results show that the proposed system could supply hot water at a temperature of 55 °C corresponding to the present resort demand of 6 m^3/h using EGH of ICE at 20% load. The proposed system could achieve the saving of 400 L/day in diesel oil (DO) fuel consumption and the payback time of new system could be evaluated as 9 months. The proposed system could produce hot water of 14 m^3/h at 25% of engine load and 29 m^3/h at full engine load which are sufficient to satisfy the regular and maximum hot water demand of resort. The presented results show the capability of the proposed system to satisfy the current hot water demand of resort and suggest the larger potential to save energy by recovering EGH of ICE. The novelty of the present work involves detailed methodology to design heat exchangers and evaluate system economics for hot water supply system based on EGH of ICE.

Keywords: economic; energy saving; hot water; heat exchanger; internal combustion engine; waste heat

Citation: Nguyen, N.-H.; Lee, D.-Y.; Garud, K.S.; Lee, M.-Y. Energy Saving and Economic Evaluations of Exhaust Waste Heat Recovery Hot Water Supply System for Resort. *Symmetry* **2021**, *13*, 624. https://doi.org/10.3390/sym13040624

Academic Editor: Mikhail Sheremet

Received: 22 March 2021
Accepted: 7 April 2021
Published: 8 April 2021

1. Introduction

In the present time, energy saving methods and techniques in various industries are very important to prevent the global energy crisis, increasing fuel consumption and oil price and decrease the fossil fuel energy use [1]. Especially, the uses of the fossil fuels have caused the ozone layer depletion and the ecological problems including the global warming and air pollution. There are two ways to overcome these international issues. One is the replacement of the old systems with low operation efficiency for the purpose of decreasing the lost energy. Another is the re-use of the waste energy from the present systems for the purpose of improving the efficiency. To reuse a part of energy exhausted to the environment after system operation as a waste heat is a general method [2]. There is lot of research interest in utilizing waste heat to satisfy the energy demand. To utilize the waste heat effectively, the attention needs to be given to the characteristics of waste heat source and the characteristics of used place. The attention aspects are type, chemical components and basic parameters (pressure, temperature and flow), stability and continuity of waste heat source, purpose and size of application, technology and waste heat recovery system diagram and economic efficiency. The waste heat source could be utilized by the direct and indirect way. If the waste heat source is clean and no-corrosive, then could be used directly otherwise indirectly through the intermediate heat exchanger. Based on the types and the physical properties of the waste heat source, the type of heat exchanger is selected. If the waste heat source is flue gas from ICE or gas turbine then the exhaust gas-to-steam heat exchanger or exhaust gas-to-hot water heat exchanger could be used [3]. If the WH source is liquid then depend on the liquid temperature, the liquid-to-steam heat exchanger or the liquid-to-hot water heat exchanger could be selected.

There are now many services as motels, hotels and resorts which were built for the purpose of travel and lodging. The tourists usually enjoy coming to the quiet places with clean environment to relax. Therefore, many resorts were built in farther places from the urban, near to the coastline or on the islands. Due to the geographic location, many resorts cannot connect with the power grid, so that the ICE for electric generation is usually chosen. Beside the power demands, hot water and steam are the important issues.

Many previous studies have focused on the waste heat recovery from the ICE to enhance the performance. The general technologies and waste heat sources of combined heat and power systems were presented in [4–6] and typical recovery of EGH from ICE were introduced in [7–9]. Separate Rankine cycle could provide additional power of 12–16% from the waste heat of diesel engine [10,11]. The recovering of waste heat of ICE stored about 10–15% of fuel power in storage system [12,13]. The waste heat recovery from ICE of vehicle was investigated by combined thermodynamic cycles and shown good effective energy savings [14–17]. The cooling capacity of absorption refrigeration system was improved by using the exhaust gas of ICE [18,19]. The exhaust gas recovery using various bottom cycles showed improvement in the engine thermal efficiency [20,21]. The Rankine steam cycle presented the maximum energy saving potentials [22,23]. Generally, depend on the load of ICE and the applied cycle, the effective energy up to 10–20% could be achieved by recovering the waste heat from ICE. Many research show that a huge amount of energy is lost in form of heat through the exhaust gas which is about 30–40% of the combustion heat [12,13,23,24]. The generated exhaust gases are about 50–70% of the fuel input and waste heat from exhaust gas could be recovered through the engine cooling [25]. Majority of the research studies are focused on the additional power for the engine by utilizing the waste heat of ICE [10,11,14,15], some of research studies were applied in cooling [18,19] and some research was focused on the residential applications [4,9,26]. However, there are only few research studies which presented the application of waste heat in residential and small commercial buildings [4,9]. Onovwiona et al. presented the assumed data and simulated results of waste heat recovery system of ICE [26]. Jia et al. have proposed novel gas-engine driven heat pump system to overcome the limitation of insufficient engine exhaust waste heat for hot water supply [27]. Liu et al. have presented the exhaust waste heat recovery system for hot water supply which comprises of solar energy collection system, drainage system and heat pump system [28]. Butrymowicz et al. have conducted experimental study on the waste heat recovery system which uses waste heat of combustion engines for heating applications [29]. Kunal et al. have investigated the utilization of waste heat of ICE for thermoelectric power generation [30,31]. Estefania et al. have developed the energy recovery-based heat pump system which uses low grade temperature source for domestic hot water supply [32]. Ahmadi et al. have proposed solar thermal energy based parallel feed water heating system for power plant units. The net energy and exergy efficiencies of the proposed system are increased by 9.5% by using solar collectors as the replacement of high-pressure feed water heaters [33]. Ahmadi et al. have also simulated the energy and exergy performances of feed water heating repowering for steam power plan under three operating modes [34]. The repowering of steam power plant with and without integration of solar energy has been investigated based on energy, energy and environmental aspects [35].

The conducted literature review reveals that there has been no study reported until now which uses the exhaust gas of ICE for the purpose of the hot water supply as an application of waste heat recovery. Therefore, this study focuses on the energy savings and economic evaluations of proposed hot water supply system which recovered the EGH from the ICE to produce hot water for resort. The recovered EGH from ICE is used to heat the water though the heat exchanger system. This will decrease the steam for heating water and reduces the oil consumption of the boiler for the existing system. The daily demand and volume flow rate of hot water at peak hour are surveyed. The fuel consumption and exhaust gas of ICE are investigated. The new heat exchanger system to recover EGH to hot water is proposed. The dew point temperature and the back pressure of exhaust gas are considered in design. The investment cost is calculated to evaluate the payback time of the proposed system. The findings clarify the effectiveness of utilizing exhaust gas heat of ICE to produce hot water, and simultaneously provide scientific data for constructing a new heat recovery system. The detailed methodology for designing the waste heat recovery heat exchanger is presented which could be reference guideline for the active researcher in the field of waste heat recovery application which is not cover in relevant articles in the open literature. In addition, the methodology for evaluating the economics of the waste recovery system is presented explicitly.

2. Methodology

2.1. Description of Engine and Hot Water Demand

The investigated luxury resort is located at the island which is a popular tourist place. It has the hotels and restaurants with five stars standard and a nice quiet beach. There are many travelers from all over the world are coming here. Beside the tourist demands, this place is used for organizing the national and international events. This resort has 300 high level rooms for 400–500 visitors at the present time, about 80–90% of the rooms are occupied daily and 100% in the holidays. It has 3 restaurants with about 1000 seats and has 01 canteen for about 200 staffs. This resort is on offshore islands, farther from the land, and the main transportation is by the ship, the electric power was generated by ICE combined with generator. The demand for hot water and steam are quite high, hot water is used mainly for bathing and steam is used for hot water heating and laundry. At the present, the hot water is supplied by the heat of boiler using DO fuel. So, if we utilize the waste heat from EGH of the ICE to replace a part or fully the steam from the boiler, the cost of fuel consumption will be decreased. The average exhaust gas temperature of ICE is about 400–500 °C. Current temperature is not enough for the engine cycle, but very suitable for another heat process. The specifications of ICE were investigated in this study as shown in Table 1. It has a close cooling system by an inside solution and outside by air cooling, so we cannot utilize this waste heat source. Thus, we only can utilize waste heat source from exhaust gas of this ICE. The used fuel of ICE is DO, the fuel consumption and exhaust gas temperature changed according to the ICE load as shown in Table 2. It is obvious that the DO fuel consumption increases as the engine load increases which results into increase in the exhaust gas temperature. This is because increase in the heat carrying capacity of exhaust gas with increase in the DO fuel consumption. The data presented in Table 2 show that the waste heat percentage is more at the higher engine load. So, increasing hot water demand at the resort could be satisfied by running the engine at the higher loads.

Table 1. Engine specifications.

Component	Specification
Manufacturer	Mitsubishi
Rated output	1530 KVA
Engine model	S16R-PTA, 4cycle, air cooler
Number of cylinders	16
Bore/stroke (mm)	170/180
Total displacement	65.4 L
Fuel consumption	273.5 L/h at 75% load
Maximum back pressure in exhaust gas system	12 in H_2O = 3000 Pa

Table 2. Fuel consumption and exhaust gas temperature according to the load.

Load (%)	Fuel DO Consumption (L/h)	Exhaust Gas Temperature (°C)
25	99.7	321
50	168.6	410
75	273.5	465
100	304.4	502
125	327.3	513

At the resort, the requirement of hot water temperature is ensured in range of 50–55 °C. The real data of daily hot water volume for 7 months including summer months (July and August), Christmas and new year (December and January) was collected. Figure 1 presents the daily hot water consumption volume of four months higher number of tourists. It can be observed that the hot water volume changed daily, monthly, and average demand is about 50–55 m^3/day. The hot water demand changes due to fluctuation in the number of tourists on daily and monthly basis. However, there are some peak days with the hot water volume increases to 80–90 m^3/day. From these collected results, the volume flow rate and its ratio with maximum hot water volume flow rate per day were determined, as shown in Table 3. The hot water demand of resort for three volumes namely, maximum, minimum and average is evaluated from the collected data of real system. Considering the average hot water demand as the baseline, the maximum and minimum hot water demands are 180% higher and 42% lower, respectively.

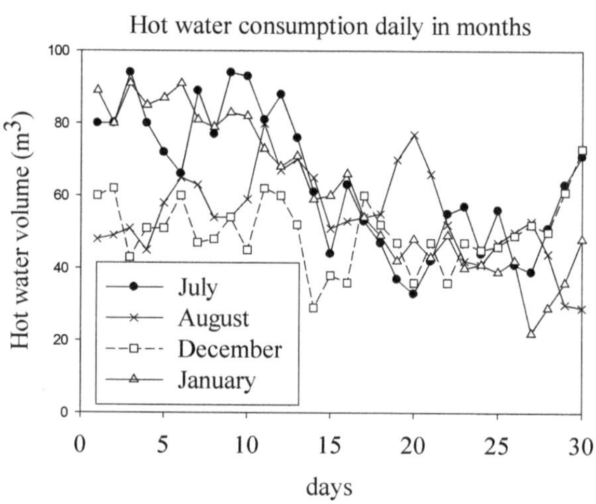

Figure 1. Daily hot water consumption volume for months with higher number of tourists.

Table 3. Volume flow rate and its ratio with maximum value per day.

Hot Water Volume	Flow (m^3/Day)	Ratio (%)
Average (V_{hwa}/d)	52	100
Maximum (V_{hwmax}/d)	94	180
Minimum (V_{hwmin}/d)	22	42

2.2. Description of Proposed System

The system diagram of proposed hot water supply system based on hot water demand and EGH recovery from ICE to produce hot water was proposed is shown in Figure 2. The EGH of ICE is not directly used to produce the hot water for resort in order to restrain the scale inside the tube exchanger. The EGH of ICE is exchanged with intermediate heat source (primary hot water), thus this system has two hot water circles. In the primary circle (closed circle), the fresh water is pumped into the primary heat exchanger and is heated by hot exhaust gas from ICE. This hot water becomes the heat source for the secondary circle. In the secondary circle (opened circle), the water is heated by primary hot water to temperature of 55 °C, then is pumped to the hot water system of the hotel and distributed to all rooms. The part of hot water with decreased temperature which is not used will be returned to the heat exchanger and reheated again. The make-up water is pumped to the system to compensate the part of used hot water.

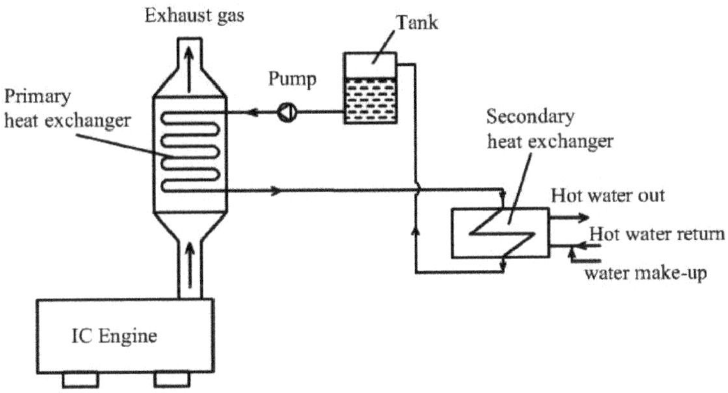

Figure 2. Proposed hot water supply system based on hot water demand and EGH of ICE.

3. Design of Proposed System

3.1. Specifications of Exhaust Gas

The load of ICE changed from 70 to 90% at the current time in a day, so the load of ICE at 80% is chosen for the heat exchanger design and correspondingly the exhaust gas temperature is $t_{eg\ in1} = 470$ °C. From the technical specifications of ICE, the mass flow rate of exhaust gas $G_{eg} = 6000$ kg/h. From the physical specifications of the used DO fuel, the sulfur concentration is $S = 0.445\%$. From the graph of flue gas dew point temperature [36], the dew point temperature of exhaust gas is $t_{dp} = 110$ °C. The reaction $SO_3 + H_2O = H_2SO_4$ in exhaust gas system which causes the corrosion in chimney or on the surface of the heat exchanger at low ICE load is prevented. We choose the minimum exhaust gas temperature $t_{eg\ out1} = 130$ °C. Thus, the total heat recovery from EGH of ICE can be determined:

$$Q_{eg} = G_{eg} \cdot c_{peg} \left(t_{eg\ in1} - t_{eg\ out1} \right) \tag{1}$$

where c_{peg} is the average specific heat of exhaust gas ($c_{peg} = 1.174$ kJ/kg °C). Thus, from Equation (1) $Q_{eg} = 665$ (kW).

3.2. Calculation of Volume Flow Rate of Hot Water

The hot water volume is always changed according to the demand. Therefore, the maximum volume flow rate of hot water per hour (V_{hwmax}/h) needs to be defined. Based on the make-up water volume into the system, the V_{hwmax}/h was collected for typical 10 days of the month with higher number of tourists as shown in the Figure 3. As can be seen, there is a time when V_{hwmax}/h is ranging from 16–20 pm daily. From these data, the ratio of average hot water volume flow rate per hour with its value per day was calculated as shown in the Figure 4. It can be observed that the ratio of V_{hwmax}/h at peak hour is nearly 6% from daily hot water volume which is suggested for other days in the year. Thus, from the maximum hot water volume as suggested in Table 3, $V_{hwmax}/d = 94$ m^3/day and the above ratio of 6% at peak hour, the V_{hwmax}/h was determined such that $V_{hwmax}/h = 94 \times 6/100 = 5.65$ m^3/h.

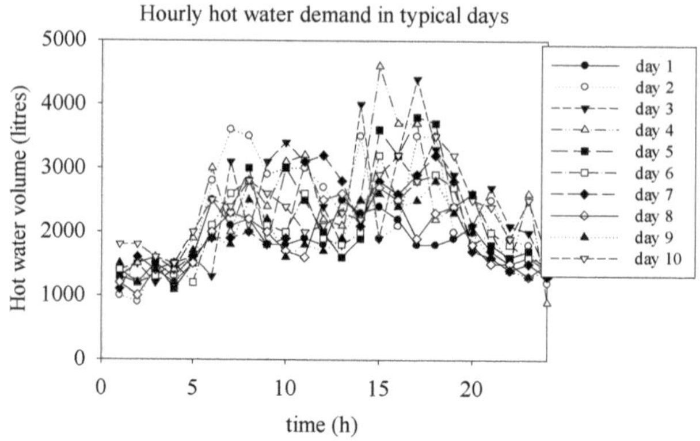

Figure 3. Hourly used hot water volume for typical days with higher number of tourists.

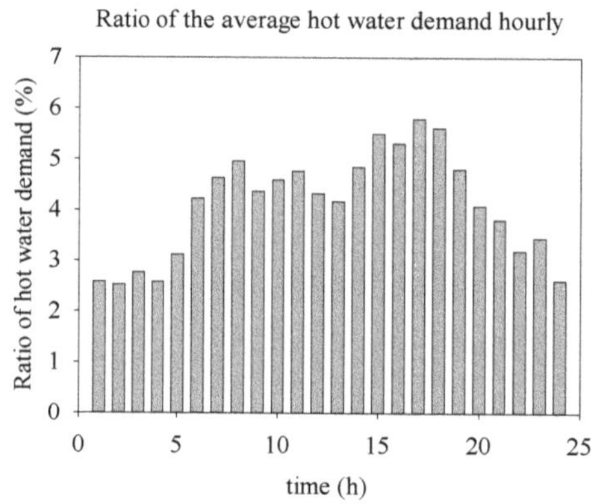

Figure 4. Ratio of average hot water volume flow rate per hour.

3.3. Maximum Production of Hot Water Volume

Figure 5 shows the diagram of separate hot water system. To ensure the continuous and adequate hot water supply, the hot water is always revolved in the system with the circulating hot water flow (V_{cc}) and requirement is $V_{cc} > V_{hwmax}/\mathrm{h}$. The circle factor ($n$) is introduced, with $n = \frac{V_{cc}}{V_{hwa}}$, we choose n = 2.75 in this study. From Table 3, the average hot water demand is V_{hwa} = 52 m^3/day = 2.2 m^3/h, the circular water flow can be defined V_{cc} = $V_{hwa} \times n$ = 6 m^3/h > V_{hwmax}/h = 5.65 m^3/h, that means the supplied hot water demand at peak hour will be ensured. From there, the volume flow rate of make-up water (V_{mu}) to the system is calculated, $V_{mu} = V_{hwa} = V_{cc} - V_{re}$, where V_{re} is the returned hot water flow rate. The temperature of returned hot water (t_{re}) is decreased because of occurred heat loss along the length of tube, this range temperature is set Δt = 15 °C. Thus, the t_{re} = 55 − 15 = 40 °C. The heat balance equation is established at the entrance of secondary heat exchanger:

$$V_{re} \cdot C_{pw}(t_{w\ re} - t_{w\ in2}) = C_{pw}(t_{w\ in2} - t_{w\ mu}) \quad (2)$$

where, C_{pw} is the specific heat of water, the water-in temperature of secondary heat exchanger ($t_{w\ in2}$), $t_{w\ mu}$ is the temperature of make-up water (t_{mu} = 25 °C at the investigated resort). Thus, from Equation (2) $t_{w\ in2}$ = 35 °C.

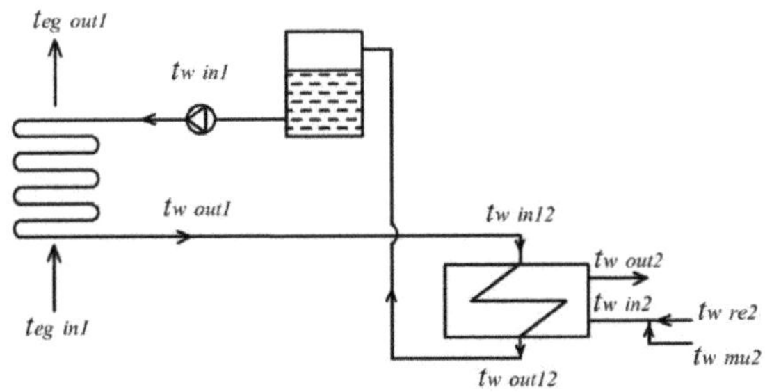

Figure 5. Diagram of separate hot water system.

The heat transfer rate from the exhaust gas to the water is presented by the balance equation:

$$\eta \cdot Q_{eg} = Q_{hw} = V_{hw} \cdot C_{pw}(t_{w\ out2} - t_{w\ in2}) \tag{3}$$

where Q_{hw} is the heat rate of water, V_{hw} is the maximum volume flow rate of hot water that could be produced, C_{pw} is the water specific heat (C_{pw} = 4.18 kJ/kg °C), η is the performance of the heat exchanger (η = 0.9).

Thus, from Equation (3) the maximum volume flow rate of hot water from the fully exhaust gas is evaluated as V_{hw} = 26 m^3/h at 80% load of ICE. This flow rate is larger than the flow rate of the circular hot water (V_{cc} = 6 m^3/h), it means that the EGH from ICE responses sufficiently to the demand of hot water at the resort.

3.4. Specifications of Heat Exchanger

The specifications of two heat exchangers were designed and chosen from the parameters of exhaust gas and hot water. Firstly, the tube diameter, tube spacing, the heat transfer coefficient (k_1), the length (L_1) and the width (W_1) are selected for the primary heat exchanger unit. Next, the area of heat exchanger (F_1), number of tube rows, number of tubes are calculated followed by calculation of the resistance. If the back pressure or resistance (Δp_1) is larger than the allowed back pressure ($[\Delta p_1]$) then L_1 and W_1 are re-adjusted. The steps are repeated and the k_1 is re-tested. The specifications for which the Δp_1 is lower than $[\Delta p_1]$ = 3000 Pa are accepted.

Calculate the resistance though the heat exchanger:

The parameters of the tube and fin such as the outside diameter of the tube (d_{out1}), horizontal spacing (s_1), fin spacing (s_f) and fin thickness (δ_f) were selected. The vertical spacing (s_2) is evaluated as $s_2 = \frac{\sqrt{3}}{2}s_1$, the slit between two fins (t_f) is calculated as $t_f = s_f - \delta_f$, the diameter of the fin (d_f) is calculated as $d_f = s_1 - t_f$, the height of the fin (h_f) is calculated as $h_f = 0.5\left(d_f - d_{out1}\right)$, number of the fin per 1 m length of tube (n_f) is evaluated as $n_f = \frac{1}{s_f}$.

The factor of the fin (CT [1]/136):

$$\varepsilon_f = 1 + \frac{d_f^2 - d_{out1}^2}{2 \times d_{in1} \times s_f} \tag{4}$$

The area without fin per 1 m length of tube (m^2):

$$F_0 = \pi \times d_{out1} \times t_f \times n_f \tag{5}$$

The area with fin per 1 m length of tube (m^2):

$$F_f = \frac{\pi}{2}\left(d_f^2 - d_{out1}^2\right)n_f \tag{6}$$

Total the outside area of fin-tube: $F_2 = F_0 + F_c$.

The equivalent diameter of the slits between tubes and fins through which gas flow passes (CT [1]/118):

$$d_{eqi} = \frac{F_0 d_{out1} + F_f\sqrt{\frac{F_f}{2n_f}}}{F_0 + F_f} \tag{7}$$

The hot water flow rate needs to be produced (V_{hw}) is known based on the hot water demand. Based on the heat balance equation of secondary heater, the primary hot water flow (V_{hw1}) is calculated as:

$$V_{hw1}c_{pw1}(t_{hw\ out1} - t_{hw\ in1})\eta = V_{hw2}c_{pw2}(t_{hw\ out2} - t_{hw\ in2}) \tag{8}$$

Based on the heat balance equation of primary heater, the exhaust gas-out temperature (t_{kr}) is calculated as:

$$G_{eg}c_{peg}(t_{eg\ in1} - t_{eg\ out1})\eta = V_{hw1}c_{pw1}(t_{hw\ out1} - t_{hw\ in1}) \tag{9}$$

The heat transfer coefficient (k_{F2}) was selected in the suitable range of k_{F2} = 40–80 $W/m^2 \cdot K$ based on common design.

Calculate the average temperature difference ($\overline{\Delta t}$) with $\Delta t_{max} = t_{egin1} - t_{hw\ out1}$ and $\Delta t_{min} = t_{egout1} - t_{hw\ in1}$:

$$\overline{\Delta t} = \frac{\Delta t_{max} - \Delta t_{min}}{\ln\frac{\Delta t_{max}}{\Delta t_{min}}} \tag{10}$$

Calculate the heat power required to heat the hot water at rate flow demand (Q_{hwd}):

$$Q_{hwd} = G_{eg}c_{peg}(t_{eg\ in1} - t_{eg\ out1}) \tag{11}$$

Calculate the waste heat power:

$$Q_{eg} = G_{eg}c_{peg}(t_{eg\ in1} - t_{air}) \tag{12}$$

Calculate the waste heat recovery efficiency:

$$\eta_{HR} = \frac{Q_{hwd}}{Q_{eg}} \times 100\ (\%) \tag{13}$$

Calculate the inside (F_1) and outside area (F_2) of fin-tube:

$$F_2 = \frac{Q_{hwd}}{k_{F2}\overline{\Delta t}} \text{ and } F_1 = \frac{F_2}{\varepsilon_f} \tag{14}$$

The total length of the heat transfer tube was calculated:

$$F_1 = \frac{F_1}{\pi d_{in1}} \tag{15}$$

The length of 1 tube (L_1) was as the width of the heat exchanger as $W_1 = L_1$. Numbers of rows of tube (z_1), number of tubes in 1 row (m_1), the height of the heat exchanger (H_1) were evaluated as, $z_1 = W_1/s_2$, $m_1 = \frac{l_1}{z_1 L_1}$, $H_1 = m_1 \cdot s_1$.

The velocity of exhaust gas flow through minimum cross section was defined (CT [1]/121):

$$\omega = \frac{\omega_0}{1 - \left(\frac{d_{out1}}{s_1} + \frac{2h\omega_f}{s_1 s_f}\right)} \tag{16}$$

where ω_0 is the velocity of exhaust gas-in, $\omega_0 = \frac{G_{eg}}{\rho f}$ and f is the area of cross section, $f = L \times W$.

From the average temperature exhaust gas (t_k) and $t_{eg} = \frac{t_{eg\ in1} - t_{eg\ out1}}{2}$, the heat conductive coefficient (λ) and the viscosity (v) of exhaust gas were known.

The number Re of exhaust gas was calculated:

$$\text{Re}_{eg} = \frac{\omega d_{eqi}}{v} \quad (17)$$

The resistance factor (ξ) for the staggered arrangement tubes with circle fins was defined (CT [1]/119):

$$\xi = 0.72 \times \text{Re}_{eg}^{-0.245} \left(\frac{s_1 - d_{out1}}{s_f} + 2 \right)^{0.9} \times \left(\frac{s_1 - d_{out1}}{d_{out1}} \right)^{0.9} \times \left(\frac{d_{eqi}}{d_{out1}} \right)^{0.9} \times \left(\frac{s_1 - d_{out1}}{s_2 - d_{out1}} \right)^{-0.1} \quad (18)$$

The resistance of friction (Δp_f), the local resistance at inlet and outlet (Δp_l), the total resistance of the system (Δp) were calculated, $\Delta p_1 = \Delta p_f + \Delta p_l$:

$$\Delta p_f = \xi \times \rho \times \frac{\omega^2}{2} \times z_1 \quad (19)$$

$$\Delta p_l = \xi \times \rho \times \frac{\omega^2}{2} \quad (20)$$

The total resistance of the system (Δp) has to be lower than the allowable resistance in the exhaust gas system of ICE [Δp] = 3000 Pa. Using the design procedure explained above, the specifications of primary and secondary heat exchangers are evaluated. The type of heat exchanger selected for primary case is tube and fin type heat exchanger. The water to be heated is flowing through tubes whereas the exhaust gas is flowing around the tubes. The flow paths for water and exhaust gas are presented in Figure 6a. The tube and fins in the primary heat exchanger are made up of steel. The inner and outer diameters of tube are 27.24 mm and 34 mm, respectively and total length of tube is 4.4 m with spacing of 7 mm. The fin thickness and spacing are designed as 1 mm and 4 mm, respectively. The arrangement and dimensions of fins are depicted in Figure 6b. The summary for specifications of primary heat exchanger is shown in Table 4. Similarly, the type of heat exchanger, the heat transfer coefficient (k_2), the area of heat exchanger (F_2), number of tube rows, number of tubes and the resistance are calculated for the secondary heat exchanger unit. The shell and tube heat exchanger are selected for secondary circle. The water heated through the exhaust gas is passing through the shell whereas, the fresh water is circulating through the tubes. The flow patterns for both water paths are shown in Figure 7. The secondary heat exchanger comprises of 3 shell passes and 36 tubes divided equally in 3 tube passes. The outer diameter and length of tube are evaluated as 19 mm and 740 mm, respectively. The summary for specifications of secondary heat exchanger is shown in Table 5.

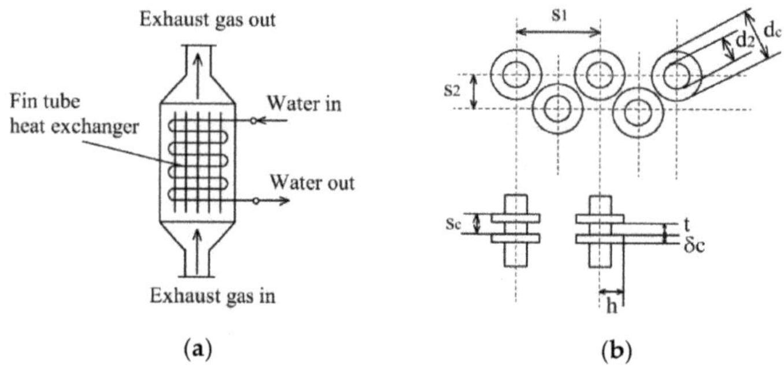

Figure 6. (**a**) Primary heat exchanger and (**b**) specifications of fin tube.

Table 4. Specifications of primary heat exchanger.

Component	Specification
Heat exchanger type	Liquid–gas (Figure 6a), liquid (water) flows inside the tube, gas flows outside the tube
Tube type	Fin–tube, staggered arrangement (Figure 6b)
Material of tube and fin	steel
Heat conductive coefficient	46 W/m^2–K
Outside–tube diameter	34 mm
Thickness of tube	3.38 mm
Horizontal spacing	7 mm
Fin spacing	4 mm
Fin thickness (δ_c)	1 mm
Total length of tube	4.4 m
Length of 1 tube	0.5 m
Fin factor	15
Resistance	1210 Pa

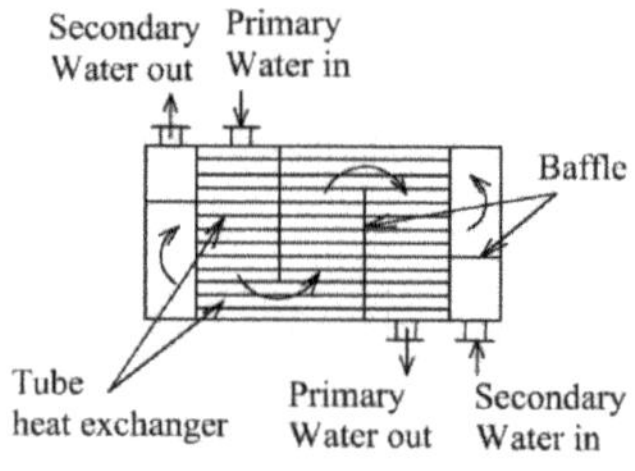

Figure 7. Secondary heat exchanger.

Table 5. Specifications of secondary heat exchanger.

Component	Specification
Heat exchanger type	shell–tube (Figure 7), the lower temperature water is heated by the higher temperature water.
Outside–tube diameter	19 mm
Length of tube	740 mm
Heat conductive coefficient	46 W/m^2–K
Number of pass in primary side	3 passes
Number of tubes	36 tubes
Number of pass in secondary side	3 passes

4. Evaluation of Proposed System

4.1. Performance Evaluation

Table 6 presents the exhaust gas temperature and heat recovered power for various engine loads and hot water flow rate. The exhaust gas temperature and the heat recovered power are changed due to the frequent change of ICE load to satisfy the demands during the actual operation. The exhaust gas temperature and heat recovered power are increasing with increase in the hot water demand and corresponding increase in engine load. To satisfy the increasing demand of hot water, it is required to increase the hot water flow rate. This means higher exhaust heat is needed to satisfy the increasing demand therefore, the engine load has increased to produce the exhaust gas with higher temperature. The higher portion of heat could be recovered by increase in the exhaust gas temperature due to increase in the heat transfer capacity. As can be seen that when the engine is running at 25% corresponding EGH of ICE is enough to produce the hot water demand of 6 m^3/h. The hot water demand changes with time at the resort which results into change in the heat recovered efficiency and the exhaust gas-out temperature. The calculated heat recovered efficiency and exhaust gas temperature corresponding to various hot water demand are shown in the Table 7. The waste recovery power and efficiency are lower at the lower hot water flow rate demand due to higher exhaust gas temperature (higher exhaust waste heat). This means larger portion of exhaust heat is wasted as the demand is not high. However, with increase in the hot water flow rate demand, the exhaust gas temperature reduces which shows significant increase in the heat recovery power and efficiency. As can be observed that the hot water demand is in range 2–6 m^3/h, corresponding to the heat recovered power is lowered by 25% and the exhaust gas-out temperature is still rather high t_{eg} > 390 °C, it means that there are still large amounts of excess heat that has not been fully used in exhaust gas of ICE. If the flow rate demand of hot water increases to V_{hw} = 25 m^3/h in the future, the exhaust gas-out temperature is t_{eg} = 140 °C still higher than the dew point temperature of the exhaust gas t_{dp} = 130 °C. Therefore, the condensation cannot occur in the exhaust gas system, the safety of operation is ensuring. The higher demand of hot water could be satisfied at the higher ICE load. The hot water flow rates of 26 m^3/h could be achieved at the ICE load of 80% and that could be further increased to 29 m^3/h at the ICE full load. The heat recovery power and efficiency of 155 kW and 23%, respectively are achieved for the proposed system at the regular hot water flow rate demand of 6 m^3/h. At the maximum hot water flow rate demand of 25 m^3/h, the highest heat recovery power of 645 kW and highest heat recovery efficiency of 97% could be achieved.

Table 6. Temperature and heat power of exhaust gas, flow rate of produced hot water change on ICE load.

Load (%)	Exhaust Gas Temperature (°C)	Exhaust Gas Heat Power (kW)	Hot Water Flow Rate (m^3/h)
25	321	374	14
50	410	548	21
75	465	655	25
80	470	665	26
100	502	728	28
125	513	749	29

Table 7. The heat recovery efficiency and the exhaust gas-out temperature change on the hot water demand.

Hot Water Demand Gnn (m^3/h)	Heat Recovered Power Q (kW)	Efficiency HS (%)	Exhaust Gas-Out Temperature t_{kr} (°C)
2	52	8	444
4	103	16	417
6	155	23	391
8	206	31	364
10	258	39	338
15	387	58	272
20	516	78	206
25	645	97	140

4.2. Economic Evaluation

Figure 8 presents the daily DO volume consumption of boiler for 7 months. Depending on the demand, the DO fuel consumption changes daily, higher DO fuel consumption corresponding to higher demand and vice-versa. From the collected data as presented in Figure 8, the average volume of DO consumption is evaluated about 600 L/day, among which about 30% was used in the production of steam for laundry, equivalent to 200 L/day. Thus, the average volume of DO consumption corresponding to the supplied hot water is 400 L/day. Based on this data, the annual savings (*E*) for reducing the volume of DO fuel consumption was evaluated in the Table 8. The hot water supply system using EGH of ICE could eliminate the existing boiler system which results into fuel savings utilized to run the existing system. The annual saving of 110,880 $/year could be achieved using the proposed system. The total cost of new system (C) at the current time was estimated as shown in Table 9. The payback time (*T*) was defined:

$$T = \frac{\ln \frac{E}{E - i * c}}{\ln(1+i)} * 12 \text{ months} \quad (21)$$

where i is the compound rate yearly, taken i = 20%/year. Thus, the payback time is evaluated as T = 09 months from Equation (21).

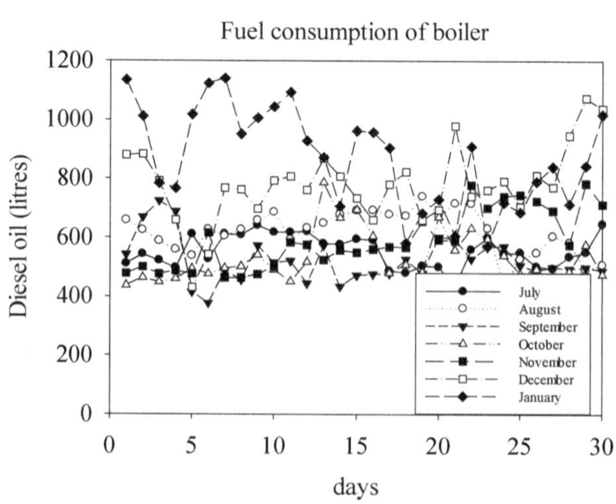

Figure 8. Daily DO volume consumption of boiler.

Table 8. Annual savings of recovered EGH.

Parameter	Result
Average DO consumption	400 L/day
Cost of DO	0.77 $/L
Cost of hot water per day	400 L/day × 0.77 $/L = 308 $/day
Monthly saving	308 $/day × 30 days = 9240 $/month
Annual saving	12 months × 9240 $/month = 110,880 $/year

Table 9. Cost of the new hot water system.

Parameter	Result
The primary heat exchanger	20,000 $/unit
The secondary heat exchanger	15,000 $/unit
The tube and valves	15,000 $
The auxiliary equipment	8000 $
Shipping and maintenance	12,000 $
Total of cost	70,000 $

5. Conclusions

In this study, the energy savings and economic performances of the EGH recovery system from ICE was analyzed and evaluated for supplying hot water to the resort. The new heat exchanger system to recover exhaust gas heat from ICE and utilize to heat the water was proposed considering the dew point temperature and back pressure of exhaust gas system. The heat recovery and economic efficiency were calculated and evaluated. The regular hot water demand for the resort is ranging from 2–6 m^3/h which could be fully satisfied by the proposed system which supplies the hot water with the flow rate of 14 m^3/h at 25% ICE load and the hot water flow rate of 26 m^3/h at 80% load. These flow rates have sufficient and excess capacity to satisfy the current hot water demand of 6 m^3/h at temperature of 55 °C to the resort. In case the demand of hot water supply increases in future due to higher number of tourists then it could be achievable by the proposed system up to maximum 29 m^3/h with increasing ICE load. Despite of increasing hot water demand and recovering larger amount of waste heat from the exhaust gas using the proposed system but still the exhaust gas temperature is lower than the dew point temperature which depicts the safe operation without condensation and corrosion in the exhaust system. At the hot water flow rate demand of 6 m^3/h, the proposed system achieves the heat recovery power of 155 kW and heat recovery efficiency of 23%. And the heat recovery power and efficiency are achieved maximum up to 645 kW and 97%, respectively at the maximum hot water flow rate demand of 25 m^3/h. The proposed system reduces the daily fuel consumption of DO up to 400 L/day which results into the annual saving of 110,880 $/year and the calculated payback time for the proposed system is 9 months. Apart from excellent performance and economics, the proposed system could reduce the CO_2 emission by saving fuel consumption which results into reduced global warming and environmental degradation. The results indicate that the potential energy savings of recovered EGH system from ICE is significantly large. The proposed system has effectively utilized the energy economically and the generated scientific data could be used in the construction of new heat recovery system. In the proposed system still larger amount of waste heat is remained un-utilized therefore, as the future direction, the designed new heat exchanger could be optimized based on heat transfer performances to make the efficiency utilization of remaining waste heat which could improve the economics of the proposed heat recovery system. In addition, various statistical approaches such as, Taguchi analysis, grey relational analysis and analysis of variance could be employed to evaluate the most influencing factors to the performances of designed waste heat recovery heat exchanger in the future works.

Author Contributions: Conceptualization, N.-H.N. and D.-Y.L.; methodology, N.-H.N. and D.-Y.L.; formal analysis, N.-H.N. and D.-Y.L.; investigation, N.-H.N. and D.-Y.L.; resources, N.-H.N. and D.-Y.L.; data curation, N.-H.N. and D.-Y.L.; writing—original draft preparation, N.-H.N., D.-Y.L., K.S.G. and M.-Y.L.; writing—review and editing, N.-H.N., D.-Y.L., K.S.G. and M.-Y.L.; visualization, N.-H.N. and D.-Y.L.; supervision, M.-Y.L. and D.-Y.L.; project administration, M.-Y.L. and D.-Y.L.; funding acquisition, M.-Y.L. and D.-Y.L. All authors have read and agreed to the published version of the manuscript.

Funding: This research received no external funding.

Institutional Review Board Statement: Not applicable.

Informed Consent Statement: Not applicable.

Data Availability Statement: The data presented in this study will be available on request to the corresponding author.

Acknowledgments: This work was supported by the Dong-A University research fund.

Conflicts of Interest: The authors declare no conflict of interest.

References

1. Lee, M.Y.; Seo, J.H.; Lee, H.S.; Garud, K.S. Power generation, efficiency and thermal stress of thermoelectric module with leg geometry, material, segmentation and two-stage arrangement. *Symmetry* **2020**, *12*, 786. [CrossRef]
2. Chung, J.H.; Jeong, S. Experimental Performance Analysis of a Small Thermoelectric System Applicable to Real-Time PCR Devices. *Symmetry* **2020**, *12*, 1963. [CrossRef]
3. Rolf, B.; Henrick, N.; Judy, W. *Combined-Cycle Gas and Steam Turbine Power Plants*; PennWell Corp: Oklahoma, OK, USA, 1999.
4. Saidur, R.; Rezaei, M.; Muzammil, W.K.; Hassan, M.H.; Paria, S.; Hasanuzzaman, M. Technologies to recover exhaust heat from internal combustion engines. *Renew. Sustain. Energy Rev.* **2012**, *16*, 5649–5659. [CrossRef]
5. Wang, T.; Zhang, Y.; Peng, Z.; Shu, G. A review of researches on thermal exhaust heat recovery with Rankine cycle. *Renew. Sustain. Energy Rev.* **2011**, *15*, 2862–2871. [CrossRef]
6. Onovwiona, H.I.; Ugursal, V.I. Residential cogeneration systems: Review of the current technology. *Renew. Sustain. Energy Rev.* **2006**, *10*, 389–431. [CrossRef]
7. Beith, R. *Small and Micro Combined Heat and Power (CHP) Systems*; Elsevier: Cambridge, UK, 2011.
8. European Commission. *Combined Heat and Power Generation CHP*; Directorate-General for Energy Save II Programme. Semantic Scholar, European Commission, 2001. Available online: https://www.semanticscholar.org/paper/DIRECTORATE-GENERAL-FOR-ENERGY-SAVE-II-Programme-by-chillers/655f1a38e94d533beba346e0f522ffecd2aceeff (accessed on 22 March 2021).
9. U.S. Department of Energy, Office of Energy Efficiency and Renewable Energy. *Review of Combined Heat and Power Technologies*; U.S. Department of Energy, Office of Energy Efficiency and Renewable Energy: Washington, DC, USA, 1999.
10. Bari, S.; Hossain, S.N. Waste heat recovery from a diesel engine using shell and tube heat Exchanger. *Appl. Therm. Eng.* **2013**, *61*, 355–363. [CrossRef]
11. Hossain, S.N.; Bari, S. Waste heat recovery from the exhaust of a diesel generator using Rankine Cycle. *Energy Convers. Manag.* **2013**, *75*, 141–151. [CrossRef]
12. Pandiyarajan, V.; Pandian, M.C.; Malan, E.; Velraj, R.; Seeniraj, R.V. Experimental investigation on heat recovery from diesel engine exhaust using finned shell and tube heat exchanger and thermal storage system. *Appl. Energy* **2011**, *88*, 77–87. [CrossRef]
13. Pandiyarajan, V.; Chinnappandian, M.; Raghavan, V.; Velraj, R. Second law analysis of a diesel engine waste heat recovery with a combined sensible and latent heat storage system. *Energy Policy* **2011**, *39*, 6011–6020. [CrossRef]
14. He, M.; Zhang, X.; Zeng, K.; Gao, K. A combined thermodynamic cycle used for waste heat recovery of internal combustion engine. *Energy* **2011**, *36*, 6821–6829. [CrossRef]
15. Kuhn, V.; Klemes, J.; Bulatov, I. MicroCHP: Overview of selected technologies, products and field test results. *Appl. Therm. Eng.* **2008**, *28*, 2039–2048. [CrossRef]
16. Peng, Z.; Wang, T.; He, Y.; Yang, X.; Lu, L. Analysis of environmental and economic benefits of integrated Exhaust Energy Recovery (EER) for vehicles. *Appl. Energy* **2013**, *105*, 238–243. [CrossRef]
17. Pertti, K.; Tuomo, E.; Lisa, W.; Jorma, H.; Juhani, L. Temperature optimisation of a diesel engine using exhaust gas heat recovery and thermal energy storage (diesel engine with thermal energy storage). *Appl. Therm. Eng.* **2010**, *30*, 631–638.
18. Manzela, A.A.; Hanriot, S.M.; Cabezas-Gomez, L.; Sodre, J.R. Using engine exhaust gas as energy source for an absorption refrigeration system. *Appl. Energy* **2010**, *87*, 1141–1148. [CrossRef]
19. Talbi, M.; Agnew, B. Energy recovery from diesel engine exhaust gases for performance enhancement and air conditioning. *Appl. Therm. Eng.* **2002**, *22*, 693–702. [CrossRef]
20. Liu, J.P.; Fu, J.Q.; Ren, C.Q.; Wang, L.J.; Xu, Z.X.; Deng, B.L. Comparison and analysis of engine exhaust gas energy recovery potential through various bottom cycles. *Appl. Therm. Eng.* **2013**, *50*, 1219–1234. [CrossRef]

21. Wang, E.H.; Zhang, H.G.; Fan, B.Y.; Ouyang, M.G.; Zhao, Y.; Mu, Q.H. Study of working fluid selection of organic Rankine cycle (ORC) for engine waste heat recovery. *Energy* **2011**, *36*, 3406–3418. [CrossRef]
22. Domingues, A.; Santos, H.; Costa, M. Analysis of vehicle exhaust waste heat recovery potential using a Rankine cycle. *Energy* **2013**, *49*, 71–85. [CrossRef]
23. Wang, T.; Zhang, Y.; Zhang, J.; Shu, G.; Peng, Z. Analysis of recoverable exhaust energy from a light-duty gasoline engine. *Appl. Therm. Eng.* **2013**, *53*, 414–419. [CrossRef]
24. Yu, C.; Chau, K.T. Thermoelectric automotive waste heat energy recovery using maximum power point tracking. *Energy Convers. Manag.* **2009**, *50*, 1506–1512. [CrossRef]
25. Shi, X.; Che, D.; Agnew, B.; Gao, J. An investigation of the performance of compact heat exchanger for latent heat recovery from exhaust flue gases. *Int. J. Heat Mass Transf.* **2011**, *54*, 606–615. [CrossRef]
26. Onovwiona, H.I.; Ugursal, V.I.; Fung, A.S. Modeling of internal combustion engine based cogeneration systems for residential applications. *Appl. Therm. Eng.* **2007**, *27*, 848–861. [CrossRef]
27. Jia, L.L.; Zhang, R.; Zhang, X.; Ma, Z.X.; Liu, F.G. Experimental analysis of a novel gas-engine-driven heat pump (GEHP) system for combined cooling and hot-water supply. *Int. J. Refrig.* **2020**, *118*, 84–92. [CrossRef]
28. Liu, L.; Fu, L.; Jiang, Y. Application of an exhaust heat recovery system for domestic hot water. *Energy* **2010**, *35*, 1476–1481. [CrossRef]
29. Butrymowicz, D.; Gagan, J.; Łukaszuk, M.; Śmierciew, K.; Pawluczuk, A.; Zieliński, T.; Kędzierski, M. Experimental validation of new approach for waste heat recovery from combustion engine for cooling and heating demands from combustion engine for maritime applications. *J. Clean. Prod.* **2021**, *290*, 125206. [CrossRef]
30. Garud, K.S.; Seo, J.H.; Cho, C.P.; Lee, M.Y. Artificial neural network and adaptive neuro-fuzzy interface system modelling to predict thermal performances of thermoelectric generator for waste heat recovery. *Symmetry* **2020**, *12*, 259. [CrossRef]
31. Garud, K.S.; Seo, J.H.; Patil, M.S.; Bang, Y.M.; Pyo, Y.D.; Cho, C.P.; Lee, M.Y. Thermal–electrical–structural performances of hot heat exchanger with different internal fins of thermoelectric generator for low power generation application. *J. Therm. Anal. Calorim.* **2021**, *143*, 387–419. [CrossRef]
32. Hervás-Blasco, E.; Navarro-Peris, E.; Corberán, J.M. Optimal design and operation of a central domestic hot water heat pump system for a group of dwellings employing low temperature waste heat as a source. *Energy* **2019**, *188*, 115979. [CrossRef]
33. Ahmadi, G.; Toghraie, D.; Akbari, O.A. Efficiency improvement of a steam power plant through solar repowering. *Int. J. Exergy* **2017**, *22*, 158–182. [CrossRef]
34. Ahmadi, G.R.; Toghraie, D. Parallel feed water heating repowering of a 200 MW steam power plant. *J. Power Technol.* **2015**, *95*, 288–301.
35. Ahmadi, G.; Toghraie, D.; Azimian, A.; Akbari, O.A. Evaluation of synchronous execution of full repowering and solar assisting in a 200 MW steam power plant, a case study. *Appl. Therm. Eng.* **2017**, *112*, 111–123. [CrossRef]
36. Muller, M.R.; Simek, M.; Mak, J.; Mitrovic, B. *Modern Industrial Assessments: A Training Manual*; Rutgers University Press: New Brunswick, NJ, USA, 2001.

symmetry

MDPI

Article

Experimental Study on Heating Performances of Integrated Battery and HVAC System with Serial and Parallel Circuits for Electric Vehicle

Taek-Kyu Lim [1,†], Kunal Sandip Garud [2,†], Jae-Hyeong Seo [2], Moo-Yeon Lee [2,*] and Dong-Yeon Lee [3,*]

1 Thermal Management R&D Center, KATECH, 303 Pungse-ro, Pungse-Myun, Cheonan 31214, Korea; tklim@katech.re.kr
2 Department of Mechanical Engineering, Dong-A University, 37 Nakdong-Daero 550, Saha-gu, Busan 49315, Korea; 1876936@donga.ac.kr (K.S.G.); cheonchw@donga.ac.kr (J.-H.S.)
3 Department of Robotics and Intelligent Machine, Yeungnam University, 280 Daehak-Ro, Gyeongsan, Gyeongbuk 38541, Korea
* Correspondence: mylee@dau.ac.kr (M.-Y.L.); dylee@ynu.ac.kr (D.-Y.L.); Tel.: +82-51-200-7642 (M.-Y.L.)
† These authors contributed equally to this work.

Abstract: The objective of the present study is to conduct experiments for investigating heating performances of integrated system with serial and parallel circuits for battery and heating ventilation and air conditioning system (HVAC) of electric vehicles under various operating conditions. In addition, the artificial neural network (ANN) model is proposed to accurately predict the heating performances of integrated system with serial and parallel circuits for battery and HVAC. A test bench of integrated system with serial and parallel circuits has been developed for establishing the trade-off between battery heating and HVAC heating. The heating performances namely, battery out temperature, battery temperature rise rate, battery heating capacity, HVAC heating capacity and total heating capacity are evaluated experimentally for the integrated system with serial and parallel circuits. The behavior of various heating performances is evaluated under influence of flow rate and heater power. Battery out temperature reaches 40 °C within 10 min with rise rate of 2.17 °C/min for the integrated system with serial circuit and that within 20 min with rise rate of 1.22 °C/min for the integrated system with parallel circuit. Integrated system with serial circuit shows higher HVAC heating capacity than integrated system with parallel circuit which are 5726.33 W and 3869.15 W, respectively. ANN model with back-propagation algorithm, Levenberg-Marquardt training variant, Tan-sigmoidal transfer function and 20 hidden neurons presents the accurate prediction of heating performances of the integrated system with serial and parallel circuits for battery and HVAC.

Keywords: battery; heating performance; HVAC; parallel; serial

check for updates

Citation: Lim, T.-K.; Garud, K.S.; Seo, J.-H.; Lee, M.-Y.; Lee, D.-Y. Experimental Study on Heating Performances of Integrated Battery and HVAC System with Serial and Parallel Circuits for Electric Vehicle. *Symmetry* **2021**, *13*, 93. https://doi.org/10.3390/sym13010093

Received: 4 December 2020
Accepted: 4 January 2021
Published: 7 January 2021

1. Introduction

In recent times, increasing demand of high energy efficiency and zero emission has resulted into shifting the major means of transportation towards the electric vehicle [1]. The low driving range and battery life are two major hurdles in the development of electric vehicles. The cabin is heated in winter using electric energy of the battery, which results into reduction in driving range of vehicle [2]. The charge-discharge performances of the batteries are degrading significantly as the temperature reduces. In cold weather conditions, to maintain the battery performance and battery life, effective battery thermal management in form of preheating of battery is essential [3]. Numerous studies have been conducted focusing on cabin heating and battery heating in cold weather conditions to improve the driving range and battery performance.

Guo et al. presented effective heating for battery cell and battery pack using echelon internal heating strategy [4]. Ruan et al. developed an optimal internal heating strategy for rapid battery heating [5]. Lei et al. suggested an intermittent self-heating lithium-ion battery method for heating of battery with temperature uniformity [6]. Shang et al. proposed buck-boost conversion based high frequency alternating current heater for battery heating in low temperature conditions. As the AC heating frequency increases, the heating speed and efficiency improve, due to lithium-ion transport and increase in heat generation of ohmic resistance [7]. Fan et al. concluded that the discharge rate has negligible effect on the heating performance of battery thermal management system compare with external heating source. In addition, a higher mass flow rate of the heating medium gives better heating performance [8]. Delos Reyes et al. investigated the behavior in driving ranges of Mitsubishi i-MiEV and Nissan Leaf for the ambient temperature variation in a range of 20 to −15 °C [9].

Positive temperature coefficient (PTC) heaters are widely used for cabin heating, however, they consume more energy. Therefore, heat pumps have been used as the replacement of PTC heater for cabin heating [10]. Zhang et al. showed coefficient of performance (COP) of 1.25 and improvement of 57.7% in heating capacity using economized vapor injection heat pump system [11]. Cho et al. proposed a coolant source heat pump which uses waste heat from electric devices for heating a passengers' compartment of electric bus [12]. Qin et al. presented that the air source heat pump with refrigerant injection shows enhancement in the heating capacity compared with the conventional air source heat pump for electric vehicles [13–15]. Ahn et al. investigated heating performances of air source heat pump, waste heat pump and dual source (air + waste heat) heat pump for electric vehicle, and concluded that the dual source heat pump shows superior heating performances compared with air source and waste heat pumps [16]. Lee et al. proposed an R744 based stack coolant heat pump, which attains a heating capacity of 5.0 kW at an ambient temperature of −20 °C [17]. Shi et al. suggested R32 based economized vapor injection heat pump system for temperature range of −2 to 15 °C, and showed higher coefficient of performance compare with conventional single stage heat pump [18]. Jung et al. showed that the single injection heat pump with optimum port angle of 440° and dual injection heat pump with optimum port angles of 535°/355° present enhancement of 7.5% and 9.8%, respectively, in coefficient of performance, compared with non-injection heat pump at ambient temperature of −10 °C [19]. Patil et al. proposed a 2.0 kW burner that shows a maximum efficiency of 96.7% for the cabin heating of an electric vehicle [20]. Zhang et al. showed that the heat pump system with desiccant reduces cabin heat load and power consumption by 42% and 38%, respectively, at an ambient temperature of −20 °C, compared to traditional heat pump system [21]. Choi et al. investigated the heating performances of vapor injection heat pump system for cabin heating of electric vehicle in cold weather conditions [22,23]. Ahn et al. showed that the dual evaporator heat pump has 62% higher heating coefficient of performance for cabin in electric vehicle compare with conventional heat pump [24]. Lee et al. proposed a mobile heat pump which uses waste heat of electric devices for heating in an electric bus. The heating coefficient of performance of proposed heat pump is evaluated as 2.4 [25]. Liu et al. investigated heating performances of propane-based heat pump system for cabin heating in electric vehicles and found that the proposed system shows superior heating performance above the ambient temperature of −10 °C [26]. Li et al. compared heating performances of an R134a based heat pump and an R1234yf based heat pump, and showed that the R1234yf based heat pump is a potential candidate for the replacement of the R134a based heat pump for cabin heating in cold weather conditions [27]. Bellocchi et al. developed a heat pump with a regenerative heat exchanger, which reduces power consumption by 17–52% and reduces the decrease in driving range up to 6% for electric vehicles [28]. Lee et al. proposed air source heat pump system with a heating coefficient of 3.26 and a heating capacity of 3.10 kW at an ambient temperature of −10 °C for cabin heating in electric vehicle [29]. Lee et al. have experimentally investigated the performance characteristics of heat pump system integrated with a high pressure side chiller under cold and hot weather conditions for light duty commercial electric vehicles [30]. Jeffs et al. integrated five different heat sources with heat pump system for efficient heating of battery and cabin. An energy saving of 14.8% was achieved with a heat pump system integrated with different heat sources [31]. Further, Jeff et al. proposed an optimal strategy for tradeoff heating between battery and cabin. The optimal strategy enables 6.2% of improvement in range for no battery heating and 5.5% of improvement in cabin comfort for full battery heating [32].

From the literature review, there are very few studies that discuss combined cabin and battery heating. Therefore, the objective of the present study is to develop an integrated system with serial and parallel circuits to enable the trade-off between battery heating and heating ventilation and air conditioning system (HVAC) (cabin) heating. The battery and HVAC heating performances, namely, battery out temperature, battery temperature rise rate, battery heating capacity, HVAC (cabin) heating capacity and total heating capacity, are compared experimentally for the integrated system with serial and parallel circuits under the influence of heater power and flow rate. In addition, an ANN model is developed in the present study, to accurately predict the battery and HVAC heating performances.

2. Experimental Method

The experimental set-up of the integrated system with serial and parallel circuits is shown in Figure 1. To control the experimental ambient conditions, the experimental set-up is housed inside the psychrometric calorimeter which is controlled using PID controller. The experimental set-up comprises of serial and parallel circuits with components namely, battery, HVAC heater core, heater, water pump, working fluid tank, radiator, temperature sensors, flow meters and valves. The specifications of various components of experimental set-up are presented in Table 1. Both the circuits are operated using three valves, named valve1, valve2 and valve3. For the integrated system with serial circuit, valve1 and valve3 are cutoff and only valve2 is operated. Whereas, for the integrated system with parallel circuit, valve2 is cutoff and valve1 and valve3 are operated. The valve1 supplies working fluid for the battery heating and valve3 supplies working fluid for HVAC heating in the integrated system with parallel circuit. The battery is not the actual battery but the mimic of GM Volt (2010) battery model whose specifications are: weight = 198 kg, specific heat = 143 J/kg K and battery capacity = 16 kWh. The total working fluid flow rate of 24 L/min is divided for battery heating and HVAC heating in the integrated system with parallel circuit using valve1 and valve3. For the integrated system with parallel circuit, the flow ratio term is defined which is the ratio of battery flow rate to HVAC flow rate. In the case of an integrated system with serial circuit, full working fluid flow rate of 24 L/min is either supplied for battery heating or HVAC heating using valve2. The working of the integrated system with serial and parallel circuits involves a supply of working fluid using water pumped to a heater, where it is heated to higher temperature. From the heater, the heated working fluid is divided between the battery and HVAC in the case of an integrated system with a parallel circuit, and totally supplied for either the battery or HVAC in the case of an integrated system with serial circuit. The collected heated working fluid in tank is transferred through the radiator where it is cooled and again transferred to the heater using a water pump. To analyze the heating performances of the battery and HVAC, two modes (serial and parallel) of operation are employed. In addition, the heater power is varied as 2 kW, 4 kW and 6 kW in both the modes, and the flow ratio is varied as 2/8, 3/7, 5/5, 7/3 and 8/2 in a parallel mode of operation, to investigate the behavior of battery and HVAC heating performances. Battery out temperature, battery temperature rise rate, battery heating capacity, HVAC heating capacity and total heating capacity are investigated under various operating conditions. The accuracy of various experimental devices and instruments is shown in Table 2. Additionally, Figure 1b shows the picture of an experimental set-up of the integrated system with serial and parallel circuits.

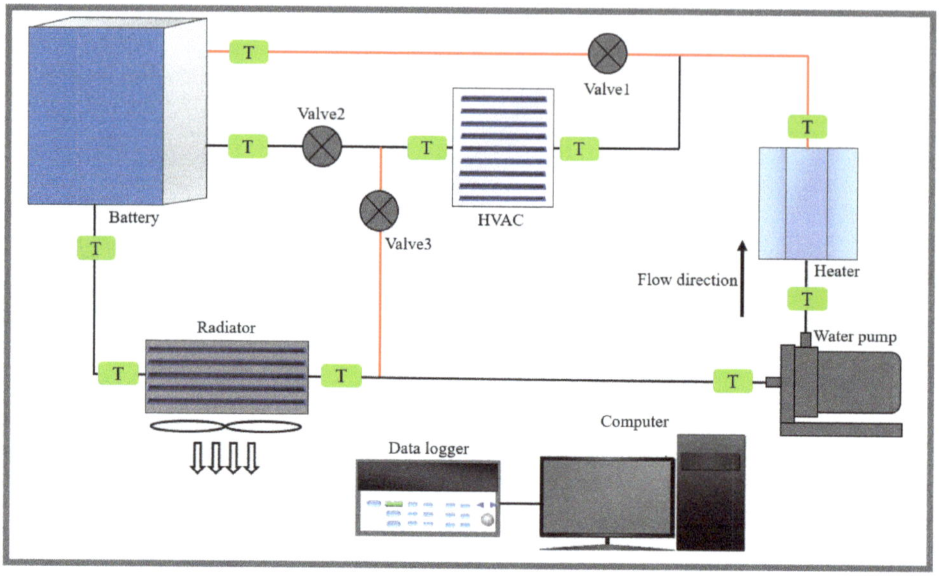

(**a**) Schematic diagram

(**b**) Picture

Figure 1. (**a**) Schematic diagram and (**b**) picture for experimental set-up of integrated system with serial and parallel circuits.

Table 1. The specifications of components of experimental set-up.

Component	Specification
Pipe	Material: stainless
Heater	Type: sheath
Water pump	Max flow: 25 L/min Max head: 25 m Power voltage: 24 VCD

Table 1. *Cont.*

Component	Specification
Radiator	Applied vehicle: GM Volt Core size: 147 × 206 × 28 mm
HVAC	Applied vehicle: Kona Core size: 152 × 222 × 26 mm
Heater	510 V, 11.8 A

Table 2. Accuracy of various experimental devices and instruments.

Device/Instrument	Accuracy
T-type thermocouple	0.75%
DAQ	-200 °C ≤ TS ≤ -100 °C, ± (0.10% of reading) -100 °C ≤ TS ≤ 400 °C, ± (0.10% of reading)
Flow rate sensor	±1.50%

The poor calibration, instrumental errors, positional errors of probes, environmental error etc., are responsible for the uncertainty in the experimental results [33]. Therefore, the uncertainty analysis is conducted on the experimental results of the integrated system with serial and parallel circuits, to ensure the accuracy and reliability of the experimental test results [34]. The errors in experimental data of temperature and flow rate measurements cause uncertainties in the heating performances of battery and HVAC of the integrated system with serial and parallel circuits. The uncertainties in the output parameter due to uncertainties in the input parameters are calculated using Equation (1) [35]. The uncertainties in various parameters for the experimental study on integrated system with serial and parallel circuits could be calculated using the concept of linear fraction approximation [36].

$$U_F = [(\frac{\partial F}{\partial X_1} U_1)^2 + (\frac{\partial F}{\partial X_2} U_2)^2 + \ldots + (\frac{\partial F}{\partial X_n} U_n)^2]^{\frac{1}{2}} \quad (1)$$

Here, $X_1, X_2, \ldots X_n$ are the input parameters, $U_1, U_2, \ldots U_n$ are the uncertainties in the input parameters, F is the output parameter and U_F is the uncertainty in the output parameter. Using Equation (1), the uncertainties in flow rate, battery out temperature and HVAC heating capacity are calculated as 1.50%, 0.75% and 1.68%, respectively.

3. Artificial Neural Network (ANN)

In recent times, ANNs are used for the performance prediction, forecasting, modeling, simulation and optimization of various physical systems. The working principle of ANN is based on a biological neural network [37]. When the input and output parameters with a larger data set is interrelated in a nonlinear relationship with each other, in those cases the ANN is the efficient tool to relate these parameters with less complexity than the conventional mathematical techniques [38]. The structure of the ANN model includes three layers namely, an input layer, a hidden layer and an output layer [39]. Each layer comprises a suitable number of neurons: the number of neurons in the input layer is equal to the number of input parameters, the number of neurons in the output layer is equal to the number output parameters and the number of neurons in the hidden layer is adjusted based on the training error [40]. Neurons in one layer are connected with the other layer's neurons using weights and two neurons are connected by a single weight value [41]. The structure of the ANN model including layers, neurons and weights is trained using a training algorithm. The construction of training algorithm includes training variant and transfer function [39]. The maximum training error and maximum number of epochs are set as stopping criteria for training. The structure of the ANN model is trained using the selected training algorithm until the desired output or permissible error is obtained [42]. In the training process, the weights assigned between neurons are adjusted to achieve the desired output. If the desired output or the permissible errors is not achieved, then the different combination of training algorithm, training variant, transfer function, number of hidden layers and hidden neurons is used for further training. The trained ANN model that shows a higher prediction accuracy, with a predicted output closer to the actual results, is suggested as the optimum ANN model.

In the present study, an ANN model with various algorithms is developed for the prediction of battery and HVAC heating performances of the integrated system with a serial circuit and with a parallel circuit. The formulated structure of an ANN model for the integrated system with serial and parallel circuits is presented in Figure 2. The aim of the development of ANN models is to predict the battery and HVAC heating performances, which are indicated by battery out temperature and HVAC temperature difference. The battery out temperature and HVAC temperature difference are most affected by heater power and flow rate, which vary in real time. Therefore, the ANN model for the integrated system with serial and parallel circuits is formulated to predict the battery out temperature and HVAC temperature difference as the output parameters, for various conditions of heater power, flow rate and time as the input parameters. The training algorithm comprises of back-propagation algorithm, Levenberg-Marquardt (LM) training variant, Tan-Sigmoidal (Tan) and Log-Sigmoidal (Log) transfer functions, one hidden layer and 10, 15 and 20 number of hidden neurons [38]. The maximum training error and maximum number of epochs are set to 10^{-6} and 1000, respectively. The ANN model is trained for the selected algorithm with various combinations, until the maximum training error and the maximum epochs are reached.

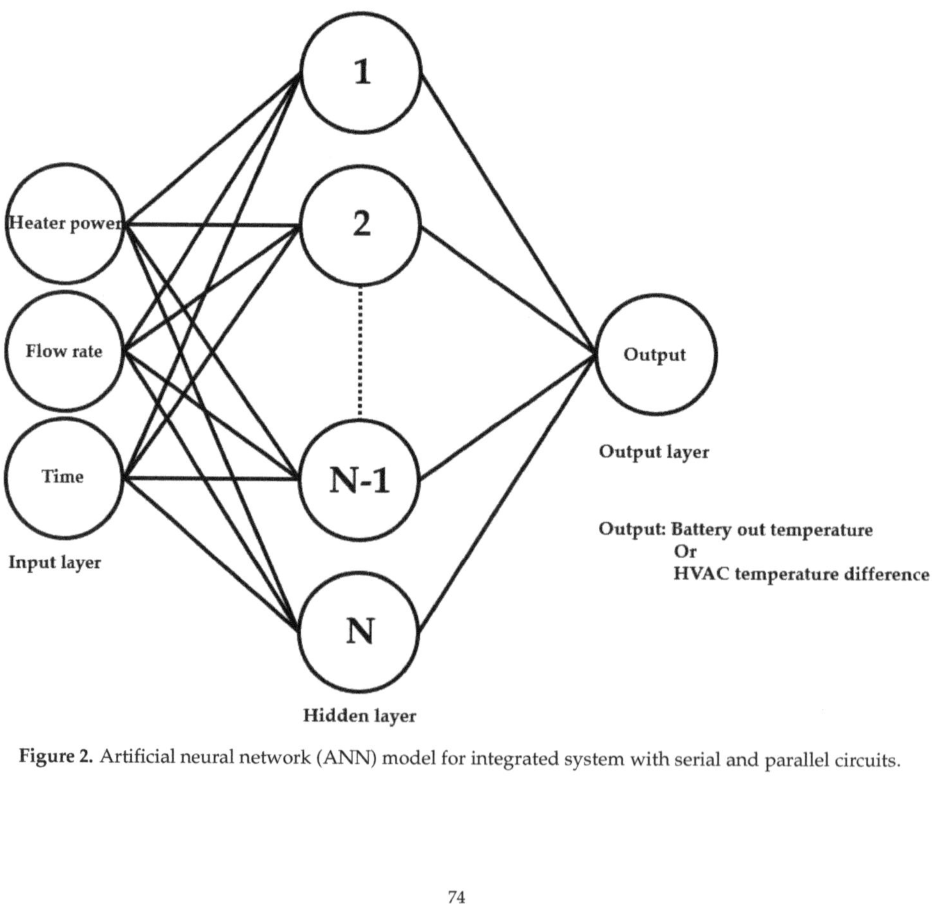

Figure 2. Artificial neural network (ANN) model for integrated system with serial and parallel circuits.

4. Data Reduction

The battery temperature rise rate is calculated using Equation (2). The battery temperature rise rate is presented in °C/min [4].

$$Battery\ temperature\ rise\ rate = \frac{End\ time\ temperature - Initial\ time\ temperature}{total\ time} \quad (2)$$

The battery heating capacity is calculated using the battery temperature rise rate, mass of battery and specific heat of battery, as presented by Equation (3).

$$Battery\ heating\ capacity = m_B \times C_{pB} \times Battery\ temperature\ rise\ rate \quad (3)$$

Here, m_B is mass of battery and C_{pB} is specific heat of battery.

The HVAC heating capacity is calculated using the temperature difference of working fluid at the inlet and outlet of HVAC, flow rate of working fluid in HVAC and specific heat of working fluid, as presented by Equation (4).

$$HVAC\ heating\ capacity = \rho \times \dot{Q}_{HVAC} \times C_{pf} \times (T_{HVAC,in} - T_{HVAC,out}) \quad (4)$$

Here, ρ is density of working fluid, $\dot{Q}_{HVAC}$ is flow rate of working fluid in HVAC, C_{pf} is specific heat of working fluid, $T_{HVAC,in}$ and $T_{HVAC,out}$ are the temperatures of the working at the inlet and outlet of HVAC, respectively.

Equations (5)–(7) are used to calculate the coefficient of determination (R^2), root mean square error (RMSE) and coefficient of variance (COV), respectively [43].

$$R^2 = 1 - \frac{\sum_{m=1}^{n} (X_{pre,m} - Y_{mea,m})^2}{\sum_{m=1}^{n} (Y_{mea,m})^2} \quad (5)$$

$$RMSE = \sqrt{\frac{\sum_{m=1}^{n} (X_{pre,m} - Y_{mea,m})^2}{n}} \quad (6)$$

$$COV = \frac{RMSE}{|\overline{Y}_{mea}|} \times 100 \quad (7)$$

Here, n presents number of data points, $X_{pre,m}$ presents predicted value of output parameter at data point m, $Y_{mea,m}$ presents experimental (actual) value of output parameter at data point m and $\overline{Y}_{mea}$ presents the average value of all experimental data points.

5. Results and Discussion

The experimental results of battery and HVAC heating performances for the integrated system with serial and parallel circuits are comprehensively discussed in this section. Battery and HVAC heating performances of the integrated system with parallel circuit are discussed in Sections 5.1 and 5.2 elaborates the battery and HVAC heating performances of the integrated system with serial circuit. The comparison of integrated system with serial and parallel circuits for battery and HVAC heating performances is presented in Section 5.3. Section 5.4 discusses the results of ANN model for prediction of battery and HVAC heating performances in the case of an integrated system with serial and parallel circuits.

5.1. Heating Performances of Battery and HVAC for Integrated System with Parallel Circuit

The experimental results for battery and HVAC heating performances of the integrated system with parallel circuit namely, battery out temperature, battery temperature rise rate, battery heating capacity and HVAC heating capacity at various heater powers and flow ratios are elaborated in this section.

5.1.1. Battery out Temperature and Battery Temperature Rise Rate

The variation of battery out temperature with time for the integrated system with parallel circuit at various heater powers and flow ratios is presented in Figure 3a. The maximum battery heating temperature is cutoff at 40 °C because battery functions effectively below this temperature [8]. Therefore, the stopping criteria for the experiment at each flow rate and various heater powers is set as 40 °C, as shown in Figure 3a. The purpose of battery heating is to reach the maximum cut off temperature (40 °C) in minimum time using the proposed integrated heating system. If the experiment reaches the cut off temperature at any flow ratio and heater power, then it is stopped. However, if the experiments fail to reach cutoff temperature or the increasing gradient of temperature curve at any flow ratio and heater power is low then the experiment is stopped. For all flow ratios, as the heater power increases from 2 to 6 kW, the battery out temperature increases. This means higher heater power enables higher and faster heating of battery. A higher flow ratio indicates higher flow rate of working fluid for battery heating. Therefore, as the flow ratio increases from 2/8 to 8/2, the battery out temperature enhances towards the higher heating temperature. In the case of heater power of 2 kW, the battery heating temperature reaches to only 35 °C after 3000 s. The increasing gradient for each temperature curve increases slowly with time, hence, experiments are stopped on or before 3000 s, because, from the trends of temperature curve, it is not expected that they could reach to cutoff temperature. Whereas the battery heating temperature reaches to 40 °C within 2000 s for heater power of 4 kW and within 1000 s for heater power of 6 kW.

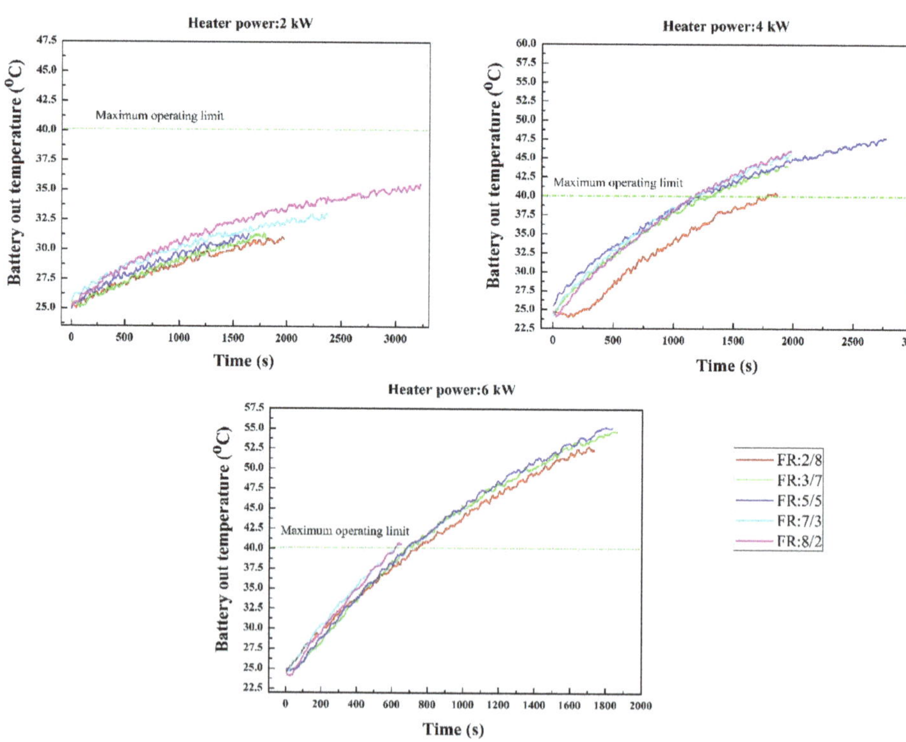

(a) Battery out temperature

Figure 3. *Cont.*

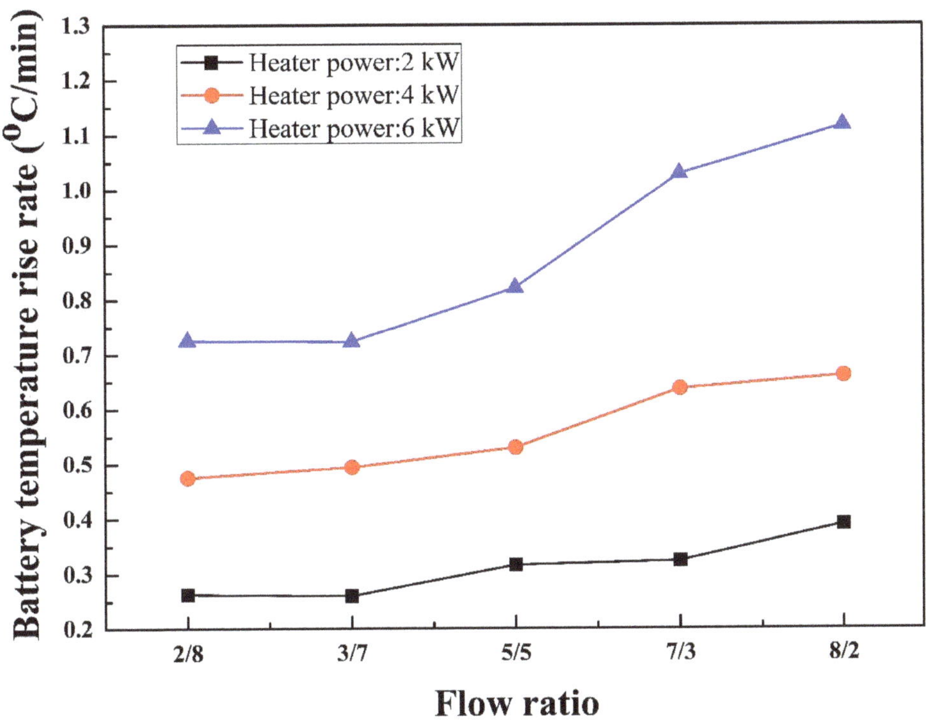

(**b**) Battery temperature rise rate

Figure 3. Variation of (**a**) battery out temperature and (**b**) battery temperature rise rate of parallel circuit for various heater power and flow ratios.

The variation of the battery temperature rise rate with flow ratio for the integrated system with parallel circuit at various heater powers is shown in Figure 3b. Due to an increase in the battery out temperature with increase in the heater power, the battery temperature rise rate enhances with heater power. In addition, the higher flow rate of working fluid for battery heating increases the battery temperature rise rate as the flow ratio increases. As the flow ratio increases from 2/8 to 8/2, the battery temperature rise rate increases by 48.20% for heater power of 2 kW, 38.90% for heater power of 4 kW and 53.90% for heater power of 6 kW. The maximum battery temperature rise rate for heater powers of 2 kW, 4 kW and 6 kW are observed 0.39 °C/min, 0.66 °C/min and 1.17 °C/min, respectively, at the flow ratio of 8/2.

5.1.2. Battery and HVAC Heating Capacities

The behavior of battery heating capacity with flow ratio for the integrated system with a parallel circuit at various heater powers is shown in Figure 4a. The battery heating capacity increases with increase in flow ratio as well as heater power, due to an increase in the battery out temperature and the battery temperature rise rate with heater power and flow ratio. The battery heating capacity curves are less steep with the flow ratio for the lower heater powers of 2 kW and 4 kW, compared with the higher heater power of 6 kW. The battery heating capacities increase from 124.14 to 183.99 W, 224.62 to 311.65 W and 456.36 to 702.43 W for heater powers of 2 kW, 4 kW and 6 kW, respectively, when the flow ratio increases from 2/8 to 8/2. The maximum battery heating capacities for heater powers of 4 kW and 6 kW are higher by 69.30% and 281.70%, respectively, than that for heater power of 2 kW.

The effect of flow ratio and heater power on HVAC heating capacity of the integrated system with parallel circuit is presented in Figure 4b. The lower flow ratio indicates higher flow rate of working fluid through HVAC and vice versa. Therefore, the HVAC heating capacity is higher at the lower flow ratio and lower at the higher flow ratio. This means HVAC heating capacity decreases as the flow ratio increases for all heater powers. The HVAC heating capacities decrease by 26.50%, 39.40% and 50.00% for heater powers of 2 kW, 4 kW and 6 kW, respectively, as the flow ratio increases from 2/8 to 8/2. The decreasing curve of HVAC heating capacity are steeper at higher heater power compared with lower heater power. The maximum HVAC heating capacities for heater powers of 2 kW, 4 kW and 6 kW are observed at flow ratio of 2/8 which are 1315.50 W, 1915.22 W and 3482.23 W, respectively. Min et al. have proposed fuzzy logic based electric vehicle thermal management system to maintain the desire temperatures of battery and cabin [44]. Seo et al. have investigated heat transfer characteristics of an integrated heating system for thermal management of cabin and battery of electric vehicle [45]. In the case of the integrated system with a parallel circuit, the sum of heating capacities for battery and HVAC are not equivalent to the supplied input for all heater powers of 2 kW, 4 kW and 6 kW. The heat losses occurred in experimental components, valves, pipes, pipe fittings/connections and ambient from the heated working fluid.

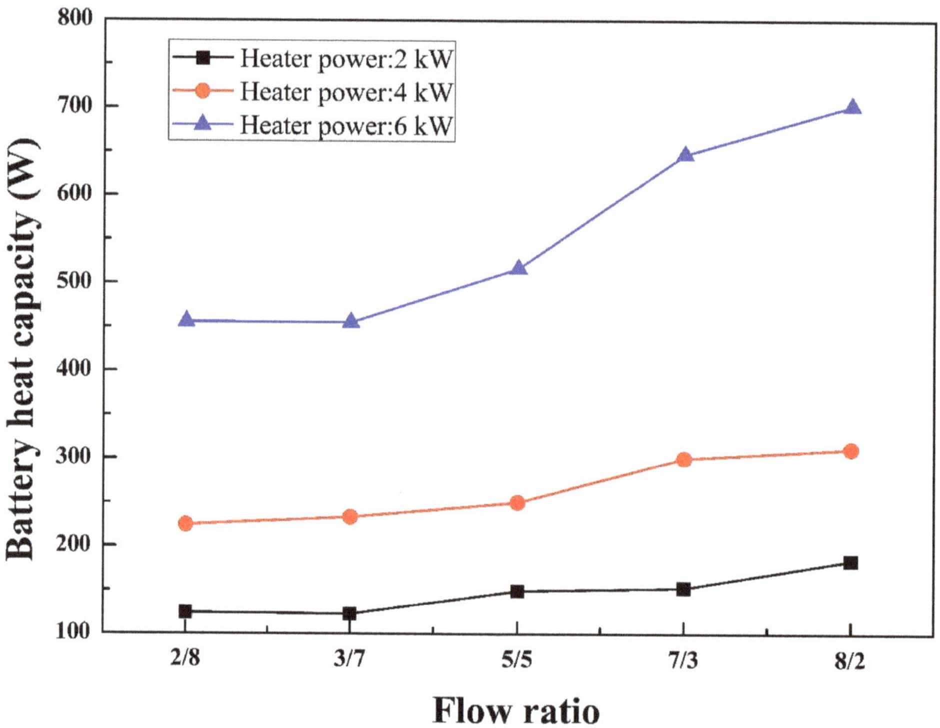

(**a**) Battery heating capacity

Figure 4. *Cont.*

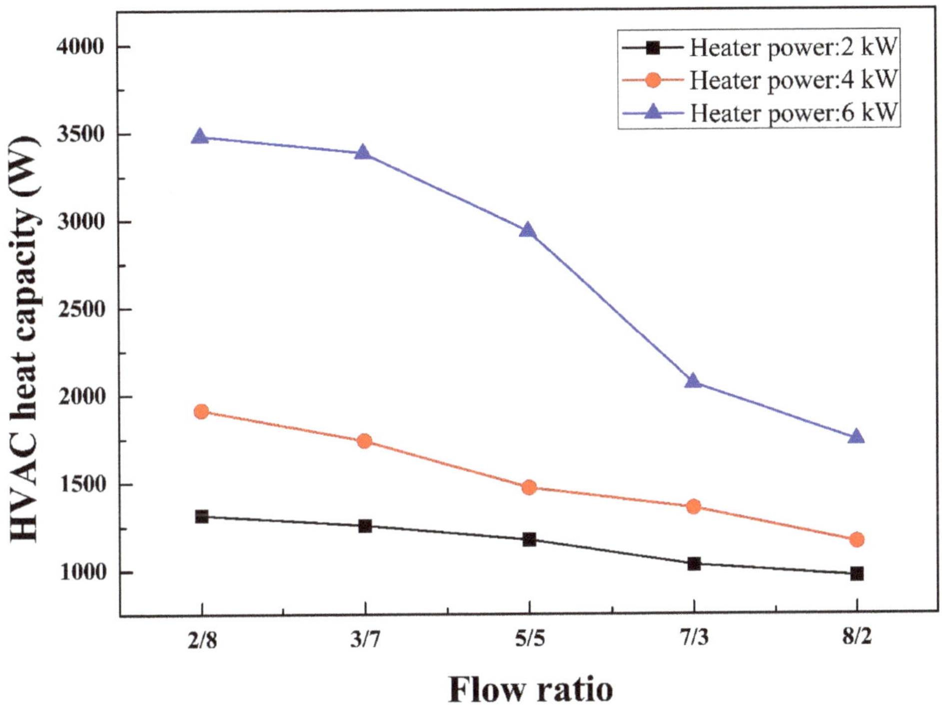

(**b**) Heating ventilation and air conditioning system (HVAC) heating capacity

Figure 4. Variation in (**a**) battery heating capacity and (**b**) HVAC heating capacity of parallel circuit with flow ratios for various heater powers.

5.2. Heating Performances of Battery and HVAC for Integrated System with Serial Circuit

The experimental results for battery and HVAC heating performances of the integrated system with serial circuit, such as battery out temperature, battery temperature rise rate, battery heating capacity and HVAC heating capacity at various heater powers are discussed in this section.

5.2.1. Battery out Temperature and Battery Temperature Rise Rate

The variation of battery out temperature with time for the integrated system with serial circuit at various heater powers is presented in Figure 5a. As the heater power increases from 2 kW to 6 kW, the battery out temperature increases. Hence, higher heater power shows faster battery heating performance compare with lower heater power. In the case of an integrated system with a serial circuit, full flow rate of working fluid is used in battery heating, therefore, the time needed to reach 40 °C is shorter than the integrated system with parallel circuit for all heater powers. The battery out temperature reaches to 40 °C within 400 s for 6 kW heater power, which is within 600 s for 4 kW heater power and that within 1200 s for 2 kW heater power. Ruan et al. have also shown the heating temperature behavior of single cell battery over the time. The battery heating from temperature of −30 °C to 2.1 °C is achieved within 103 s [5].

The variation of battery temperature rise rate of the integrated system with serial circuit for various heater powers is shown in Figure 5b. As a result of the increase in the battery out temperature with increase in the heater power, the battery temperature rise rate is higher at higher heater power. The battery temperature rise rates are 102.90% and

204.60% higher for heater powers of 4 kW and 6 kW, respectively, compared to the battery temperature rise rate at heater power of 2 kW. The battery temperature rise rate is higher for the integrated system with serial circuit compare to the integrated system with parallel circuit for all heater powers because full flow rate of working fluid is used for battery heating in the integrated system with serial circuit which significantly enhances the battery out temperature for the integrated system with serial circuit compared with the integrated system with parallel circuit. The battery temperature rise rates in the case of an integrated system with serial circuit are higher by 83.20%, 119.20% and 94.60% for heater powers of 2 kW, 4 kW and 6 kW, respectively, compared to the integrated system with parallel circuit. Guo et al. have achieved the battery cell heating from temperature of −20.30 °C to 10.02 °C within 13.70 min at an average temperature rise of 2.21 °C/min and battery pack heating from a temperature of −20.84 °C to 10 °C is achieved within 12.40 min at an average temperature rise of 2.47 °C/min [4].

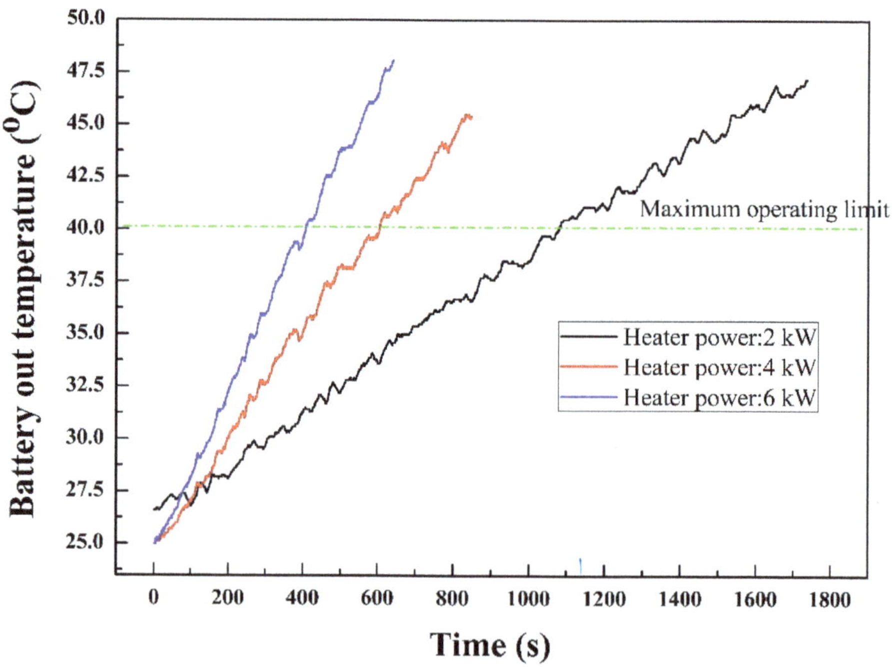

(a) Battery out temperature

Figure 5. *Cont.*

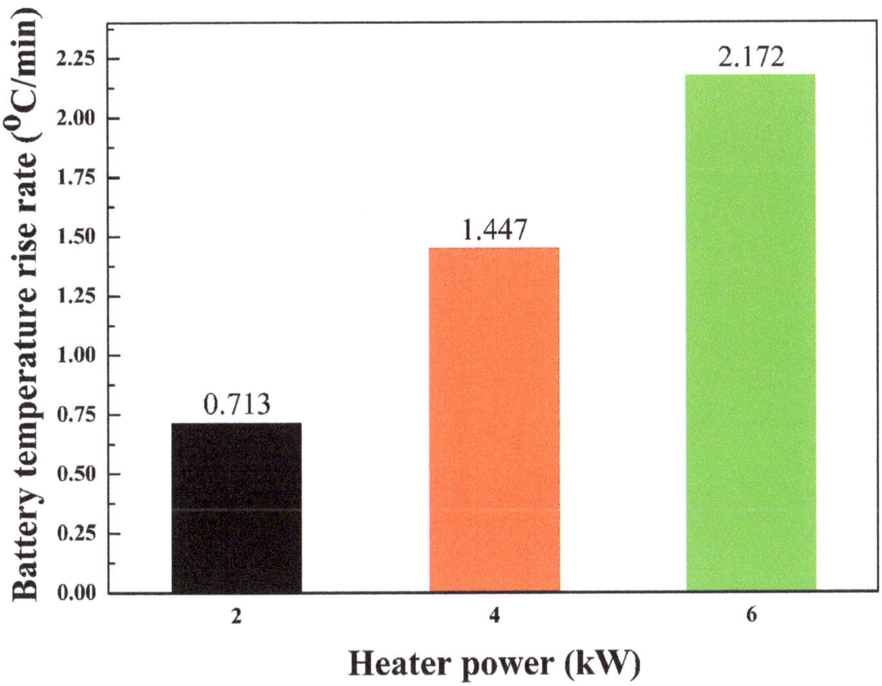

(b) Battery temperature rise rate

Figure 5. Variation of (a) Battery out temperature and (b) Battery temperature rise rate of serial circuit for various heater power.

5.2.2. Battery and HVAC Heating Capacities

The effect of heater power on battery heating capacity of the integrated system with serial circuit is shown in Figure 6a. Due to higher battery heating performance at higher heater power, the battery heating capacity increases with increase in the heater power. The battery heating capacities at heater powers of 2 kW, 4 kW and 6 kW are 336.37 W, 682.75 W and 1025.16 W, respectively. As a result of higher battery heating temperature of the integrated system with serial circuit compare to integrated system with parallel circuit for all heater powers, the battery heating capacities of integrated system with serial circuit for the heater powers of 2 kW, 4 kW and 6 kW are higher by 82.80%, 119.10% and 45.90%, respectively, than those of integrated system with parallel circuit.

The effect of heater power on HVAC heating capacity of the integrated system with serial circuit is shown in Figure 6b. As whole flow rate of working fluid is used for HVAC heating in the case of an integrated system with serial circuit, the HVAC heating capacity of integrated system with serial circuit is higher than that of integrated system with parallel circuit for all heater powers. The HVAC heating capacities of integrated system with serial circuit for heater powers of 2 kW, 4 kW and 6 kW are higher by 46.50%, 93.60% and 64.40%, respectively, than those of integrated system with parallel circuit. The HVAC heating capacity of integrated system with serial circuit also increases as the heater power increases. The HVAC heating capacities at heater powers of 4 kW and 6 kW are enhanced by 92.30% and 196.90%, compared to that of 2 kW. Zhang et al. have shown maximum cabin heating capacity of 2097 W using an economized vapor injection heat pump system whereas, Patil et al. have presented maximum cabin heating capacity of 2171.5 W using a 2.0 kW burner [11,20].

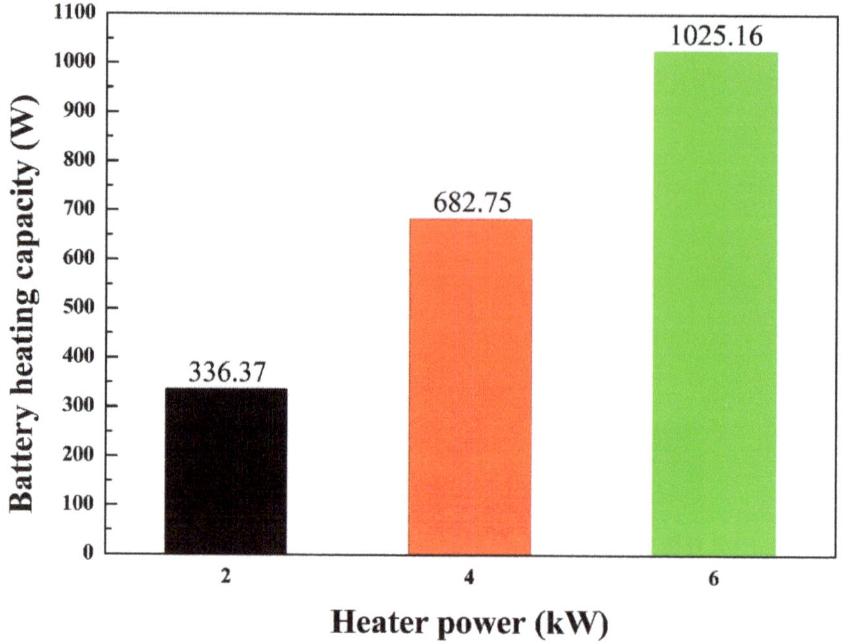

(a) Battery heating capacity

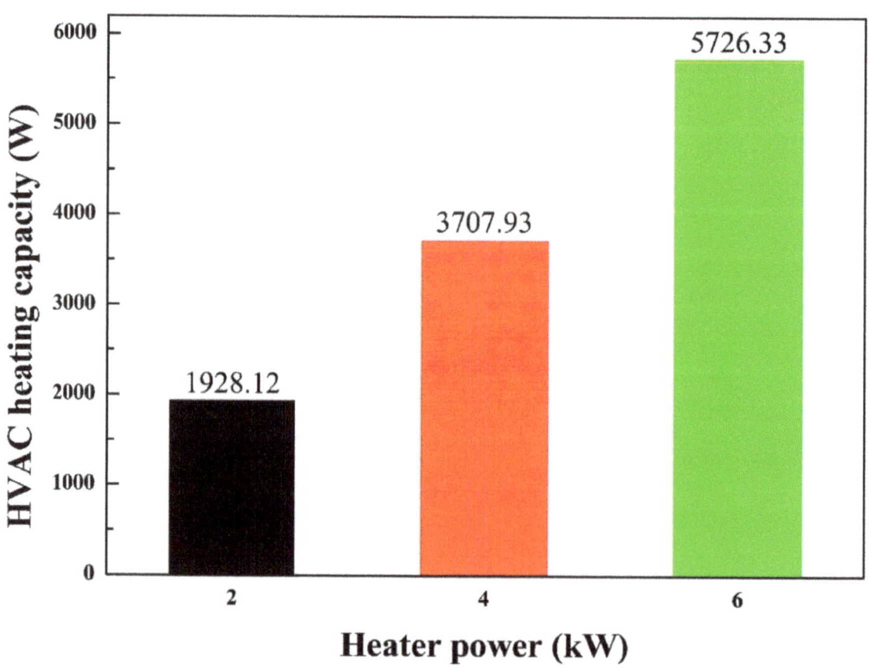

(b) HVAC heating capacity

Figure 6. Effect of heater power on (**a**) battery heating capacity and (**b**) HVAC heating capacity of serial circuit.

5.3. Total Heating Capacity of Integrated System with Serial and Parallel Circuits

The total heating capacities comparison of the integrated system with serial and parallel circuits for various heater powers is presented in Figure 7. In addition, in the case of an integrated system with parallel circuit, the effect of flow ratio on total heating capacity is also included. As presented in Figure 7, the total heating capacity for parallel circuit is the sum of heating capacities of battery and HVAC at all heater powers and flow ratios, whereas the total heating capacity for the serial circuit is the heating capacity of the battery when the battery only heated using an entire flow rate, and that is the heating capacity of HVAC when only HVAC is heated using a full flow rate of working fluid. In the integrated system with parallel circuit, the total flow rate of working fluid (24 L/min) is divided to heat up the battery and HVAC therefore, the total heating capacity is the sum of heating capacities of battery and HVAC. While, in the case of an integrated system with serial circuit, the total flow rate of working fluid is used for heating of battery or heating of HVAC, therefore, the heating capacity of the battery or heating capacity of HVAC is only the total heating capacity. In total, the heating capacity of the integrated system with parallel circuit, the heating capacity of HVAC is more dominant compare with heating capacity of battery. Therefore, the total heating capacity of the integrated system with parallel circuit decreases as the flow ratio increases for all heater powers because of decrease in the heating capacity of HVAC. The maximum total heating capacities for the integrated system with parallel circuit are found at flow ratio of 2/8 for heater powers of 2 kW, 4 kW and 6 kW, which are 1439.66 W, 2139.60 W and 3938.59 W, respectively. For the integrated system with serial and parallel circuits, the total heating capacity increases as the heater power increases. The total heating capacity increases by 173.50% for the integrated system with parallel circuit, 204.80% for battery heating of integrated system with serial circuit and 197% for HVAC heating of the integrated system with serial circuit, as the heater power increases from 2 kW to 6 kW. For heater power of 2 kW, the maximum total heating capacity of the integrated system with parallel circuit is 327.90% higher than total battery heating capacity of the integrated system with serial circuit, and that is 25.30% lower than total HVAC heating capacity of the integrated system serial circuit. At heater power of 4 kW, the total heating capacity of the integrated system with parallel circuit is 212.30% higher and 42.30% lower than the total battery heating capacity and total HVAC heating capacity of the integrated system with serial circuit, respectively. In the same way, at a heater power of 6 kW, the total heating capacity of integrated system with parallel circuit is 284.20% higher and 31.20% lower than the total battery heating capacity and total HVAC heating capacity of the integrated system with serial circuit, respectively. As shown in Figure 7, the output as the sum of heating capacities of battery and HVAC in the case of parallel circuit and heating capacity of battery or HVAC in the case of serial circuit are lower than the input as heater power. The heat losses for the integrated system with serial and parallel circuits are occurred into experimental components. as well as three valves, pipes and pipe fittings/connections. In addition, some heat loss occurs from working fluid to ambient. The integrated system with serial circuit shows rapid heating performance compared with the integrated system with parallel circuit, but the tradeoff heating between battery and HVAC is not possible in the case of the integrated system with serial circuit. However, the integrated system with parallel circuit enables the tradeoff between the heating of HVAC and that of the battery. Despite of slow heating performance of the integrated system with parallel circuit, the desired heating performance could be achieved.

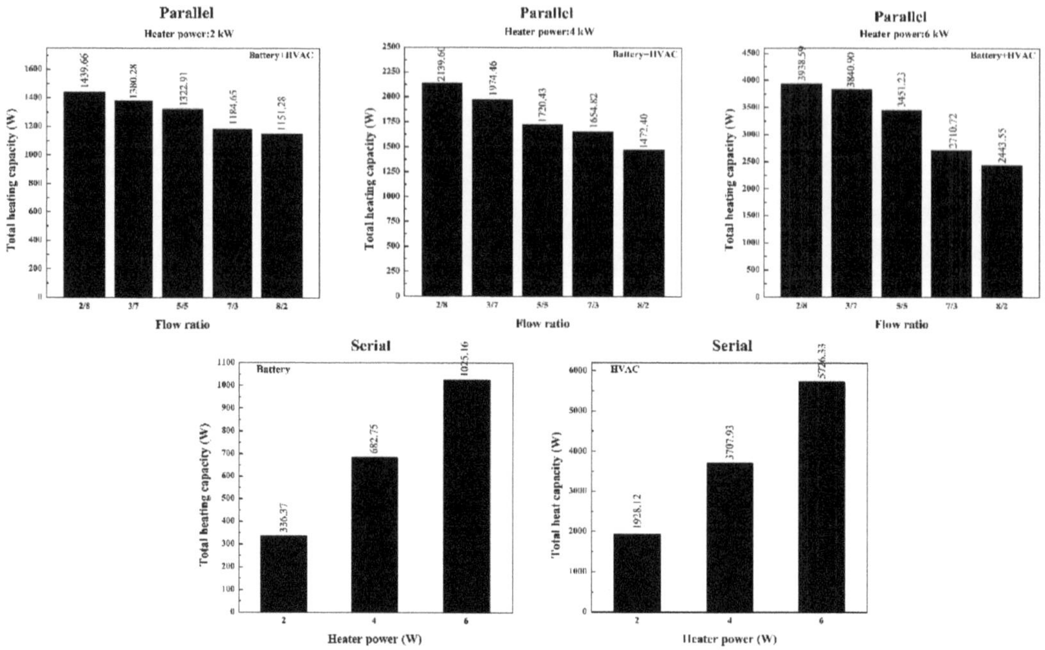

Figure 7. Comparison of total heating capacities of parallel and serial circuits for various heater powers.

5.4. ANN Model for Battery and HVAC Heating Performances

The battery and HVAC heating performances of the integrated system with serial and parallel circuits predicted by ANN model with various algorithm are compared with corresponding experimental (actual) results. The comparison of predicted results and actual results is done in terms of three statistical parameters namely, coefficient of determination (R^2), root mean square error (RMSE) and coefficient of variance (COV). The algorithm with highest value of R^2 and lowest values of RMSE and COV are suggested as the optimum algorithm for ANN model.

5.4.1. Integrated System with Parallel Circuit

The battery out temperature of the integrated system with parallel circuit for various heater powers, flow ratios and time is predicted using various algorithms of ANN models. The experimental data of battery out temperature of the integrated system with parallel circuit at various time, flow rates and heater powers are used for the training, testing and validation of ANN models with various algorithms. The prediction accuracy of various algorithms of ANN model for battery out temperature at flow ratio of 5/5 and all heater powers is presented in Table 3. A higher prediction accuracy is obtained for other flow ratios as well. For all heater powers and hidden neurons, LM with Tan training variant shows better prediction accuracy compared to LM with Log training variant. In addition, prediction accuracy increases as the number of hidden neurons increases for all heater powers. Therefore, in respective case of each heater power, LM-Tan with 20 hidden neurons shows highest prediction accuracy for battery out temperature of the integrated system with parallel circuit. The prediction accuracy of ANN model with LM-Tan-20 algorithm for battery out temperature of the integrated system with parallel circuit is R^2 with 0.999971, 0.999979 and 0.999979, RMSE with 0.154166, 0.183539 and 0.201043, as well as COV with 0.536230, 0.460489 and 0.472796 for heater powers of 2 kW, 4 kW and 6 kW, respectively. The comparison of battery out temperature of the integrated system with parallel circuit for experiment and ANN model with LM-Tan-20 algorithm at various heater powers is also presented in Figure 8a. Figure 8a shows the prediction capability of suggested algorithm and closeness of predicted results by suggested algorithm with the experimental results.

The same algorithms of the ANN model are used to predict the HVAC temperature difference of the integrated system with parallel circuit. The ANN model with various algorithms is trained, tested and validated for the experimental HVAC temperature difference data at various flow rates, heater powers and time. The prediction accuracy of various algorithms of ANN models is presented at flow ratio of 5/5 and all heater powers for HVAC temperature difference in Table 3. For other flow ratios, a higher prediction accuracy is also obtained. In the case of HVAC temperature difference of the integrated system with parallel circuit, the LM with Tan training variant also shows superior prediction accuracy than the LM with Log training variant for all heater powers. Additionally, higher number of hidden neurons shows higher prediction performance for all heater powers. For the HVAC temperature difference of the integrated system with parallel circuit, the LM-Tan-20 shows R^2 of 0.803624, 0.956737 and 0.994077, RMSE of 0.631449, 0.549005 and 0.245819, as well as COV of 44.75612, 21.27336 and 7.909049 for heater powers of 2 kW, 4 kW and 6 kW, respectively. The prediction accuracy of the LM-Tan-20 algorithm is better for the battery out temperature than the HVAC temperature difference, because of linear trends of battery out temperature and zigzag trends of HVAC temperature difference for all heater powers. Figure 8b shows the closeness of LM-Tan-20 algorithm predicted HVAC temperature difference and corresponding actual results for all heater powers. For some points deviation between the predicted and actual results is larger because of zigzag trend of HVAC temperature difference curves at all heater powers, the suggested algorithm is not able to follow all points accurately. However, the minimum prediction accuracy is 0.8, which is an acceptable prediction performance for the suggested ANN model. Mohanraj et al. have suggested the Levenberg-Marquardt training algorithm as the optimum for the accurate performance prediction with maximum R^2 of 0.999 and lowest values of RMSE and COV [39].

Table 3. Prediction accuracy of ANN models for battery heating performance and HVAC heating performance of integrated system with parallel circuit.

	Heater Power	Algorithm	Number of Hidden Neurons	R^2	RMSE	COV
Battery heating performance	2 kW	LM-Tan	10	0.999964	0.172015	0.598317
			15	0.999965	0.170036	0.591433
			20	0.999971	0.154166	0.536230
		LM-Log	10	0.999962	0.176631	0.614369
			15	0.999965	0.170120	0.591725
			20	0.999968	0.163048	0.567126
	4 kW	LM-Tan	10	0.999976	0.197036	0.494352
			15	0.999978	0.189969	0.476623
			20	0.999979	0.183539	0.460489
		LM-Log	10	0.999976	0.198585	0.498239
			15	0.999976	0.196935	0.494101
			20	0.999978	0.187425	0.470238
	6 kW	LM-Tan	10	0.999975	0.218650	0.514204
			15	0.999976	0.211399	0.497152
			20	0.999979	0.201043	0.472796
		LM-Log	10	0.999931	0.360343	0.847426
			15	0.999976	0.212566	0.499896
			20	0.999977	0.208056	0.489290

Table 3. *Cont.*

	Heater Power	Algorithm	Number of Hidden Neurons	R^2	RMSE	COV
HVAC heating performance	2 kW	LM-Tan	10	0.786929	0.657742	46.61972
			15	0.798038	0.640366	45.38813
			20	0.803624	0.631449	44.75612
		LM-Log	10	0.785127	0.660519	46.81651
			15	0.794651	0.645714	45.76716
			20	0.799947	0.637332	45.17311
	4 kW	LM-Tan	10	0.955619	0.556058	21.54665
			15	0.956040	0.553408	21.44398
			20	0.956737	0.549005	21.27336
		LM-Log	10	0.954056	0.565763	21.92271
			15	0.955687	0.555626	21.52991
			20	0.956573	0.550047	21.31374
	6 kW	LM-Tan	10	0.993304	0.261362	8.409131
			15	0.993737	0.252775	8.132854
			20	0.994077	0.245819	7.909049
		LM-Log	10	0.991877	0.287864	9.291802
			15	0.993562	0.256273	8.245384
			20	0.993748	0.252538	8.125237

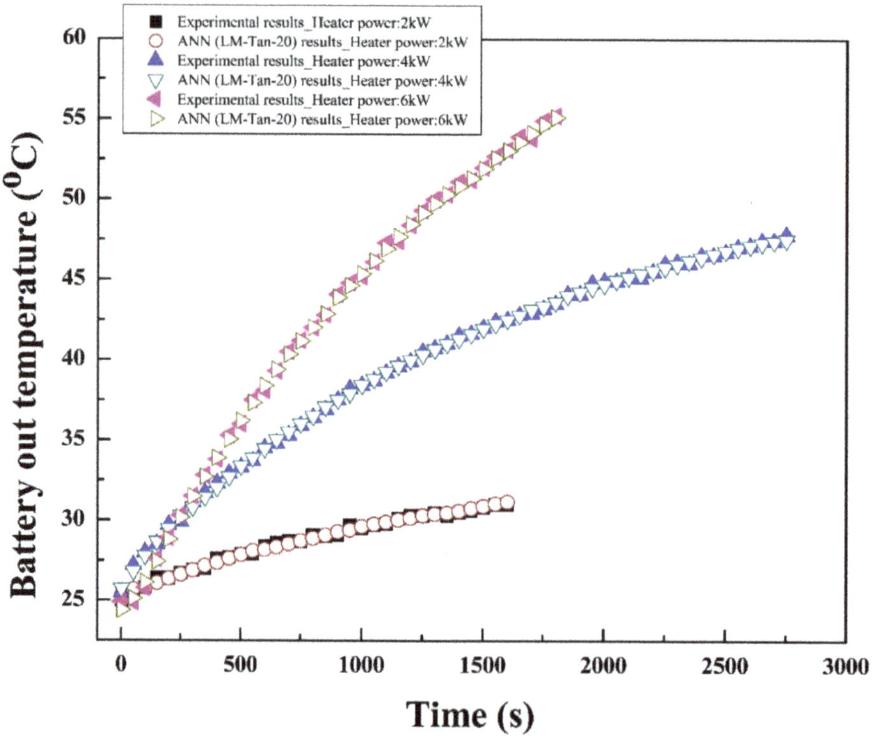

(**a**) Battery out temperature

Figure 8. *Cont.*

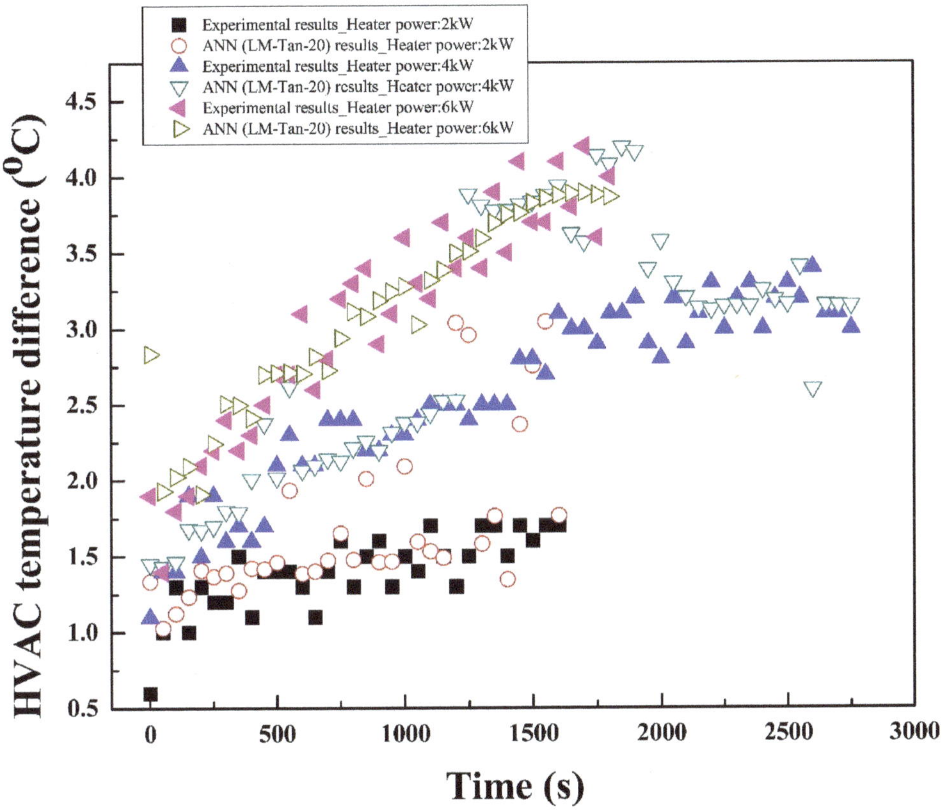

(**b**) HVAC temperature difference

Figure 8. Comparison of (**a**) battery out temperature and (**b**) HVAC temperature difference of parallel circuit for experimental and ANN model with LM-Tan-20 algorithm at various heater powers.

5.4.2. Integrated System with Serial Circuit

The battery out temperature of the integrated system with serial circuit is predicted for various algorithms of ANN model using heater power, flow rates and time as the input conditions. The experimental battery out temperature, heater powers and flow rates at various time are considered to train, test and validate the various algorithms of ANN model. The prediction accuracy of ANN model with various algorithms for battery out temperature of the integrated system with serial circuit at various heater powers is presented in Table 4. The prediction accuracy of ANN model with LM-Tan-20 algorithm is superior for all heater powers. LM-Tan-20 algorithm shows R^2, RMSE and COV values of 0.999970, 0.205704 and 0.554591, respectively, at a heater power of 2 kW, those values of 0.999980, 0.170831 and 0.479862, respectively, at a heater power of 4 kW, and those values of 0.999982, 0.158077 and 0.431001, respectively, at heater power of 6 kW. The battery out temperatures predicted by the LM-Tan-20 algorithm for various time and heater powers are compared with the corresponding experimental values in Figure 9a. A higher degree of closeness between the actual and predicted results could be observed for the suggested algorithm of ANN model.

The experimental results of HVAC temperature difference, heater power, flow rates and time are used to train, test and validate the same algorithms of the ANN models. Trained ANN models are used to predict the HVAC temperature difference for various heater powers with time. Table 4 shows the prediction accuracy of various algorithms of the ANN model for HVAC temperature difference of the integrated system with serial circuit at various heater powers. ANN model with LM-Tan-20 algorithm is suggested as the optimum model, which shows higher value of R^2 and lowest values of RMSE and COV. ANN model with LM-Tan-20 algorithm shows R^2, RMSE and COV values of 0.987539, 0.128786 and 11.30589, respectively, at 2 kW heater power, those of 0.998338, 0.132884 and 4.164997, respectively, at 4 kW heater power and those of 0.998081, 0.134271 and 4.444122, respectively, at 6 kW heater power. Figure 9b shows the comparison of experimental HVAC temperature difference and LM-Tan-20 algorithm predicted HVAC temperature difference for various heater powers. For all heater powers, the predicted results show closer agreement with corresponding experimental results.

Table 4. Prediction accuracy of ANN models for battery heating performance and HVAC heating performance of integrated system with serial circuit.

	Heater Power	Algorithm	Number of Hidden Neurons	R^2	RMSE	COV
Battery heating performance	2 kW	LM-Tan	10	0.999963	0.227245	0.612664
			15	0.999966	0.220835	0.595383
			20	0.999970	0.205704	0.554591
		LM-Log	10	0.999961	0.234787	0.632999
			15	0.999964	0.226566	0.610835
			20	0.999967	0.216158	0.582773
	4 kW	LM-Tan	10	0.999979	0.164997	0.463474
			15	0.999978	0.197546	0.470634
			20	0.999980	0.170831	0.479862
		LM-Log	10	0.999969	0.201567	0.566196
			15	0.999978	0.170591	0.479186
			20	0.999980	0.161314	0.453130
	6 kW	LM-Tan	10	0.999972	0.198939	0.542413
			15	0.999981	0.164124	0.447487
			20	0.999982	0.158077	0.431001
		LM-Log	10	0.999967	0.214850	0.585792
			15	0.999975	0.188060	0.512750
			20	0.999981	0.161996	0.441684
HVAC heating performance	2 kW	LM-Tan	10	0.986886	0.132118	11.59836
			15	0.987196	0.130549	11.46062
			20	0.987539	0.128786	11.30589
		LM-Log	10	0.986881	0.132142	11.60054
			15	0.987086	0.131105	11.50950
			20	0.987226	0.130393	11.44697
	4 kW	LM-Tan	10	0.998091	0.142413	4.463665
			15	0.998137	0.140689	4.409637
			20	0.998338	0.132884	4.164997
		LM-Log	10	0.998029	0.144687	4.534949
			15	0.998106	0.141871	4.446673
			20	0.998229	0.137182	4.299688
	6 kW	LM-Tan	10	0.997789	0.144090	4.769143
			15	0.997864	0.141669	4.688989
			20	0.998081	0.134271	4.444122
		LM-Log	10	0.997779	0.144445	4.780861
			15	0.997815	0.143260	4.741652
			20	0.997894	0.140671	4.655938

Based on the results discussed in Sections 5.4.1 and 5.4.2, ANN model with LM-Tan-20 algorithm is suggested to accurately predict the heating performances of battery and HVAC for the integrated system with serial and parallel circuits. Kunal et al. proposed an ANN model with the optimum structure as the Levenberg-Marquardt training algorithm and Tan-sigmoidal transfer function for the accurate performance prediction [43].

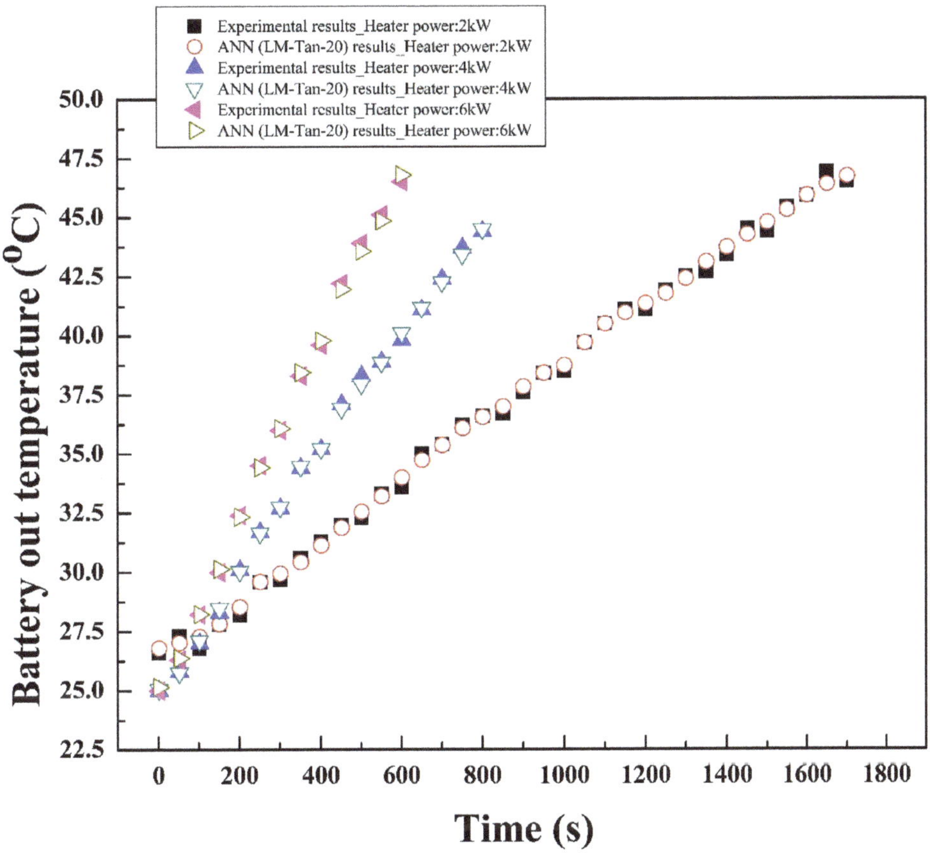

(**a**) Battery out temperature

Figure 9. *Cont.*

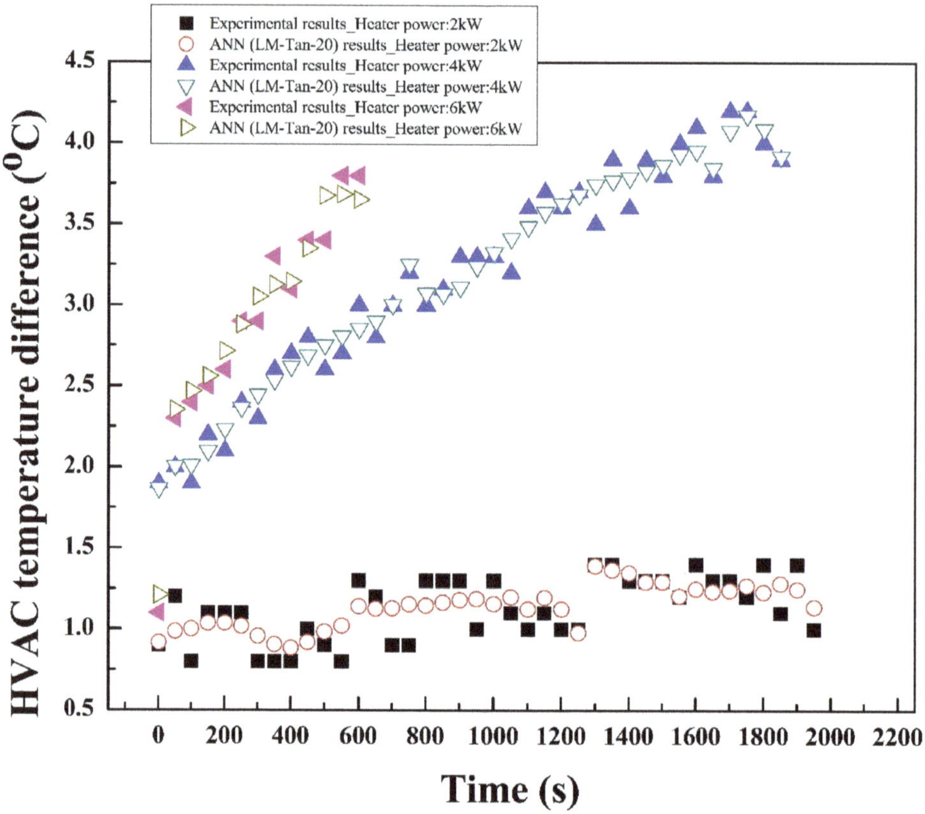

(**b**) HVAC temperature difference

Figure 9. Comparison of (**a**) battery out temperature and (**b**) HVAC temperature difference of serial circuit for experimental and ANN model with LM-Tan-20 algorithm at various heater powers.

6. Conclusions

An experimental study is conducted on the integrated system with serial and parallel circuits to investigate the heating performances of battery and HVAC for an electric vehicle. In addition, ANN models are developed to predict the battery and HVAC heating performances of the integrated system with serial and parallel circuits. The following findings are summarized from the present study.

(a) The effect of heater power on heating performances of the integrated system with serial and parallel circuits and effect of flow ratio on heating performances of the integrated system with parallel circuit are analyzed. As the heater power increases, the heating performances increases for the integrated system with serial and parallel circuits. With an increase in the flow ratio to the battery, battery heating performance enhances, whereas HVAC heating performance decreases.

(b) In the case of integrated system with parallel circuit, battery out temperature reaches 40 °C within 20 min at the rate of 1.22 °C/min. Battery heating capacity is evaluated as 764.99 W and HVAC heating capacity is evaluated as 3869.15 W.

(c) The battery out temperature reaches to 40 °C within 10 min at the rate of 2.17 °C/min for the integrated system with serial circuit. The battery and HVAC heating capacities for the integrated system with serial circuit are evaluated as 1025.16 W and 5726.33 W, respectively.

(d) The integrated system with serial circuit enables faster heating performance than the integrated system with parallel circuit for both battery and HVAC. However, the integrated system with parallel circuit enables the tradeoff heating between battery and HVAC at the desired level with the slower rate.

(e) The battery and HVAC heating performances of the integrated system with serial and parallel circuits are accurately predicted using the developed ANN models with back-propagation training algorithm, Levenberg-Marquardt training variant, Tan-sigmoidal transfer function and 20 hidden neurons.

(f) The effects of various operating conditions on the heating performances of battery and HVAC using the proposed integrated system with serial and parallel circuits could be investigated and optimized, to find the optimum point for tradeoff heating of battery and HVAC under various conditions. The extracted results could be used in practical applications such as under cold weather conditions, the extracted optimum point for tradeoff heating of battery and HVAC could successfully achieve the efficient battery and HVAC heating performances of full size commercial electric vehicles with increased driving range and improved battery performance and life. The integrated system with serial circuit could be used for applications where rapid heating of battery or HVAC is needed, whereas the integrated system with parallel circuit could be used for applications where tradeoff simultaneous heating of battery and HVAC is needed. Thus, the proposed integrated system with serial and parallel circuits has the practical applicability to enable rapid, as well as tradeoff heating for both battery and HVAC in electric vehicles.

Author Contributions: Conceptualization, T.-K.L.; K.S.G.; J.-H.S. and M.-Y.L.; methodology, T.-K.L. and K.S.G.; formal analysis, T.-K.L. and K.S.G.; investigation, T.-K.L. and K.S.G.; resources, T.-K.L.; K.S.G. and J.-H.S.; data curation, T.-K.L.; K.S.G. and J.-H.S.; writing—original draft preparation, T.-K.L.; K.S.G.; J.-H.S. and M.-Y.L.; writing—review and editing, T.-K.L.; K.S.G., J.-H.S.; M.-Y.L. and D.-Y.L.; visualization, T.-K.L.; K.S.G. and J.-H.S.; supervision, M.-Y.L. and D.-Y.L.; project administration, T.-K.L.; M.-Y.L. and D.-Y.L.; funding acquisition, T.-K.L.; M.-Y.L. and D.-Y.L. All authors have read and agreed to the published version of the manuscript.

Funding: This research received no external funding.

Institutional Review Board Statement: Not applicable.

Informed Consent Statement: Not applicable.

Data Availability Statement: The data presented in this study will be available on request to the corresponding author.

Acknowledgments: This work was supported by the Technology Innovation Program (20011622, Development of battery pack system applied high efficiency heat control polymer and part component) funded by the Ministry of Trade, Industry & Energy (MOTIE, Korea), the National Research Foundation of Korea (NRF) grant funded by the Korea government (MSIT) (No. 2020R1A2C1011555) and the 2020 Yeungnam University Research Grant.

Conflicts of Interest: The authors declare no conflict of interest.

References

1. Wu, S.; Xiong, R.; Li, H.; Nian, V.; Ma, S. The state of the art on preheating lithium-ion batteries in cold weather. *J. Energy Storage* **2020**, *27*, 101059.
2. Kang, B.H.; Lee, H.J. A review of recent research on automotive HVAC systems for EVs. *Int. J. Air-Cond. Refrig.* **2017**, *25*, 1730003.
3. Lei, Z.; Zhang, Y.; Lei, X. Temperature uniformity of a heated lithium-ion battery cell in cold climate. *Appl. Therm. Eng.* **2018**, *129*, 148–154.
4. Guo, S.; Xiong, R.; Wang, K.; Sun, F. A novel echelon internal heating strategy of cold batteries for all-climate electric vehicles application. *Appl. Energy* **2018**, *219*, 256–263.

5. Ruan, H.; Jiang, J.; Sun, B.; Su, X.; He, X.; Zhao, K. An optimal internal-heating strategy for lithium-ion batteries at low temperature considering both heating time and lifetime reduction. *Appl. Energy* **2019**, *256*, 113797.
6. Lei, Z.; Zhang, Y.; Lei, X. Improving temperature uniformity of a lithium-ion battery by intermittent heating method in cold climate. *Int. J. Heat Mass Transf.* **2018**, *121*, 275–281.
7. Shang, Y.; Zhu, C.; Lu, G.; Zhang, Q.; Cui, N.; Zhang, C. Modeling and analysis of high-frequency alternating-current heating for lithium-ion batteries under low-temperature operations. *J. Power Sources* **2020**, *450*, 227435.
8. Fan, R.; Zhang, C.; Wang, Y.; Ji, C.; Meng, Z.; Xu, L.; Chin, C.S. Numerical study on the effects of battery heating in cold climate. *J. Energy Storage* **2019**, *26*, 100969.
9. Reyes, J.R.M.D.; Parsons, R.V.; Hoemsen, R. Winter happens: The effect of ambient temperature on the travel range of electric vehicles. *IEEE Trans. Veh. Technol.* **2016**, *65*, 4016–4022.
10. Qi, Z. Advances on air conditioning and heat pump system in electric vehicles–A review. *Renew. Sust. Energ. Rev.* **2014**, *38*, 754–764.
11. Zhang, Z.; Li, W.; Shi, J.; Chen, J. A study on electric vehicle heat pump systems in cold climates. *Energies* **2016**, *9*, 881.
12. Cho, C.W.; Lee, H.S.; Won, J.P.; Lee, M.Y. Measurement and evaluation of heating performance of heat pump systems using wasted heat from electric devices for an electric bus. *Energies* **2012**, *5*, 658–669.
13. Qin, F.; Zhang, G.; Xue, Q.; Zou, H.; Tian, C. Experimental investigation and theoretical analysis of heat pump systems with two different injection portholes compressors for electric vehicles. *Appl. Energy* **2017**, *185*, 2085–2093.
14. Qin, F.; Xue, Q.; Zhang, G.; Zou, H.; Tian, C. Experimental investigation on heat pump for electric vehicles with different refrigerant injection compressors. *Energy Procedia* **2015**, *75*, 1490–1495.
15. Qin, F.; Xue, Q.; Velez, G.M.A.; Zhang, G.; Zou, H.; Tian, C. Experimental investigation on heating performance of heat pump for electric vehicles at− 20 C ambient temperature. *Energy Convers. Manag.* **2015**, *102*, 39–49.
16. Ahn, J.H.; Kang, H.; Lee, H.S.; Jung, H.W.; Baek, C.; Kim, Y. Heating performance characteristics of a dual source heat pump using air and waste heat in electric vehicles. *Appl. Energy* **2014**, *119*, 1–9.
17. Lee, H.S.; Won, J.P.; Cho, C.W.; Kim, Y.C.; Lee, M.Y. Heating performance characteristics of stack coolant source heat pump using R744 for fuel cell electric vehicles. *J. Mech. Sci. Technol.* **2012**, *26*, 2065–2071.
18. Shi, Y.; Guo, X.; Zhang, X. Study on economized vapor injection heat pump system using refrigerant R32. *Int. J. Air-Cond. Refrig.* **2016**, *24*, 1650006.
19. Jung, J.; Jeon, Y.; Lee, H.; Kim, Y. Numerical study of the effects of injection-port design on the heating performance of an R134a heat pump with vapor injection used in electric vehicles. *Appl. Therm. Eng.* **2017**, *127*, 800–811.
20. Patil, M.S.; Cho, C.P.; Lee, M.Y. Numerical study on thermal performances of 2.0 kW burner for the cabin heater of an electric passenger vehicle. *Appl. Therm. Eng.* **2018**, *138*, 819–831.
21. Zhang, L.; Hashimoto, K.; Hasegawa, H.; Saikawa, M. Performance analysis of a heat pump system with integrated desiccant for electric vehicles. *Int. J. Refrig.* **2018**, *86*, 154–162. [CrossRef]
22. Choi, Y.U.; Kim, M.S.; Kim, G.T.; Kim, M.; Kim, M.S. Performance analysis of vapor injection heat pump system for electric vehicle in cold startup condition. *Int. J. Refrig.* **2017**, *80*, 24–36. [CrossRef]
23. Kwon, C.; Kim, M.S.; Choi, Y.; Kim, M.S. Performance evaluation of a vapor injection heat pump system for electric vehicles. *Int. J. Refrig.* **2017**, *74*, 138–150. [CrossRef]
24. Ahn, J.H.; Kang, H.; Lee, H.S.; Kim, Y. Performance characteristics of a dual-evaporator heat pump system for effective dehumidifying and heating of a cabin in electric vehicles. *Appl. Energy* **2015**, *146*, 29–37. [CrossRef]
25. Lee, D.Y.; Cho, C.W.; Won, J.P.; Park, Y.C.; Lee, M.Y. Performance characteristics of mobile heat pump for a large passenger electric vehicle. *Appl. Therm. Eng.* **2013**, *50*, 660–669. [CrossRef]
26. Liu, C.; Zhang, Y.; Gao, T.; Shi, J.; Chen, J.; Wang, T.; Pan, L. Performance evaluation of propane heat pump system for electric vehicle in cold climate. *Int. J. Refrig.* **2018**, *95*, 51–60. [CrossRef]
27. Li, W.; Liu, R.; Liu, Y.; Wang, D.; Shi, J.; Chen, J. Performance evaluation of R1234yf heat pump system for an electric vehicle in cold climate. *Int. J. Refrig.* **2020**, *115*, 117–125. [CrossRef]
28. Bellocchi, S.; Guizzi, G.L.; Manno, M.; Salvatori, M.; Zaccagnini, A. Reversible heat pump HVAC system with regenerative heat exchanger for electric vehicles: Analysis of its impact on driving range. *Appl. Therm. Eng.* **2018**, *129*, 290–305. [CrossRef]
29. Lee, H.S.; Lee, M.Y. Steady state and start-up performance characteristics of air source heat pump for cabin heating in an electric passenger vehicle. *Int. J. Refrig.* **2016**, *69*, 232–242. [CrossRef]
30. Lee, M.Y.; Garud, K.S.; Jeon, H.B.; Lee, H.S. A Study on Performance Characteristics of a Heat Pump System with High-Pressure Side Chiller for Light-Duty Commercial Electric Vehicles. *Symmetry* **2020**, *12*, 1237. [CrossRef]
31. Jeffs, J.; McGordon, A.; Picarelli, A.; Robinson, S.; Tripathy, Y.; Widanage, W.D. Complex heat pump operational mode identification and comparison for use in electric vehicles. *Energies* **2018**, *11*, 2000. [CrossRef]
32. Jeffs, J.; Dinh, T.Q.; Widanage, W.D.; McGordon, A.; Picarelli, A. Optimisation of Direct Battery Thermal Management for EVs Operating in Low-Temperature Climates. *Energies* **2020**, *13*, 5980. [CrossRef]
33. Arun, K.R.; Kunal, G.; Srinivas, M.; Kumar, C.S.; Mohanraj, M.; Jayaraj, S. Drying of untreated Musa nendra and Momordica charantia in a forced convection solar cabinet dryer with thermal storage. *Energy* **2020**, *192*, 116697. [CrossRef]

34. Garud, K.S.; Seo, J.H.; Patil, M.S.; Bang, Y.M.; Pyo, Y.D.; Cho, C.P.; Lee, M.Y. Thermal–electrical–structural performances of hot heat exchanger with different internal fins of thermoelectric generator for low power generation application. *J. Therm. Anal. Calorim.* **2020**, 1–33. [CrossRef]
35. Seo, J.H.; Garud, K.S.; Lee, M.Y. Grey Relational Based Taguchi Analysis on Thermal and Electrical Performances of Thermoelectric Generator System with Inclined Fins Hot Heat Exchanger. *Appl. Therm. Eng.* **2020**, 116279. [CrossRef]
36. Raj, A.K.; Kunal, G.; Srinivas, M.; Jayaraj, S. Performance analysis of a double-pass solar air heater system with asymmetric channel flow passages. *J. Therm. Anal. Calorim.* **2019**, *136*, 21–38. [CrossRef]
37. Yazdani-Chamzini, A.; Zavadskas, E.K.; Antucheviciene, J.; Bausys, R. A model for shovel capital cost estimation, using a hybrid model of multivariate regression and neural networks. *Symmetry* **2017**, *9*, 298. [CrossRef]
38. Garud, K.S.; Jayaraj, S.; Lee, M.Y. A review on modeling of solar photovoltaic systems using artificial neural networks, fuzzy logic, genetic algorithm and hybrid models. *Int. J. Energy Res.* **2020**, *45*, 6–35. [CrossRef]
39. Mohanraj, M.; Jayaraj, S.; Muraleedharan, C. Performance prediction of a direct expansion solar assisted heat pump using artificial neural networks. *Appl. Energy* **2009**, *86*, 1442–1449. [CrossRef]
40. Islam, K.T.; Raj, R.G.; Mujtaba, G. Recognition of traffic sign based on bag-of-words and artificial neural network. *Symmetry* **2017**, *9*, 138. [CrossRef]
41. Moya-Rico, J.D.; Molina, A.E.; Belmonte, J.F.; Tendero, J.C.; Almendros-Ibanez, J.A. Characterization of a triple concentric-tube heat exchanger with corrugated tubes using Artificial Neural Networks (ANN). *Appl. Therm. Eng.* **2019**, *147*, 1036–1046. [CrossRef]
42. Şahin, A.Ş. Performance analysis of single-stage refrigeration system with internal heat exchanger using neural network and neuro-fuzzy. *Renew. Energy* **2011**, *36*, 2747–2752. [CrossRef]
43. Garud, K.S.; Seo, J.H.; Cho, C.P.; Lee, M.Y. Artificial Neural Network and Adaptive Neuro-Fuzzy Interface System Modelling to Predict Thermal Performances of Thermoelectric Generator for Waste Heat Recovery. *Symmetry* **2020**, *12*, 259. [CrossRef]
44. Min, H.; Zhang, Z.; Sun, W.; Min, Z.; Yu, Y.; Wang, B. A thermal management system control strategy for electric vehicles under low-temperature driving conditions considering battery lifetime. *Appl. Therm. Eng.* **2020**, *181*, 115944. [CrossRef]
45. Seo, J.H.; Patil, M.S.; Cho, C.P.; Lee, M.Y. Heat transfer characteristics of the integrated heating system for cabin and battery of an electric vehicle under cold weather conditions. *Int. J. Heat Mass Transf.* **2018**, *117*, 80–94. [CrossRef]

symmetry

MDPI

Article

Experimental Performance Analysis of a Small Thermoelectric System Applicable to Real-Time PCR Devices

Jae Hoon Chung [1] and Sehoon Jeong [2,3,4,*]

1 Power Service BG, Doosan Heavy Industries & Construction, Changwon 51711, Korea; jaehoon.chung@doosan.com
2 Department of Healthcare Information Technology, Inje University, Gimhae 50834, Korea
3 Department of Information Communication System, Inje University, Gimhae 50834, Korea
4 Paik Institute for Clinical Research, Inje University, Busan 47392, Korea
* Correspondence: jeongsh@inje.ac.kr; Tel.: +82-55-320-4131

Received: 26 October 2020; Accepted: 21 November 2020; Published: 27 November 2020

check for updates

Abstract: At the International Space Station (ISS), not only observation of the space environment, but also biological and medical research under weightlessness has been conducted. The efficient use of energy from waste heat from the ISS away from the Earth is very important to the efficient operation of the ISS. To develop a thermoelectric module that can be used for real-time polymerase chain reaction (PCR) machinery used in biological and medical research, we simulated and evaluated the thermoelectric waste heat recovery system. Specifically, the thermoelectric module was attached to a stainless steel duct, and a hot air blower was faced with the duct inlet. The power of the thermoelectric system was measured by controlling the temperature of the hot air inlet. Additionally, the thermoelectric performance was evaluated according to the heat sink attached to the cold side of the thermoelectric module. Here, we also found the optimal heat exchange factors to improve the power and efficiency of the thermoelectric module. In this regard, it is expected that the thermoelectric module development and analysis study using waste heat will play an important role in the biological and medical research that is being conducted at ISS by developing a real-time PCR utilizing it.

Keywords: thermoelectric system; waste heat recovery; real-time polymerase chain reaction

1. Introduction

The increasing presence of humans in orbit over the last 50 years has shown that humans can adapt to short-term space flight. However, we still know very little about the long-term exposure to the space flight environment and its health-related consequences. Future missions to other planets and space objects, such as Mars and the moon and asteroids, provide a remarkable scientific opportunity for space biologists to explore life's ability to adapt to the environment of space flight during long-term missions.

More specifically, the study of animals in space allows us to investigate the effect of gravity on biological development, an area of research that is not open to humans. The ISS era promises an opportunity to observe and test various features of animal development during long-term exposure to microgravity, as well as access to centrifuges where specimens can be exposed to partial gravity loads. Space probes such as Neurolab provided evidence that frogs and rats need critical periods of gravity for biological development. The ability to store animals on the ISS for multiple life cycles allows scientists to determine exactly how, when, where, and why these gravitational dependencies exist [1]. By studying experimental animals aboard the space probe, scientists can better understand the adaptive response of animals or humans to long-term space flight. The results of these missions can also help determine the requirements for optimal human health in space [2,3].

There is no doubt that the Earth's gravitational field affects the physiology, morphology, and the behavior of life in almost every phenomenon. Space biology research covers a wide range of biological subfields, including gravity, developmental, and even radiation biology. It also focuses on advanced techniques, including molecular technology, genomics, DNA arrays, gene arrays, cell culture technology, and the study of related habitat systems [1]. Here, the most essential technique in life science research is polymerase chain reaction (PCR). PCR is used to make millions of copies of target DNA fragments, which have a wide range of specialized applications and are used by scientists in all fields of biology [4–6]. This is because PCR allows the identification and quantification of specific target species, even when very small numbers are present. Real-time PCR is based on the revolutionary method of PCR, an advanced laboratory technique in molecular biology, which monitors the amplification of target DNA molecules in real-time during PCR [7–9]. Thus, real-time PCR is an indispensable tool in modern molecular biology and has transformed scientific research and diagnostic medicine.

The PCR allows the polymerase to select a gene to amplify from a mixed DNA sample by adding small pieces of DNA that are complementary to the gene of interest. These tiny pieces of DNA are known as primers because the polymerase binds and prepares a DNA sample ready to begin copying the gene of interest. During PCR, temperature changes inside the PCR unit are of utmost importance, which is used to control the activity of the polymerase and the binding of primers. To start the reaction, the temperature is raised to 95 °C so that all double-stranded DNA is melted into a single strand. Then lower the temperature to 60 °C, which allows the primer to bind to the gene of interest. Accordingly, the polymerase has a place to bind and can start copying strands of DNA. The optimal temperature for the polymerase to work is 72 °C, so at this point, the temperature is raised to 72 °C to allow the enzyme to work faster. This temperature change is repeated over about 40 cycles. One copy continues until billions of copies have been made [10–12].

The efficient use of waste heat energy from ISS far from Earth is critical to the efficient operation of ISS. Since a large amount of thermal energy generated from spacecraft systems in space is released to the outside, applying a thermoelectric power generation system as a waste heat recovery device to utilize waste heat can increase the overall efficiency of the existing system. In this regard, we explored the applicability of the method of using thermoelectric power modules to utilize the temperature changes in the PCR process performed on spacecraft.

The main advantages of thermoelectric generators are direct energy conversion, no moving parts, and a relatively simple structure. However, the application of thermoelectric generators has a disadvantage that it is difficult to use variously due to the limitation of low conversion efficiency. Recently, the idea of using the TEG (thermoelectric generator) system as a waste heat recovery device has been under serious consideration in the automobile and incinerator industry. For waste heat recovery, it is unnecessary to consider the cost of the thermal energy input. The high conversion efficiency is not crucial. Rather, achieving higher power generation may be considered. Several types of thermoelectric generators for waste heat recovery were analyzed and developed by researchers with various approaches. Yu et al. developed a numerical model for a thermoelectric generator with a parallel-plate heat exchanger [13]. In this study, the optimization of the heat exchanger and thermoelectric module geometry was simultaneously performed by the numerical procedure. Crane et al. validated numerical heat exchanger models integrated with thermoelectric modules against experimental data [14]. The work suggests that the operating temperature range of the thermoelectric device should be delicately controlled to maximize the effectiveness of the system. Nuwayhid et al. developed a thermoelectric generator with natural convection cooling which recovered waste heat from a domestic woodstove [15]. In particular, the design factors that could achieve a low cost per watt were demonstrated. Esarte et al. researched the optimum heat exchanger system to maximize the performance of a thermoelectric generator [16]. In particular, theoretical expressions for heat exchangers were developed to compare with the experimental results and matched quite well in the low flow rate of the cooling medium. Saqr et al. reviewed the thermal design of thermoelectric generators for automobile exhaust waste heat recovery [17]. Four main factors that control the thermal efficiency of the thermoelectric generator were presented: heat exchanger geometry, heat exchanger materials, the installation site of the thermoelectric generator within the car, and the coolant system of the thermoelectric generator.

Regarding the above works, the present study tried to hold focus on the other aspects of optimized conditions. As an example, thermoelectric generation performance was evaluated depending on the compressive force applied to the module. The compressive force is, in specific, a crucial means not only to lock TEG device to the heat source, but also to minimize the heat resistance, so TEG can perform maximum power. However, some optimal force exists where, putting more compressive force can reduce the performance because of heat transfer convergence to its maximum level, and putting more stress on the module can reduce the power because of the material's deformation. Moreover, putting more stress may result in mechanical failure. Additionally, the effect of the thermoelectric module surface temperature uniformity on the overall power generation performance was evaluated quantitatively. On the other hand, the experiment on the thermoelectric module power performance depending on the heat sink was conducted in a similar but partially different way from what Esarte et al. did [16]. The main difference was that in the present study, various types of heat sinks with different geometry (fin height, fin array) were applied whereas in Esarte's work, heat sinks were almost similar to one other with the same size except that fin thicknesses and numbers were varied from one another. After all, the main purpose of the present study was to simulate the waste heat recovery TEG system in automobiles and small-sized industrial facilities and thus, find the design factor that enhances the overall performance of the TEG system.

2. Materials and Methods

Before discussing the experiment, thermoelectric generation phenomena will be briefly explained. The solid-state direct energy conversion is based on the thermoelectric phenomena which cover the Seebeck, Peltier, and Thompson effects. For the thermoelectric generation case, the dominating thermoelectric phenomenon is the Seebeck effect of which the main principle is as follows: When the junctions of two dissimilar semiconductors, as shown in Figure 1, are maintained at different temperatures, the current flows through the closed-loop. The temperature difference between the two junctions and the electromotive force has the following linear relation where is the Seebeck coefficient.

$$E = \alpha_{AB}\Delta T \tag{1}$$

where. E—open circuit voltage (V), α—Seebeck coefficient (V/K), ΔT—hot/cold side temperature difference [17]. The figure of merit ZT (figure of merit) determines the generation performance of a specific thermoelectric material and is characterized by the thermal/electrical conductivity, Seebeck coefficient, and the hot/cold side average temperature as described by Equation (2).

$$\mathrm{ZT} = \frac{\alpha^2\sigma}{\kappa}\overline{T} \tag{2}$$

where: ZT—figure of merit (-); κ—thermal conductivity (W/m^2·K); σ—electrical conductivity (Ω); $\overline{T}$—mean temperature (K) [17].

ZT is a measure that determines the performance of thermoelectric material itself. Higher ZT is desirable in general. Looking at Equation (2), we can find why higher ZT is desirable. Low thermal conductivity between the hot and cold side of the module ensures ΔT is relatively well preserved so that electricity is well-produced according to Equation (1). High electrical conductivity and the Seebeck coefficient implies the material has more capability to produce electricity. Higher mean operating temperature implies the material is capable of operating at high temperature, and consequently higher ΔT.

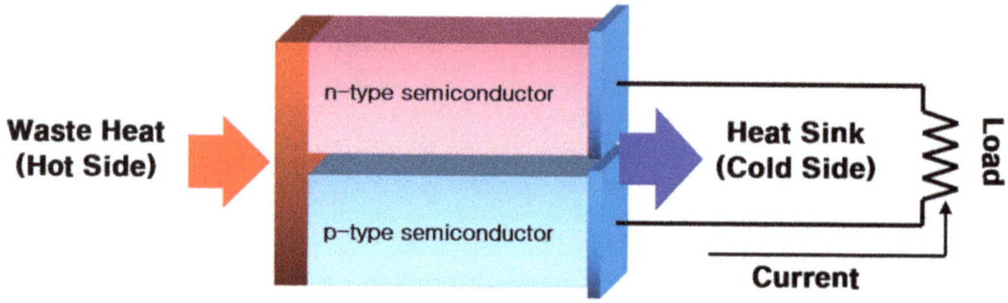

Figure 1. Schematic of the basic thermoelectric generation circuit. A temperature difference between two dissimilar semiconductors produces a voltage difference between the two substances, which can be applied to thermal-to-electrical energy conversion.

The conversion efficiency, described by Equation (3) could be derived by using energy balance and is shown as a function of ZT and the hot/cold side temperature in the thermoelectric material.

$$\varepsilon = \frac{T_H - T_C}{T_H} \frac{\sqrt{ZT+1} - 1}{\sqrt{ZT+1} + \frac{T_C}{T_H}} \tag{3}$$

where: T_H—hot side temperature (K); T_C—cold side temperature (K) [17].

As seen in Equation (3), T_H and T_C are fixed with varying ZT. We can see conversion efficiency increases with higher ZT.

Current commercial thermoelectric materials usually have ZT of 0.8–1 and conversion efficiency of 4–5%. However, thermoelectric material Bi_2Te_3, GeTe, PbTe are being developed continuously with each material being doped with Bi and In, and due to the recent developments, it has been reported ZT as high as 2.4 and that conversion efficiency as high as 15% can be achieved [18–20].

Now, looking at the thermoelectric generation system for exhaust gas waste heat recovery, the overall system efficiency cannot exactly match the thermoelectric conversion efficiency due to heat loss and thermal contact resistance which occurs at the interfaces. Considering such irreversible effects, the overall system efficiency of the TEG can be analyzed as follows:

$$\eta_{ov} = \varepsilon \times \eta_{HX} \times \rho, \tag{4}$$

where: ε—conversion efficiency (-); η_{HX}—heat exchanger efficiency; ρ—heat flux efficiency [16].

η_{HX} and ρ indicates the heat exchanger efficiency and heat flux efficiency, respectively. Here, heat exchanger efficiency, η_{HX}, is defied the ratio of $\Delta T_{\text{co-ci}}$ to $\Delta T_{\text{hi-ci}}$, which depends on fin geometry, convection type, and flow condition. Heat flux efficiency can be defined as a ratio of actual heat flux to maximum theoretical heat flux. Thermal resistance indeed exists within the interface region of the TEG system such as heat source to a thermoelectric module, and heat sink. Since the conversion efficiency, ε is an uncontrollable value at this moment, the focus should be on improving the other two values, by either designing a highly efficient heat exchanger or reducing thermal contact resistance.

In the present study, three different experiments were conducted, independently. In the first experiment, the thermoelectric module generation performance was evaluated using a high-performance thermoelectric module with heat source and sink block. The general specifications of the thermoelectric module are specified in Table 1. In the second experiment, the TEG system for exhaust gas waste heat recovery was constructed and the power performance characteristics were compared to those of the first experiment. In the last experiment, various types of heat sinks were applied to the thermoelectric module to determine the power generation performance. For the measurement devices, a K-Type thermocouple was used to measure the hot/cold side temperature of the thermoelectric module as well as the inlet/outlet gas temperature inside the TEG system. The voltage was measured by extending the lead wire of the thermoelectric module to the multimeter. The airflow rate in the TEG system was measured using the TESTO portable sensing probe. Overall, such data were converted into digitalized formats by using the Agilent 34970A data logger.

Table 1. Specifications for thermoelectric (TE) modules.

Model	Material	Max Oper. Temp.	Dimension	Electric Resistance
HZ-20	Bi_2Te_3	230 °C	75 × 75 mm^2	0.3 Ω

2.1. Module Performance Experimental Setup

As described in Table 1, the thermoelectric module is originally produced from Hi-Z Inc., USA, and was applied to the experimental device to evaluate its performance as seen in Figure 2. The HZ20 model was specified as having a maximum power performance of 19 W. A copper plate with four cartridge heaters inserted inside the plate was used as a heating device to ensure the temperature uniformity of the hot side surface of the thermoelectric module. An aluminum water cooling jacket with a thickness of 2 mm, was attached on the cold side surface of the thermoelectric module. Using a PID temperature controller, the module hot side temperature was controlled from 40–230 °C for the HZ20 model. On the other hand, the cooling jacket was constantly supplied with water of 8LPM and an approximate temperature of 17–18 °C. The hot side temperature was increased by 10 °C for each step, and the open-circuit voltage was measured. When each module achieved its maximum operating temperature, the electric power was measured with variable load resistance which ranges from 0.1–100 Ω. The reason for such measurement was to find the matching load in which the maximum power could be achieved. Additionally, by putting the load cell above the heat sink, the whole pile was compressed by the C-clamp. After so, the module generation performance depending on the compressive force was measured as well.

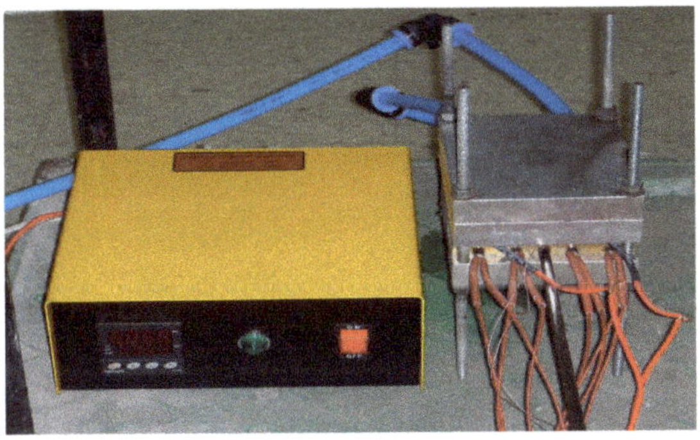

Figure 2. Experiment setup for measuring module performance.

2.2. TEG Setup and Power Performance Testing

A waste-heat recovering TEG system was constructed by the assembly of a duct, a heat exchanger, a water cooling jacket, and a thermoelectric module. The details of such assembly are described in Figure 3. The exhaust gas was simulated by a hot air blower which was capable of discharging air at a temperature ranging from 25–250 °C. Making indirect contact with hot air, the module was attached to the upper wall of the duct. Unlike the previous experiments, the module hot side temperature was controlled by using a hot air blower which was attached to the duct inlet. Figure 4 shows the schematic of the TEG system experimental setup. Strictly, it was the center of the module hot side temperature which was measured, but it will be abbreviated as just module hot side temperature as a convenience from next on. Additionally, the hot air inlet/outlet temperature and air flowrate at the normal temperature were measured. The module hot side temperature was controlled from 40–230 °C for the HZ20 model. As the whole system stabilized, the power performance was measured.

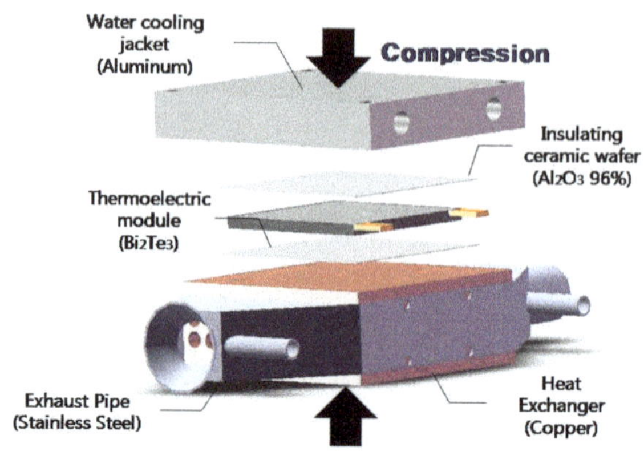

Figure 3. Assembly of waste heat recovering thermoelectric generator (TEG) system.

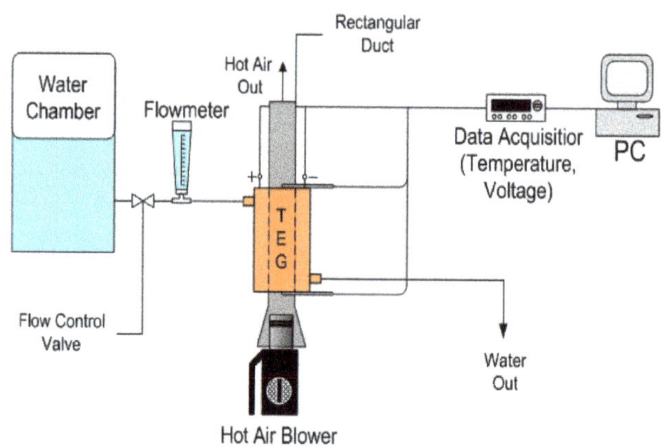

Figure 4. Schematic of a TEG experimental setup. The experimental setup is divided into two parts: (1) hardware for the experiment—hot air blower, exhaust pipe, a thermoelectric system with heat exchanger, water chamber to supply cold water; (2) data acquisition system: desktop computer, sensors to measure electricity, temperature, data logger.

Imitating exhaust pipe systems capable of waste heat recovery, the porous copper heat exchanger is stored inside the pipe system to maximize the heat transfer. The water jacket attached to the other side of the thermoelectric module was designed and aluminum is selected due to anti-corrosion characteristics and high heat conductivity. Between the water cooling jacket and exhaust pipe system, a commercial thermoelectric module is attached.

2.3. Power Generation Performance of TEM with Various Heat sink Types

For the present experiment, a total of six types of heat sinks were applied to the module, and each heat sink had its belonging case as shown in Figure 5. The experiment was conducted in two steps. For natural convection cooling, only the heat sink was attached to the cold side of the module in the first step. In the second step, a cooling fan was attached to the heat sink, for forced convection cooling. The cooling fan had a maximum flowrate of 10 CFM and consumed 5 W of electrical power. Additionally, it was attached to the heat sink in a way so that the air could flow in a parallel direction to the fins. As the module performance experiment, the module hot side temperature was controlled by the copper heating plate from 50–230 ° C and was increased by 30 °C for each step. To decrease the thermal contact, the thermal compound (silicone grease) was injected at the interface and the compressive force was applied on the whole pile by using the C-clamp. As the hot/cold side temperature of the module stabilized, the power was soon measured.

CASE	1	2	3
Shape	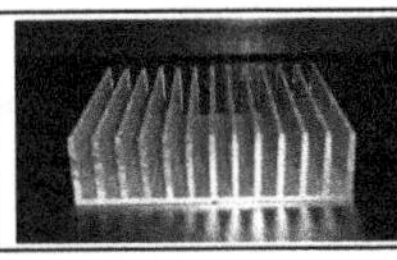	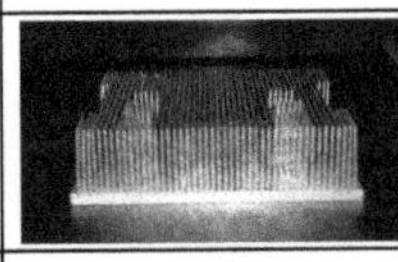	
Type	Extruded	Fin-bonded	Pin
Size	125 X 125 X 45 (mm^3)	125 X 125 X 45 (mm^3)	125 X 125 X 45 (mm^3)
Fin Number	13	46	169
Thickness (Fin/Pitch)	3/7 (mm)	1/2 (mm)	3/7 (mm)
CASE	4	5	6
Shape	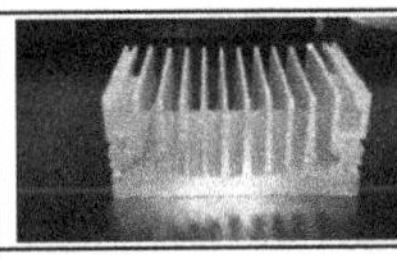		
Type	Extruded	Extruded	Extruded
Size	125 X 125 X 70 (mm^3)	125 X 125 X 120 (mm^3)	125 X 125 X 120 (mm^3)
Fin Number	12	25	9
Thickness (Fin/Pitch)	4/11 (mm)	2/8 (mm)	2/10 (mm)

Figure 5. Heat sink specifications. Basic specification of the six types of heat sinks used in the thermoelectric experiment (size, fin numbers, and thickness of fin/pitch).

3. Results and Discussion

3.1. Module Performance Experimental Results

Figure 6a describes the open-circuit voltage and electrical power of the HZ20 model, depending on the module's hot side temperature. Since ΔT has a linear relationship with voltage according to Equation (1), the power has a double square relationship with ΔT accordingly. For the hot side temperature condition, the model achieved the maximum power output of 15.5 W. Figure 6b shows the closed-circuit voltage and electrical power depending on the load resistance value. The measurements were done when the hot side temperature of the module was at the highest, which is 230 °C. It could be seen that the electric power of the thermoelectric module achieved its maximum at the load resistance value of 0.3 Ω. From such fact, it could be induced those load resistance value matched with the internal resistances of the thermoelectric module. Additionally, the values were the same as the internal resistances labeled in Table 1. In Figure 6c, the open-circuit voltage was measured depending on the compressive force applied to the thermoelectric module at hot side temperature 230 °C. From the graphs, it could be known that there existed an optimal point of the compressive force where the maximum generation performance could be achieved. The open-circuit voltage was maximized when 200 kgf of compressive forces had been applied. Applying more compressive force resulted in the generation performance decrease. One possible reason could be the internal crack within the thermoelectric module because of the excess compressive force.

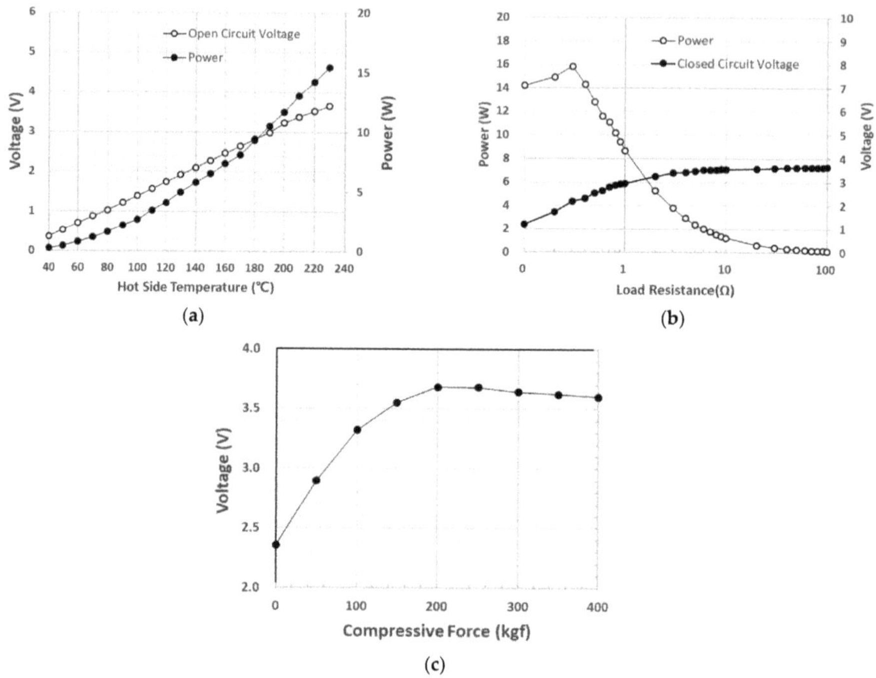

Figure 6. (**a**) Open circuit voltage and electrical power of HZ20 corresponding to the hot side temperature. (**b**) Closed-circuit voltage and electrical power corresponding to load resistance (@T_H = 230 °C). (**c**) Measurement of the open-circuit voltage depending on the compressive force applied to the thermoelectric module (@T_H = 230 °C) (n = 6).

3.2. Experimental Results of TEG Power Generation Performance

Based on the experimental results of the thermoelectric module performance test, the experimental results of the present TEG test were analyzed. Figure 7 describes the power performance of the TEG model respectively when attached to the exhaust gas duct. As mentioned earlier, since ΔT has a linear relationship with voltage according to Equation (1), it is reasonable that power came out as double square relation with ΔT. Although Figure 7 reveals quite a similar pattern with Figure 6a from the module performance test, it could be seen that the overall performance of the TEG system was declined. The maximum electric power was 8.2 W which was 2.3 W less than the previous module performance test. This was due to the presence of temperature gradient along the axial direction of the exhaust gas duct which affected the module surface temperature distribution as well. As a result, the possibility of the performance decline arises from the uniform temperature distribution on the module surface.

Since the hot side of the TEG is capable of operating up to 230 °C, and it is a device that only needs to be adjusted within 100 °C when developing a PCR device, it will not be difficult to get such a heat source in the ISS environment, whether it can be radiation energy or other heat devices. As mentioned earlier, in the process of selecting the gene to be amplified from the DNA sample of interest through PCR, the temperature change inside the unit between 60 °C and 95 °C is important; the temperature is raised to 95 °C so that all double-stranded DNA would melt into a single strand, and then lowered to 60 °C so that the primers could bind to the gene of interest. Additionally, the optimal temperature for the polymerase to operate the replication of DNA strands is 72 °C. The TEG may have great potential not only for providing power to the real-time PCR device, but also for using it as a heat. Preheating the real-time PCR device can be possible by utilizing the TEG cold side as a heat source. By using an adequate heat-resisting ceramic material, PCR can be controlled within 100 °C. This would effectively save energy for operating real-time PCR in the ISS environment where electric supply may be limited.

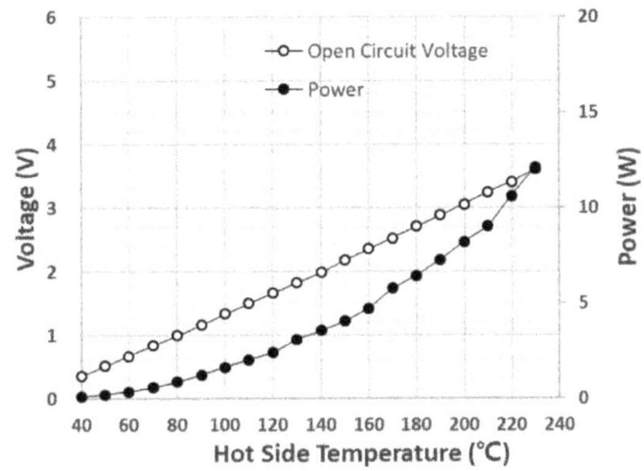

Figure 7. Open circuit voltage and power for the TEG depending on the hot side temperature of the thermoelectric module (n = 6).

3.3. Experimental Results of Power Generation Performance with Various Heat Sinks

The generation performance revealed different characteristics depending on the fin shapes of the heat sinks labeled in Figure 5. Figure 8 represents the hot/cold side temperature difference of the thermoelectric module depending on the hot side temperature for natural and forced convection cooling, respectively. Similarly, Figure 9 shows the electric power performance depending on the hot side temperature. It could be seen that the distributions of the six cases show a resemblance between Figures 6 and 7. Higher temperature difference showed higher power generation performance. For the natural convection cooling condition, as shown in Figures 8a and 9a, Case 6 (extruded fin of height 120 mm) showed the best power generation performance whereas Case 3 (bonded fin, height 45 mm) showed the lowest performance. This could be interpreted as the heat sink, having higher fins and wider fin pitches with less thermal contact resistance were best suited for the natural convection. It would have been relatively easy for natural convection induced buoyancy flows to pass over the path with enough space with reduced boundary layer blockage (Case 6) [21]. On the contrary, the heat sinks, having relatively low fins, narrow fin pitches with high thermal contact resistance, showed poor performance during the natural convection. The main reason for the low performance could be thought of as insufficient space between fins for the air to be circulated. To summarize, Case 6 achieved the highest electric power value of 2.4 W at the hot side temperature of 230 °C in natural convection mode. In the forced convection cooling case, however, the case distributions were relatively different from the previous natural convection cooling case. Case 3 showed the highest power generation performance and the most significant increase than the previous case, although Case 6 achieved relatively higher performance than others as well. This is because unlike the previous natural convection cooling condition, the air circulation between the fins in Case 3 would have been relatively easy by external means, in this case, a cooling fan, and with the contribution of a higher heat transfer area as well due to many numbers of fins [21]. Additionally, forced convection can have enough power to overcome boundary layer blockage within the passage. The highest electric power value at hot side temperature 230 °C was measured 10.5 W in this case which is 15 times the previous power value at the natural convection cooling. The heat sinks with relatively lower fins than others, except for Case 3, showed relatively low performance as expected. It could be concluded that heat sinks with higher fins, low fin pitches, and high heat transfer areas were well suited for forced convection cooling.

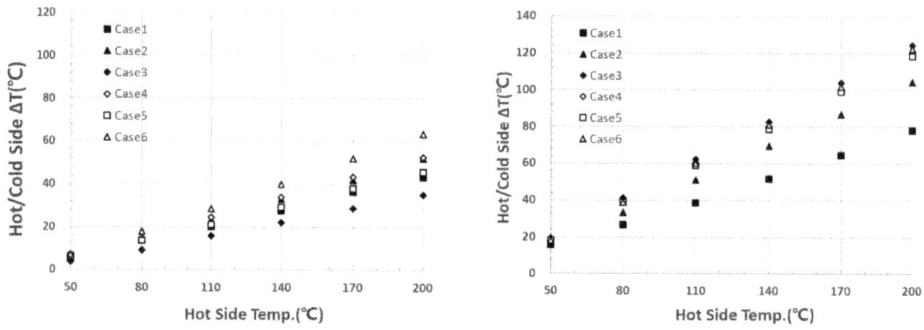

(**a**) (**b**)

Figure 8. Hot/cold temperature difference side by case corresponding to the hot side temperature for (**a**) natural convection; (**b**) forced convection ($n = 6$).

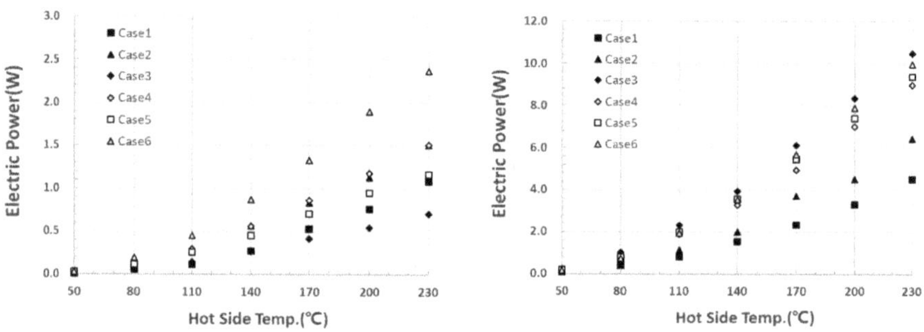

(**a**) (**b**)

Figure 9. Electric generation performance based on the electric power of the thermoelectric module side by case depending on the hot side temperature for (**a**) natural and (**b**) forced convection cooling ($n = 6$).

4. Conclusions

From the present study, the potential for adopting a thermoelectric system in the ISS environment, especially in the biomedical field, has been investigated by conducting various experiments. In conclusion, the following facts could be found out through the experiments. First, there was a maximum point where the thermoelectric power performance deteriorated when more compression was applied. Second, one of the important factors of a TEG system performance was the temperature uniformity of a thermoelectric module surface and the necessity of a heat spreader was important in a TEG system. Third, when using fins as a heat sink, forced convection insured much higher thermoelectric generation performance than that of natural convection, in general. However, each heat sink revealed different performance in each condition. With fulfilling optimum force, high-performance heat sink, and temperature uniformity in the ISS environment, the PCR system will be provided with stable electric energy to operate.

Author Contributions: Conceptualization, J.H.C. and S.J.; methodology, J.H.C.; software, J.H.C.; validation, J.H.C. and S.J.; formal analysis, J.H.C. and S.J.; investigation, J.H.C. and S.J.; data curation, J.H.C. and S.J.; writing—original draft preparation, J.H.C. and S.J.; writing—review and editing, J.H.C. and S.J.; visualization, J.H.C.; supervision, S.J.; funding acquisition, S.J. All authors have read and agreed to the published version of the manuscript.

Funding: This research received no external funding.

Acknowledgments: This work was partially supported by the National Research Foundation of Korea (NRF) grant funded by the Korea government (MSIT) (No. NRF-2018R1C1B5086455).

Conflicts of Interest: The authors declare no conflict of interest.

References

1. Clément, G.S.K. *Fundamentals of Space Biology: Research on Cells, Animals, and Plants in Space*; Springer: New York, NY, USA, 2006; Volume 18, p. 376.
2. Goswami, N.; Roma, P.G.; De Boever, P.; Clément, G.; Hargens, A.R.; Loeppky, J.A.; Evans, J.M.; Peter Stein, T.; Blaber, A.P.; Van Loon, J.J.W.A.; et al. Using the Moon as a high-fidelity analogue environment to study biological and behavioral effects of long-duration space exploration. *Planet. Space Sci.* **2012**, *74*, 111–120. [CrossRef]
3. De la Torre, G.G. Cognitive Neuroscience in Space. *Life* **2014**, *4*, 281–294. [CrossRef] [PubMed]
4. Arnheim, N.; White, T.; Rainey, W.E. Application of PCR: Organismal and Population Biology. *BioScience* **1990**, *40*, 174–182. [CrossRef]
5. Reynolds, R.; Sensabaugh, G.; Blake, E. Analysis of genetic markers in forensic DNA samples using the polymerase chain reaction. *Anal. Chem.* **1991**, *63*, 2–15. [CrossRef] [PubMed]
6. David, N.F.; Relman, D.A. Application of Polymerase Chain Reaction to the Diagnosis of Infectious Diseases. *Clin. Infect. Dis.* **1999**, *29*, 475–486.
7. Maurer, J.J. Rapid Detection and Limitations of Molecular Techniques. *Annu. Rev. Food Sci. Technol.* **2011**, *2*, 259–279. [CrossRef] [PubMed]
8. Boonham, N.; Kreuze, J.; Winter, S.; van der Vlugt, R.; Bergervoet, J.; Tomlinson, J.; Mumford, R. Methods in virus diagnostics: From ELISA to next generation sequencing. *Virus Res.* **2014**, *186*, 20–31. [CrossRef] [PubMed]
9. Umesha, S.; Manukumar, H.M. Advanced molecular diagnostic techniques for detection of food-borne pathogens: Current applications and future challenges. *Crit. Rev. Food Sci. Nutr.* **2018**, *58*, 84–104. [CrossRef] [PubMed]
10. Bej, A.K.; Mahbubani, M.H.; Atlas, R.M. Amplification of Nucleic Acids by Polymerase Chain Reaction (PCR) and Other Methods and their Applications. *Crit. Rev. Biochem. Mol. Biol.* **1991**, *26*, 301–334. [CrossRef] [PubMed]
11. Au Lorenz, T.C. Polymerase Chain Reaction: Basic Protocol Plus Troubleshooting and Optimization Strategies. *JoVE* **2012**, e3998. [CrossRef] [PubMed]
12. Adessi, C.; Matton, G.; Ayala, G.; Turcatti, G.; Mermod, J.-J.; Mayer, P.; Kawashima, E. Solid phase DNA amplification: Characterisation of primer attachment and amplification mechanisms. *Nucleic Acids Res.* **2000**, *28*, e87. [CrossRef] [PubMed]
13. Yu, J.; Zhao, H. A numerical model for thermoelectric generator with the parallel-plate heat exchanger. *J. Power Sources* **2007**, *172*, 428–434. [CrossRef]
14. Crane, D.T.; Jackson, G.S. Optimization of cross flow heat exchangers for thermoelectric waste heat recovery. *Energy Convers. Manag.* **2004**, *45*, 1565–1582. [CrossRef]
15. Nuwayhid, R.Y.; Shihadeh, A.; Ghaddar, N. Development and testing of a domestic woodstove thermoelectric generator with natural convection cooling. *Energy Convers. Manag.* **2005**, *46*, 1631–1643. [CrossRef]
16. Esarte, J.; Min, G.; Rowe, D.M. Modelling heat exchangers for thermoelectric generators. *J. Power Sources* **2001**, *93*, 72–76. [CrossRef]
17. Saqr, K.M.; Mansour, M.K.; Musa, M.N. Thermal design of automobile exhaust based thermoelectric generators: Objectives and challenges. *Int. J. Automot. Technol.* **2008**, *9*, 155–160. [CrossRef]
18. Suwardi, A.; Cao, J.; Zhao, Y.; Wu, J.; Chien, S.W.; Tan, X.Y.; Hu, L.; Wang, X.; Wang, W.; Li, D.; et al. Achieving high thermoelectric quality factor toward high figure of merit in GeTe. *Mater. Today Phys.* **2020**, *14*, 100239. [CrossRef]
19. Suwardi, A.; Cao, J.; Hu, L.; Wei, F.; Wu, J.; Zhao, Y.; Lim, S.H.; Yang, L.; Tan, X.Y.; Chien, S.W.; et al. Tailoring the phase transition temperature to achieve high-performance cubic GeTe-based thermoelectrics. *J. Mater. Chem. A* **2020**, *8*, 18880–18890. [CrossRef]

20. Perumal, S.; Samanta, M.; Ghosh, T.; Shenoy, U.S.; Bohra, A.K.; Bhattacharya, S.; Singh, A.; Waghmare, U.V.; Biswas, K. Realization of High Thermoelectric Figure of Merit in GeTe by Complementary Co-doping of Bi and In. *Joule* **2019**, *3*, 2565–2580. [CrossRef]
21. Incropera, F.P.; DeWitt, D.P.; Bergman, T.L.; Lavine, A.S. *Fundamentals of Heat and Mass Transfer*; John Wiley & Sons: Hoboken, NJ, USA, 2006.

symmetry

Article

Numerical Investigations on Magnetohydrodynamic Pump Based Microchannel Cooling System for Heat Dissipating Element

Jae-Hyeong Seo [1,†], Mahesh Suresh Patil [1,†], Satyam Panchal [2] and Moo-Yeon Lee [1,*]

1 Department of Mechanical Engineering, Dong-A University, 37 Nakdong-Daero 550, Saha-gu, Busan 49315, Korea; cheonchw@donga.ac.kr (J.-H.S.); 1576457@donga.ac.kr (M.S.P.)

2 Department of Mechanical and Mechatronics Engineering, University of Waterloo, 200 University Avenue West, Waterloo, ON N2L 3G1, Canada; satyam.panchal@uwaterloo.ca

* Correspondence: mylee@dau.ac.kr; Tel.: +82-51-200-5560

† Equal contribution.

Received: 17 September 2020; Accepted: 12 October 2020; Published: 16 October 2020

Abstract: Numerical investigations are performed on the magnetohydrodynamic (MHD) pump-based microchannel cooling system for heat dissipating element. In the present study, the MHD pump performance is evaluated considering normal current density, magnetic flux density, volumetric Lorentz force, shear stress and pump flow velocity by varying applied voltage and Hartmann number. It is found that for a low Hartmann number, the Lorentz force increases with an increase in applied voltage and Hartmann number. The velocity distribution along dimensionless width, the shear stress distribution along dimensionless width, the magnetic flux density along the dimensionless width and radial magnetic field distribution showed symmetrical behavior. The MHD pump-based microchannel cooling system performance is evaluated by considering the maximum temperature of the heat dissipating element, heat removal rate, efficiency, thermal field, flow field and Nusselt number. In addition, the influence of various nanofluids including Cu-water, TiO_2-water and Al_2O_3-water nanofluids on heat transfer performance of MHD pump-based microchannel is evaluated. As the applied voltage increased from 0.05 V to 0.35 V at Hartmann number 1.41, the heat removal rate increased by 39.5%. The results reveal that for low Hartmann number, average Nusselt number is increasing function of applied voltage and Hartmann number. At the Hartmann number value of 3.74 and applied voltage value of 0.35 V, average Nusselt numbers were 12.3% and 15.1% higher for Cu-water nanofluid compared to TiO_2-water and Al_2O_3-water nanofluids, respectively. The proposed magnetohydrodynamic microcooling system is effective without any moving part.

Keywords: cooling; Lorentz force; magnetohydrodynamics; microchannel; MHD pump

1. Introduction

Magnetohydrodynamic (MHD) pumps have been focus of research owing to various advantages over traditional pumps in many specific areas of application including biological fields, solar applications and heat transfer systems [1]. The major advantage of such pumps is that they are free of any moving parts. Additionally, the miniaturization of such pumps due to their simple structure, can be utilized in microfluidic systems, microcooling systems and microelectromechanical system (MEMS) applications [2,3]. In a few applications, where it is difficult to use conventional pumps such as molten metal pumping, these pumps are more useful and efficient. Moreover, the applications requiring no moving sections, for example, in spaceships and biological applications like blood pumping, these pumps can be used [4]. Out of various applications, one of the promising usages of MHD pumps is cooling of heat dissipating element. The coolant flow is generated by MHD pumps and can be made to flow in the microchannel where the dissipated heat from the heat dissipating element is taken away. Use of microchannels in a cooling system is one of the efficient ways of dissipating heat [5,6]. In such instances, heat transfer effectiveness and the thermal behavior of a cooling system with its influencing factors need to be investigated.

Lemoff et al. [7] developed and presented one of the first MHD micropumps with AC current using Lorentz force to pump electrolytic solution in microchannel. The authors showed that the continuous flow without any pulse can be produced. Rivero and Cuevas [8] studied MHD micropumps in one and two-dimensional flow models for laminar flows in parallel plates and rectangular ducts by considering the influence of slip condition which can be used to design MHD micropumps and characterize the flow behavior in these microfluidic devices. The 2D model presented by the authors showed more accuracy with results of experimentation as compared to 1D model [8]. Zhao et al. [9] conducted an analytical study by using the separation of variables method for generalized Maxwell fluids in a MHD rectangular micropump operated under the AC electric field and found that for given oscillating Reynolds number, large Hartmann number leads to large amplitudes of velocity. Yousofvand et al. [10] investigated heat transfer and pumping performance of electromagnetic pump considering Cu-water nanofluid as working fluid and found that for low Hartmann numbers, body force increases whereas for $Ha > 200$, the opposite trend is observed. Moghaddam analytically investigated the MHD micropump performance considering circular channel. The author found that average dimensionless velocity initially increases with increase in Hartmann number and dimensionless radius. However, after attaining peak, the average dimensionless velocity decreases with increase in Hartmann number and dimensionless radius [11]. Miroshnichenko et al. [12] studied MHD natural convection in a partially open trapezoidal cavity under the influence of various magnetic field orientations and found that an increase in uniform magnetic field value decreases the rate of heat transfer. A comprehensive study of power-law fluids in MHD natural convection has been conducted by Kefayati [13,14]. Shirvan et al. [15] conducted numerical investigations on MHD flow in a square cavity with different inlet and outlet ports. The authors presented optimization of mean Nusselt number using orthogonal array optimization. Kiyasatfar et al. [16] investigated thermal behavior and fluid motion in direct current (DC) MHD pump by varying magnetic flux density, applied current and channel size. The authors found that the maximum velocity increases with increase in applied current and as Hartmann number increases the velocity profile becomes flatter. Larimi et al. [17] studied the effect of non-uniform transverse magnetic field arrangements with a different Reynolds number for magnetic nanofluids on heat transfer and found that applying external magnetic fluid is strongly effective in fluid cooling at low Reynolds number. Kolsi et al. [18] performed a numerical study for 3D MHD natural convection inside a cubical enclosure with an inclined plate and found an optimal inclination angle of 180° for the plate. Kefayati considered various flow types including non-Newtonian nanofluids [19], blood flow [20] and power-law fluids in an internal flow [21] with focus of investigation on the effects of the power-law index, Reynolds number on thermal behavior by varying magnetic field to find optimized conditions. Further research has been conducted to understand the flow behavior of MHD considering different cases [22,23].

The MHD pump involves two types of heat transfer mechanism: forced convection and mixed convection. The micro-cooling of the heat dissipating element is a case of mixed convection owing to its microstructure and very low flow rate. Mixed convection heat transfer has attracted significant research attention of heat transfer engineers owing to various application fields including heat exchangers, electronic cooling [24], heat dissipating element cooling [25], micro-cooling, MEMS applications, solar energy applications and metal casting [26]. Micro-cooling application is one of the critical research areas which has gained importance due to recent trends of miniaturization of devices as well as high power applications, which results in large amount of heat generation in compact volume. The various cooling methods previously suggested, including direct fan cooling [27] and thermoelectric cooling, suffer from low efficiency and high-power consumption. In addition, the presence of moving components makes conventional cooling methods less desirable [28]. Therefore, in the present study, MHD pump-based microchannel cooling for a heat dissipating element is investigated. The MHD pump performance is evaluated by varying the applied voltage and Hartmann number, and its effect on various parameters including normal current density, magnetic flux density, volumetric Lorentz force, shear stress and pump flow velocity is reported. The heat transfer performance of the MHD pump-based microchannel cooling for a heat dissipating element is reported by considering the heat removal rate, efficiency, thermal field, flow field and Nusselt number. In addition, three different nanofluids, including Cu-water, TiO_2-water and Al_2O_3-water, are considered, and their influence on heat transfer performance is compared. The comparative heat transfer performance and potentials of various nanofluids in MHD pump application for microchannel cooling have not been realized. This study provides a comprehensive understanding of MHD pump performance, heat transfer performance of MHD pump-based microchannel cooling systems, and the influence of various nanofluids on heat transfer performance.

2. Method

2.1. Numerical Modeling

A schematic view of an MHD pump for cooling a heat dissipating element is presented in Figure 1. A heat dissipating element can be any microsystem including microfluidic devices, micro-batteries, electronic chips, light emitting diodes (LED), etc. The basic principle of operation of MHD pumps is based on the Lorentz force in which magnetic and electrical fields are kept perpendicular, which forces conducting fluids in a perpendicular direction to both electric currents and magnetic fields, creating an MHD pump effect. The magnetic field strength and applied current both affect the flow velocity. The magnetic field is created by keeping two small permanent magnets. The origin of the coordinate system lies between two magnets and it is equidistance from the magnets. The origin of the coordinate system lies exactly at the center of the MHD pump without considering the microchannel dimensions (Figure 1). The origin of the coordinate system has been chosen specifically at the center of the MHD pump (without considering microchannel dimensions) for simplicity in the calculations. The width of the MHD pump is chosen as a characteristic length of the system considering width as an important dimension of the MHD pump system along which various parameters are evaluated. Due to the Lorentz force, the coolant flows in the positive X-axis direction (i.e., from the MHD pump and towards the microchannel) as shown in Figure 1. The microchannel consists of four slots. Details of the MHD pump dimensions are provided in Table 1.

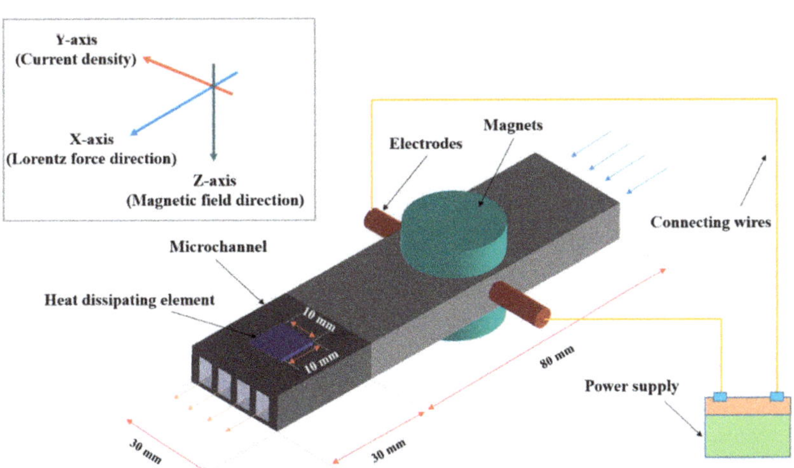

Figure 1. Schematic view of the magnetohydrodynamic (MHD) pump microchannel cooling system for a heat dissipating element.

Table 1. MHD pump and microchannel dimensions.

Item	Parameter	Values
MHD pump	Length × Height (mm)	80 × 10
Microchannel	Length × Width × Height (mm)	30 × 30 × 10
	Number of channel slots (ea)	4
Single channel	Width × Height (mm)	4 × 7
Magnet radius	Radius × Height (mm)	15 × 7.5
Heat dissipating element	Length × Width × Height (mm)	10 × 10 × 1

2.2. Governing Equations and Boundary Conditions

The modeling of the MHD phenomena involves a multiphysics problem with coupled equations between fluid flow, heat transfer, current flow, and magnetic fields, which are solved numerically. The different fields of physics involved are expressed by partial differential equations, which can be solved via the finite element method. In the present study, the numerical modeling of the MHD phenomena is conducted using COMSOL. The partial differential equations involving multiphysics behavior with coupling between fluid flow, heat transfer, electric current and magnetics are solved using the finite element method. The fluid flow and heat transfer are governed by the Navier–Stokes equation as shown below [29]. Equations (1)–(3) show continuity, momentum and energy conservation, respectively [30], where $\vec{V}$ is velocity, ρ is density, p is pressure and α is thermal diffusivity.

$$\nabla\cdot\vec{V} = 0 \tag{1}$$

$$(\vec{V}\cdot\nabla)\vec{V} = \frac{1}{\rho}\nabla P + \nabla^2\vec{V} + \frac{1}{\rho}\vec{F} \tag{2}$$

$$(\vec{V}\cdot\nabla)T = \alpha\nabla^2 T \tag{3}$$

$$\vec{F} = \vec{J}\times\vec{B} \tag{4}$$

$$\vec{J} = \sigma\left(\vec{E} + \vec{V}\times\vec{B}\right) \tag{5}$$

$\vec{F}$ is the body force due to Lorentz forces which causes fluid motion as shown in Equation (4) [9]. The electric current density which is defined by Ohm's law is shown in Equation (5) [31], where $\vec{J}$ is the electric current in y-direction and $\vec{B}$ is the magnetic field in the z-direction. The electric current and magnetic field are perpendicular which creates a Lorentz force in the x-direction.

The working fluid is Newtonian fluid with flow considered as steady and laminar based on the low Reynolds number. The thermo-physical properties of working fluid, nanoparticle and boundary conditions are presented in Table 2. The heat dissipating element that is acting on the pump's wall is assumed to be a constant volumetric heat generation source. The applied electric voltage is varied from 0.05 V to 0.35 V with an interval of 0.05 V. The Hartmann number is varied from 1.41 to 3.74. The cylindrical type permanent magnets are used for providing the magnetic field intensity. Three different types of nanofluids are considered including Cu-water, TiO_2-water, and Al_2O_3-water nanofluids. The base fluid for all the nanofluids is water. The boundary condition of opening at atmospheric pressure is applied at the coolant inlet and coolant outlet. The density of water is considered as 997.0 kg/m^3 at 25 °C and assumed as an incompressible fluid. The thermal conductivity of water is considered as 0.6069 W/m-K at 25 °C. The specific heat of water is considered as 4181.7 J/kg-K. The details about the boundary conditions and thermophysical properties of water and nanoparticles are presented in Table 2.

Table 2. Boundary conditions and thermophysical properties.

Specifications	Values			
	Boundary conditions			
Inlet coolant temperature (°C)	25			
Applied Voltage (V)	0.05, 0.10, 0.15, 0.20, 0.25, 0.30, 0.35			
Volumetric heat generation rate (W/m^3)	1.0×10^8			
Coolant inlet	Opening at atmospheric pressure			
Coolant outlet	Opening at atmospheric pressure			
	Thermophysical properties			
	Water	**Cu**	**TiO_2 [32]**	**Al_2O_3 [33]**
Density (kg/m^3)	997	8954	4260	3970
Thermal conductivity (W/m-K)	0.6069	400	8.9	25
Specific heat (J/kg-K)	4181.7	383	686.2	765

2.3. Nanofluid Relations

The density of nanofluid with various nanoparticle volume fraction is predicted by the Pak et al. [34] as shown in Equation (6). Zhong et al. [35] experimentally measured the density of TiO_2-water nanofluid and compared the predictions using the Equation (6) within 0.54%. Therefore, in the current study, the density of various nanofluids with different volume fraction is calculated using Equation (6), where ρ denotes density and ϕ denotes volume fraction.

$$\rho_{nf} = \phi\rho_n + (1-\phi)\rho_f \tag{6}$$

The viscosity of the nanofluid (μ_{nf}) with various nanoparticle volume fraction (ϕ) using viscosity of fluid (μ_f) is predicted by various researchers including Batchelor [36], Vand [37], Wang et al. [38], Duangthongsuk et al. [39] and Bobbo et al. [40], as shown in Equations (7)–(11), respectivelyBased on the model prediction accuracy with experimental data [35], in the present study, the model proposed by Want et al. [38] is used for calculating the viscosity of nanofluid.

$$\mu_{nf} = \mu_f\left(1 + 2.5\phi + 6.5\phi^2\right) \tag{7}$$

$$\mu_{nf} = \mu_f\left(1 + 2.5\phi + 7.349\phi^2\right) \tag{8}$$

$$\mu_{nf} - \mu_f\left(1 + 7.3\phi + 123\phi^2\right) \tag{9}$$

$$\mu_{nf} = \mu_f\left(1.013 + 0.092\phi - 0.015\phi^2\right) \tag{10}$$

$$\mu_{nf} = \mu_f\left(1 + 0.36838\phi + 0.25271\phi^2\right) \tag{11}$$

The thermal conductivity and specific heat of nanofluid for various volume fractions are calculated using Equation (12) [41] and Equation (13) [42], respectively. The effect of the nanoparticle volume fraction on the mixture properties is presented in Figure 2.

$$k_{nf} = k_f \frac{k_p + 2k_f + 2\phi\left(k_p - k_f\right)}{k_p + 2k_f - 2\phi\left(k_p - k_f\right)} \tag{12}$$

$$\left(\rho C_p\right)_{nf} = (1-\phi)\left(\rho C_p\right)_f + \phi\left(\rho C_p\right)_n \tag{13}$$

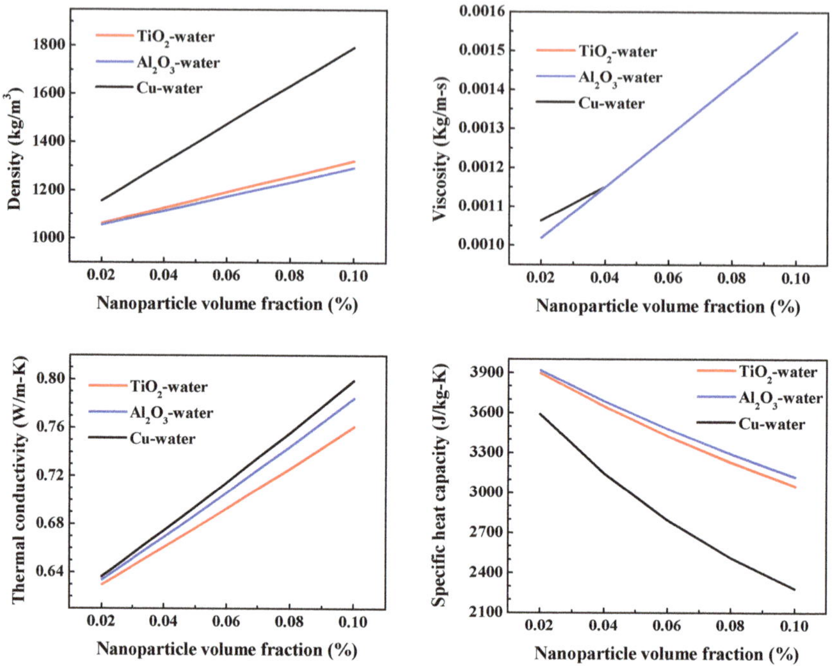

Figure 2. Effect of nanoparticle volume fraction on the mixture properties.

2.4. Mesh Independency

Figure 3 shows the details of the mesh independency test. The Lorentz force and average velocity are considered as parameters to evaluate the mesh independency. In the present study, the mesh type is defined as the number of elements in the generated mesh. Mesh type 1 contains 5.43×10^4 elements, which is a coarse mesh, whereas mesh type 5 contains 1.45×10^6 elements, which is a finer mesh. As the mesh elements increased from 9.65×10^5 to 1.45×10^6, the Lorentz force and average velocity varied only 0.008% and 0.166%, respectively. Considering the computational cost and accuracy of the numerical simulations, mesh type 4 with 9.65×10^5 elements, is selected for carrying out numerical simulations as shown in Table 3.

Table 3. Mesh details.

Mesh Type	Number of Elements
Type 1	5.43×10^4
Type 2	1.56×10^5
Type 3	6.09×10^5
Type 4	9.65×10^5
Type 5	1.45×10^6

2.5. Data Reduction

The MHD pump flow is generated by the application of electric and magnetic field, which interacts with the conducting fluid. The developed flow is described as Hartmann flow and the non-dimensional number, known as the Hartmann number (Ha), is defined as shown in Equation (14), where B is magnetic flux intensity, L is characteristics length, σ is electrical conductivity and μ is dynamic viscosity [21]. The Hartmann number gives an estimation of the magnetic forces compared to viscous force [9].

$$Ha = BL(\sigma / \mu)^{0.5} \tag{14}$$

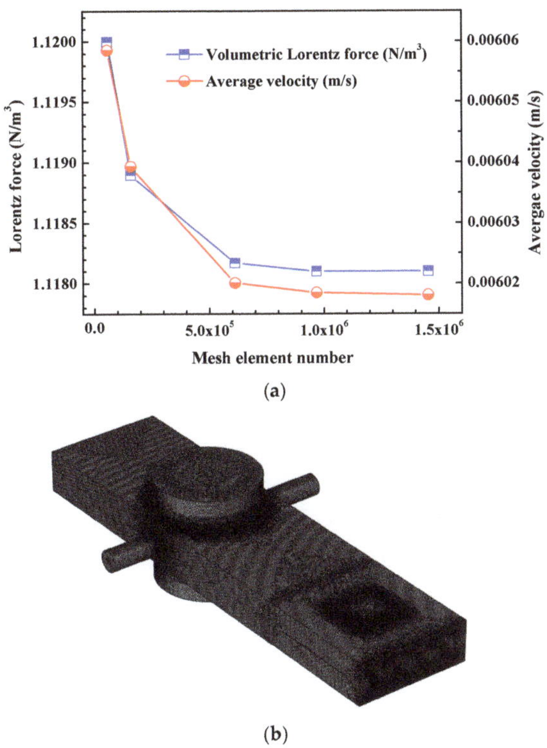

Figure 3. Mesh details (**a**) Mesh independency test (**b**) Meshing of magnetohydrodynamic (MHD) pump microchannel cooling system for heat dissipating element.

The convective heat transfer rate is used to obtain heat transfer coefficient and calculate average Nusselt number (Nu_{avg}). The heat transfer rate is evaluated as shown in Equation (15) [43].

$$Q_{conv} = m_{in} C_p \left(T_{bulk,out} - T_{bulk,in}\right) \tag{15}$$

The average heat transfer coefficient is evaluated from Equation (16). The numerator is convective heat transfer from wall to fluid and the denominator is a combined term consisting of the wall convective surface area and logarithmic mean temperature difference of the wall-and-bulk fluid [25].

$$h_{avg} = \frac{Q_{conv}}{A_{wall}\,(T_{wall} - T_{bulk})_{LMTD}} \tag{16}$$

$$(T_{wall} - T_{bulk})_{LMTD} = \frac{\Delta T_{wall-buk,in} - \Delta T_{wall-buk,out}}{\log(\Delta T_{wall-buk,in} / \Delta T_{wall-buk,out})} \tag{17}$$

where $\Delta T_{wall-buk,in}$ and $\Delta T_{wall-buk,out}$ indicate the differences between the wall temperature and bulk fluid temperature at the inlet and outlet of the channel, respectively (Equation (17)). The average

Nusselt number is calculated as shown in Equation (10) where D_h represents the hydraulic diameter and k_f represents the thermal conductivity of the fluid.

$$Nu_{avg} = \frac{h_{avg} \times D_h}{k_f} \tag{18}$$

3. Results and Discussion

The results of the numerical study on the MHD pump subjected to the mentioned boundary conditions are presented in terms of normal current density, magnetic flux density, volumetric Lorentz force, shear stress and pump flow velocity by varying applied voltage and Hartmann number. For evaluating the MHD pump performance, Cu-water nanofluid with 0.1% volume fraction was considered. In the subsequent sub-sections, the performance of the MHD pump-based microchannel cooling system is presented considering various parameters, including the maximum temperature of a heat dissipating element, heat removal rate, efficiency, thermal field, flow field and Nusselt number. In addition, the heat transfer performance of Cu-water nanofluid is compared with TiO_2-water nanofluid and Al_2O_3-water nanofluid. The study provided an in-depth understanding of the MHD pump functioning and its application in micro-cooling systems.

3.1. Validation

The numerical study is validated with the previously published literature. The server workstation with an Intel (R) Xenon(R) CPU E5-2620 v3 @2.40 GHz including 24 cores and 64 GB computation memory is used to run the simulations. To ensure the accuracy of the numerical study method, numerically predicted velocity is compared with previously published experimental data [7] and numerical data [10] as shown in Figure 4. It is demonstrated that the predicted velocity closely matches with the linear fit to the experimental data and numerical data. Thus, the validation of the numerical model is confirmed.

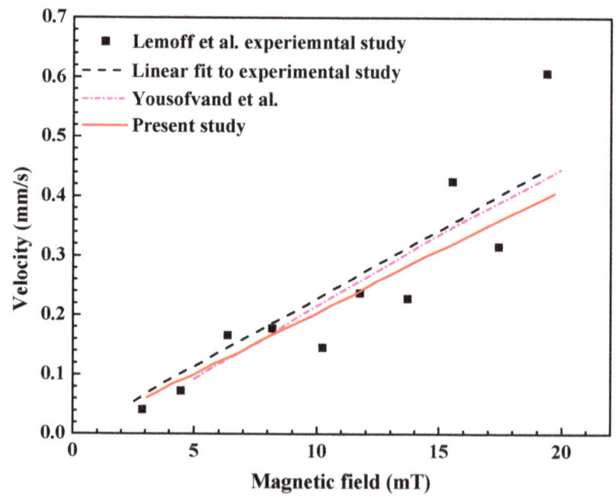

Figure 4. Velocity comparison between present study and the Lemoff et al. [7] experimental study and Yousofvand et al. [10] numerical study.

3.2. Magnetohydrodynamic Pump (MHD) Performance

Figure 5a shows the variation of normal current density with the applied voltage and Hartmann number. The normal current density increased with the increase in applied voltage. For example, as the applied voltage increased from 0.05 V to 0.35 V, at a Hartmann number value of 2.0, the normal current density increased 600%, or 6 times. For the same applied voltage, a higher normal current density is observed for the higher Hartmann number. As the Hartmann number increased from 1.41 to 3.76 at a constant applied voltage of 0.35 V, the normal current density increased 600%, or 6 times. The combined influence of a higher applied voltage and higher Hartmann number are visible with a significant increase in the normal current density.

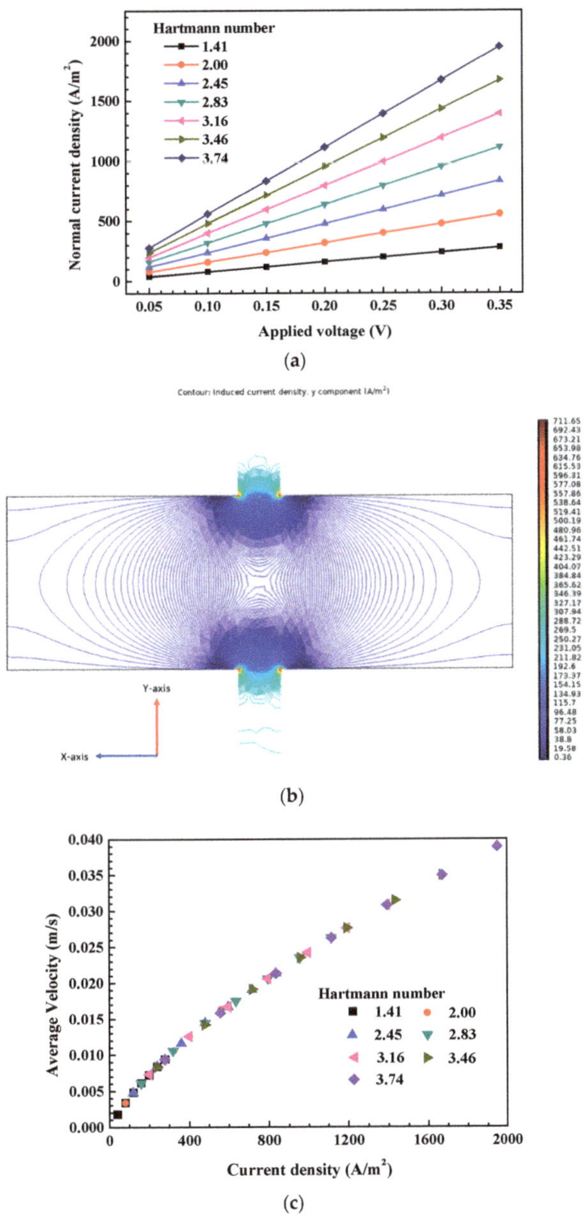

Figure 5. Current density and velocity (**a**) Normal current density variation for different applied voltage and different Hartmann number (**b**) Induced current density distribution (**c**) Variation of average velocity with current density.

Figure 5b shows the spatial variation of induced current and it can be seen that the induced current density is higher near electrode area. Figure 5c shows the variation of the average velocity with respect to current density. The average velocity increased linearly with the increase in current density. The flow rate can be increased either by increasing applied current, keeping magnetic flux constant or by increasing magnetic flux while keeping the applied current constant to enhance the pump performance. Similar trends have been observed by previously conducted studies [16]. For low Hartmann numbers, the velocity increased with an increase in the Hartmann number. However, the high Hartmann number can have a negative effect on the velocity as well as volumetric flow rate [11]. For a low Hartmann number, forced convection dominates with higher velocities which is useful for enhancing the pump performance.

Figure 6a shows the variation of magnetic flux along the dimensionless width in the *Y*-axis at the center of the magnetohydrodynamic pump. The maximum value of the magnetic flux attained is about 0.25 T at the center of the MHD pump channel. However, the value of the magnetic flux density near the conducting electrode is found to be in the order of 0.11 T. Similar results have been obtained by Aoki et al. [44]. The magnetic flux showed axisymmetric behavior for the axis passing through the center of the dimensionless width. Figure 6b shows the magnetic field distribution for the MHD pump on the *XY*-plane. As in the present study, the cylindrical permanent magnet is considered for the MHD pump application and the circular magnetic field pattern is observed. The maximum magnetic field value of the order of 100 kA/m is observed. The magnetic field showed radial symmetric behavior for the axis passing through the center of the magnet.

Figure 7 shows the volumetric Lorentz force variation for applied voltage and Hartmann number. The volumetric Lorentz force increased with increase in applied voltage. For example, as the applied voltage increased from 0.05 V to 0.35 V at a Hartmann number value of 2.0, the volumetric Lorentz force increased 600%, or 6 times. For the same applied voltage, a higher volumetric Lorentz force is observed for a higher Hartmann number. As the Hartmann number increased from 1.41 to 3.76, at a constant applied voltage of 0.35 V, the volumetric Lorentz force increased 600%, or 6 times. In the study conducted by Moghaddam on MHD micropumps, the volumetric flow rate increased owing to an increase in the Hartmann number to a value of 40, then volumetric flow rate started to decrease [11]. Similarly, the volumetric flow rate increased until a Hartmann number of 200, and then decreased in the study conducted on the MHD pump by Yousofvand et al. [10]. The present study is focused on low Hartmann numbers (Ha < 4) where the volumetric flow rate and Lorentz force increases with the increase in Hartmann number, as the defined Hartman number compares the magnetic force with the viscous force. At low Hartmann numbers, the viscous forces dominate giving a higher volumetric flow rate. As a result, the lower Hartmann number is favorable for the enhancement of heat transfer. However, a higher Hartmann number can have an adverse effect on heat transfer [10].

Figure 8a shows the shear stress variation along the non-dimensional width at the center of the magnetohydrodynamic pump. The shear stress values for all the Hartmann numbers are compared in the middle section of the channel. Regions of higher shear stress are observed near the wall for all the Hartmann numbers. The values of shear stress in the region near the walls of the channel increased as the Hartmann number increased. Shear stress is directly proportional to the rate of change of velocity. The increase in shear stress at the walls for a higher Hartmann number is observed due to the typical velocity profile of the MHD pump flow inside the channel, where the velocity profile becomes flatter at the center, and a large velocity change is seen near the walls. As the Hartmann number increased from 1.41 to 3.74, the shear stress value near the channel walls increased around 7 times, or 714%. The shear stress variation showed axisymmetric behavior for the axis passing through the center of the dimensionless width. Figure 8b shows the pressure contours for the flow cross-sectional area at the center of the pump in the *YZ*-plane, and it could be seen that higher pressure regions are observed near the wall owing to the Hartmann effect.

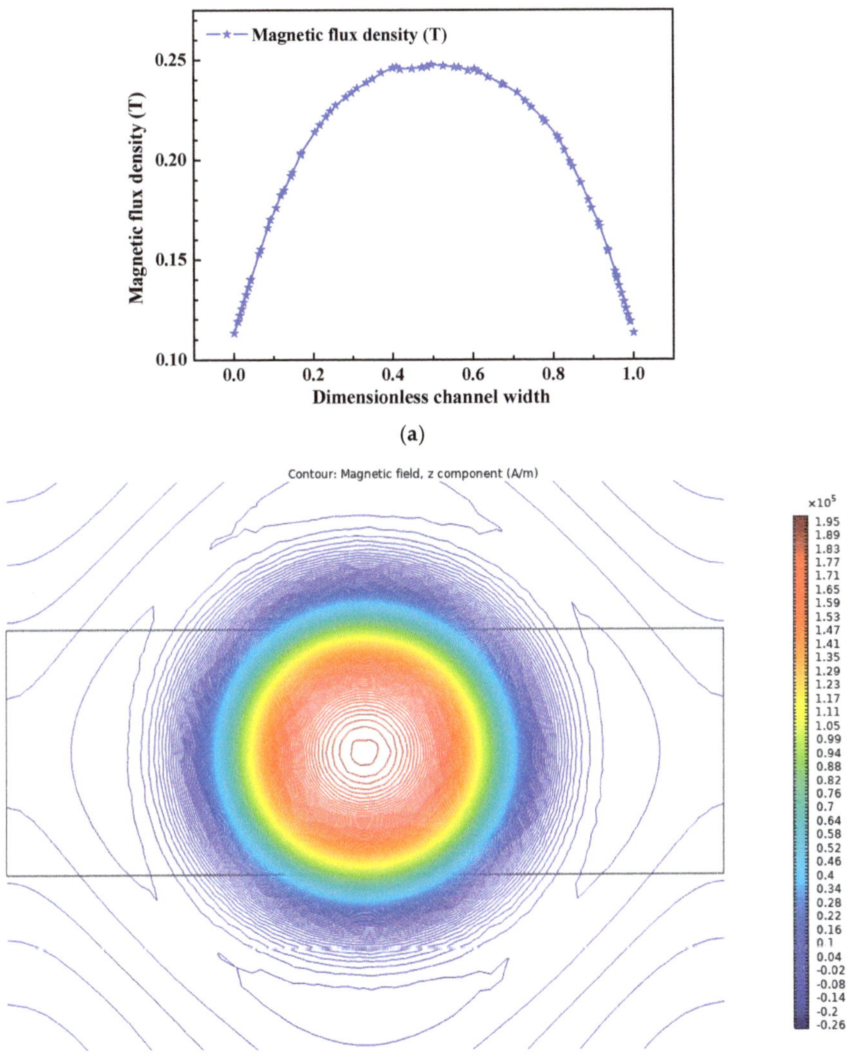

(a)

(b)

Figure 6. Magnetic flux density and magnetic field (**a**) Magnetic flux density variation with dimensionless width for different applied voltages and different Hartmann number (**b**) Magnetic field distribution at the center of the MHD pump in the *XY*-plane.

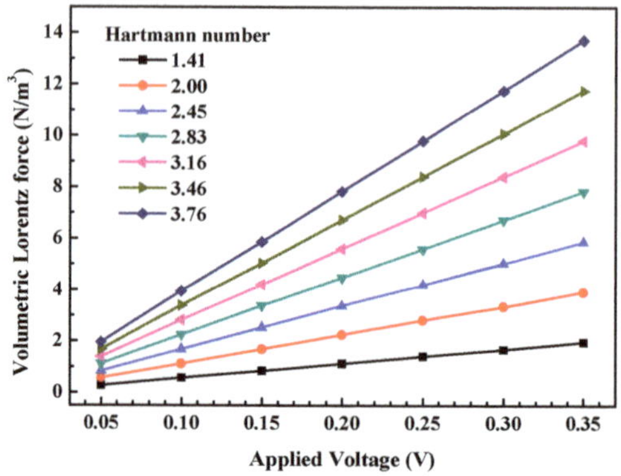

Figure 7. Volumetric Lorentz force variation for different applied voltages and different Hartmann number.

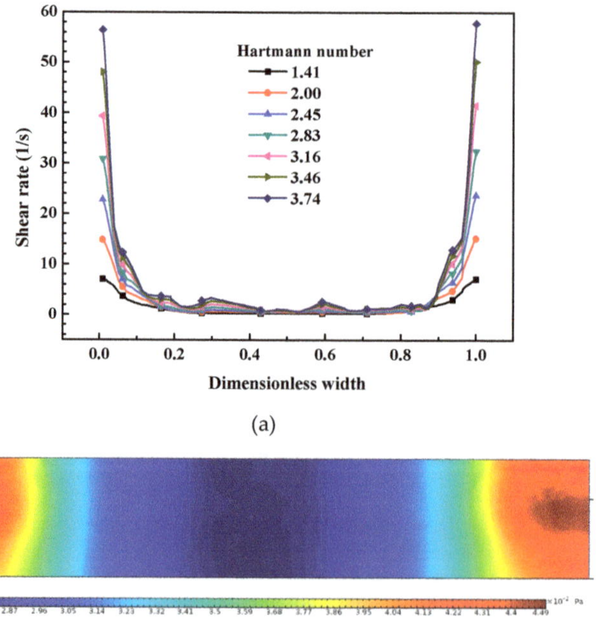

(a)

(b)

Figure 8. Shear stress and Pressure (**a**) Shear stress variation with dimensionless width for different applied voltages and different Hartmann numbers (**b**) Pressure contours for the flow cross-sectional area at the center of the pump in the *YZ*-plane

Figure 9 shows the variation of the velocity profile along the dimensionless width in the *Y*-axis imposed by the Lorentz force at the center of the magnetohydrodynamic pump. The velocity profiles show maximum values near the walls and lower values in the center of the channel owing to the Lorentz force distribution [44]. The velocity variation showed axisymmetric behavior for the axis passing through the center of a dimensionless width. The M-shape velocity profiles as observed in Figure 9 are present in many MHD pumps. This can be attributed to the position of conducting electrodes on the two opposite walls to provide the DC power supply. Moreover, the different fluids have responded with a similar velocity profile indicating that it is a geometrically affected phenomenon with the position of the electrode [45]. It could be seen from Figure 9 that as the Hartmann number increased, the velocity increased. Moreover, as the value of the Hartmann number increased, the velocity profile became flatter. The plug-like shape remained constant for a large portion of the channel width [46,47]. The current flowing in the closed loop generated a non-uniform negative small electromagnetic Lorentz force which counteracted the conducting fluid flow in the magnetic field creating a flat velocity boundary layer [45]. This phenomenon is called the Hartmann effect. For example, as the value of the Hartmann number increased from 1.41 to 3.74, the maximum velocity increased by 280% at the center of the magnetohydrodynamic pump. Moreover, as the value of the Hartmann number increased, its effect on velocity change was slightly reduced. This is evident from Figure 9, as the change in maximum velocity for the Hartmann number variation from 3.46 to 3.74 is less as compared to the variation from 1.41 to 2.00.

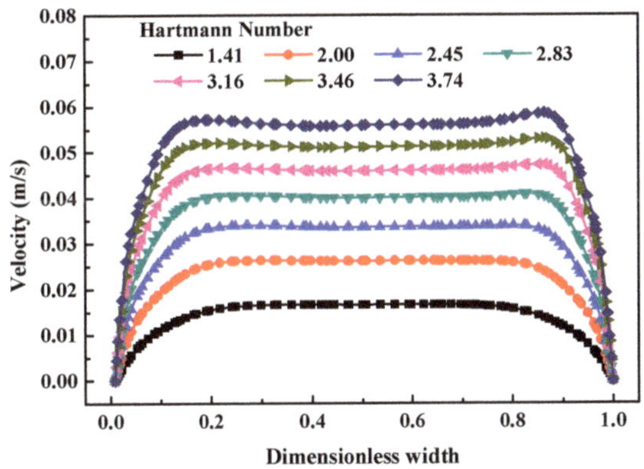

Figure 9. Velocity variation with dimensionless width.

Figure 10 shows the velocity field variation in the X-axis along the width at the center of the magnetohydrodynamic pump. The velocity at the center of the channel is higher compared to the channel wall, owing to the high shear stress observed along the channel wall. The average velocity of 0.0034, 0.0061, 0.0085, 0.0106, 0.0126, 0.0145 and 0.0164 m/s are developed for the applied voltage of 0.05, 0.10, 0.15, 0.25, 0.30 and 0.35 V at Hartmann number value of 2.0, respectively.

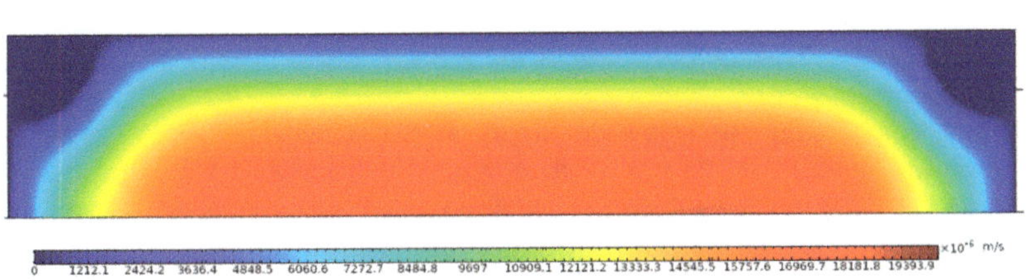

Figure 10. Velocity variation with dimensionless width.

The increase in average velocity with increase in the applied voltage is attributed to development of higher Lorentz force. It is obvious from Equation (4) that the Lorenz force will increase if the cross product of current density and magnetic field increases.

3.3. MHD-Based Microchannel Cooling System

The magnetohydrodynamic pump has various advantages over traditional pumps including low cost, low electric field and no moving parts. The Lorentz force developed by the interaction between the electric current and magnetic field can be used to propel, stir or manipulate the flow behavior in the channel. This section provided the details of the MHD micropump performance considering the applied voltage and Hartmann number.

Figure 11 shows the variation of the maximum temperature of the heat dissipating element for the varied applied voltage and Hartmann number with Cu-water with volume fraction of 0.1% as coolant. As the applied voltage is increased, the maximum temperature of the heat dissipating element decreased. For example, as the applied voltage increased from 0.05 V to 0.35 V at a Hartmann number value of 2.0, the maximum temperature of the heat dissipating element decreased by 7.7%. For the same applied voltage, lower maximum temperatures of the heat dissipating element are observed for a higher Hartmann number. As the Hartmann number increased from 1.41 to 3.76 at a constant applied voltage of 0.05 V, the maximum temperature of the heat dissipating element decreased by 11.0%. The combined influence of higher applied voltage and higher Hartmann number are visible with significant decrease in the maximum temperature of the heat dissipating element. These findings show that the applied voltage and Hartmann number have a significant effect on maintaining and controlling the maximum temperature of the heat dissipating element.

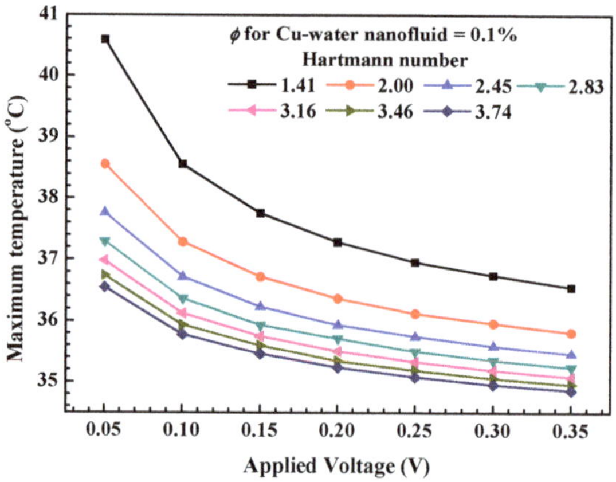

Figure 11. Maximum temperature.

Figure 12 shows variation of the heat removal rate for the varied applied voltage and Hartmann number with Cu-water with the volume fraction of 0.1% as coolant. As the applied voltage is increased, the heat removal rate increased. For example, as the applied voltage increased from 0.05 V to 0.35 V at a Hartmann number value of 2.0, the heat removal rate increased by 34.5%. For the same applied voltage, higher heat removal rates are observed for a higher Hartmann number. As the Hartmann number increased from 1.41 to 3.76 at a constant applied voltage of 0.05 V, the heat removal rate increased by 39.5%. The combined influence of a higher applied voltage and higher Hartmann number are visible with significant increase in heat removal rate. The increase in heat removal rate with a higher applied voltage is attributed to an increase in the volumetric Lorentz force as shown in Figure 7, which subsequently results in the higher volumetric flow rate. It can be seen that for a lower Hartmann number, the rate of change heat removal rate is large, whereas for a higher Hartmann number, the rate of change of heat removal rate is small. This is because the dominance of the magnetic force increased as the Hartmann number increased [10].

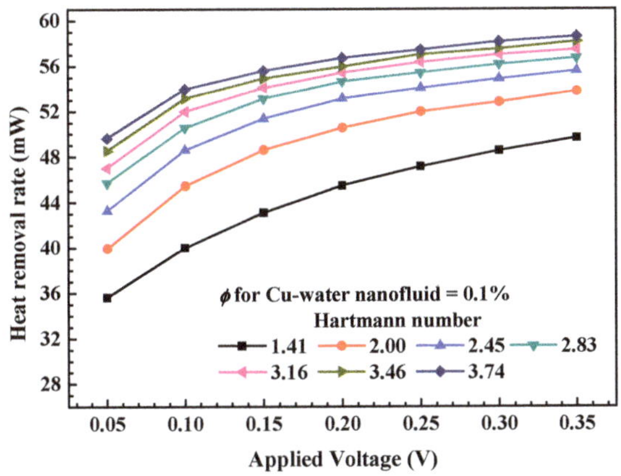

Figure 12. Heat removal rate variation.

Figure 13 shows the variation of efficiency for the varied applied voltage and Hartmann number with Cu-water with volume fraction of 0.1% as coolant. The efficiency is defined as shown in Equation (19).

$$Efficiency = \frac{Heat\ removal\ rate}{Input\ power} \quad (19)$$

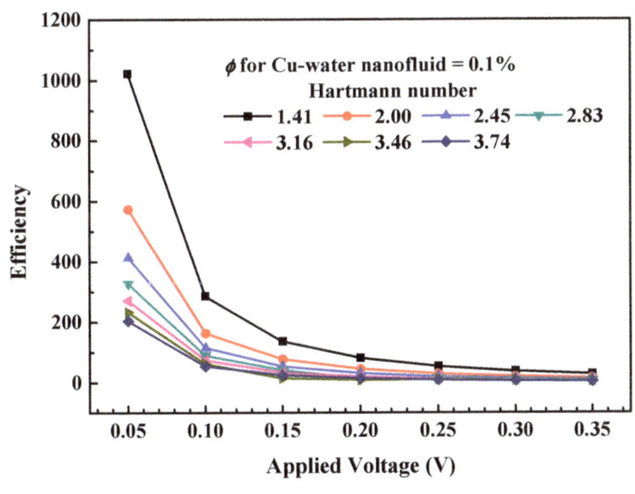

Figure 13. Variation of efficiency with applied voltage.

As shown, the efficiency decreased continuously with increase in applied voltage. This shows that, even though for higher applied voltage the heat removal rate is higher, and the temperature of the heat dissipating element is minimum, the heat removal process is less efficient. Therefore, an optimum operating range considering the heat removal rate, temperature of heat dissipating element and efficiency could be considered. As the applied voltage increased from 0.05 to 0.35 V at a Hartmann number value of 3.46, the efficiency decreased from 204.4 to 4.9. For the same applied voltage, a lower efficiency is observed for a higher Hartmann number. As the Hartmann number increased from 1.41 to 3.76 at a constant applied voltage of 0.35 V, the efficiency decreased from 29.1 to 4.9. The combined influence of the higher applied voltage and higher Hartmann number are visible with a significant decrease in efficiency. These findings show that the applied voltage and Hartmann number have a significant effect on efficiency.

Figure 14 shows the velocity and temperature distribution in the MHD pump microchannel cooling system with Cu-water with volume fraction of 0.1% as coolant. As shown in Figure 14a, the velocity is uniformly distributed in the microchannel throughout, which makes it an attractive method for the cooling heat dissipating element, especially where space and noise are constraints such as electronic devices. The rate of increase of the developed flow velocity in the magnetohydrodynamic pump cooling system is an indication of cooling performance as a higher velocity development leads to higher cooling performance. However, the increase in flow velocity has limitations owing to applied voltage and applied magnetic field. As expected, the flow velocity in the thin microchannel increased as it passed through the narrow duct of microchannel cooling system [6]. This is desirable as the heat dissipating element is placed exactly at the center of the microchannel. As shown in Figure 14b, the temperature of the coolant increased as it passed through microchannel. In the present study, the square microchannel design is investigated considering the manufacturing simplicity of the square duct. The future scope of the study involves the use of different shapes of microchannel including circular and trapezoidal. The temperature field distribution for the MHD pump microchannel at the center plane showed that heat transfer occurred along the edges of the microchannel and heat is taken away as the flow proceeded [48]. The geometry based microchannel optimization for effective thermal performance could be carried out considering the requirement of cooling performance and these findings can be used to design an effective cooling by optimizing influencing parameters.

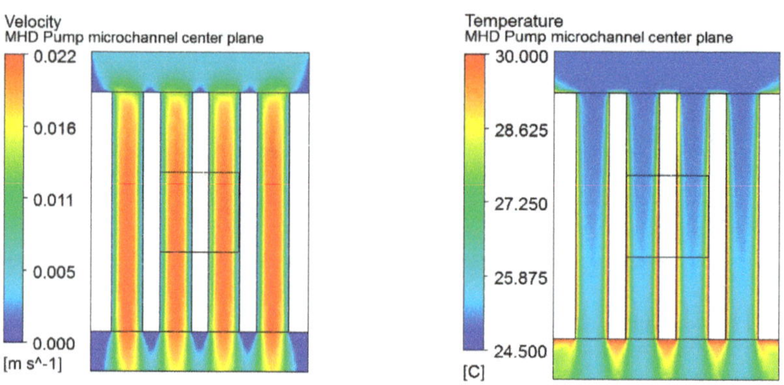

(a) (b)

Figure 14. Velocity and temperature distribution in MHD pump microchannel cooling system.

Figure 15 shows the variation of the average Nusselt number for the applied voltage and Hartmann number with Cu-water with a volume fraction of 0.1% as the coolant. The Nusselt number is an indication of enhanced heat transfer due to convection as compared to conduction [49]. The higher Nusselt number indicates the effectiveness of magnetohydrodynamic cooling systems for the heat dissipating element. The average Nusselt number increased with the applied voltage. For example, as the applied voltage increased from 0.05 V to 0.35 V at a Hartmann number value of 2.0, the average Nusselt number increased by 112.6%. For the same applied voltage, a higher average Nusselt number is observed for higher Hartmann numbers. As the Hartmann number increased from 1.41 to 3.76 at a constant applied voltage of 0.25 V, the heat removal rate increased by 100.0%. The combined influence of a higher applied voltage and higher Hartmann number are visible with a significant increase in the average Nusselt number. However, the rate of increase of the average Nusselt number decreased as the applied voltage and Hartmann number increased. The heat transfer performance slightly deteriorated as the value of the Hartmann number increased due to suppression of convection due to the magnetic field [10,50]. These findings show that the applied voltage and Hartmann number have a significant effect on the heat transfer performance of MHD micropumps.

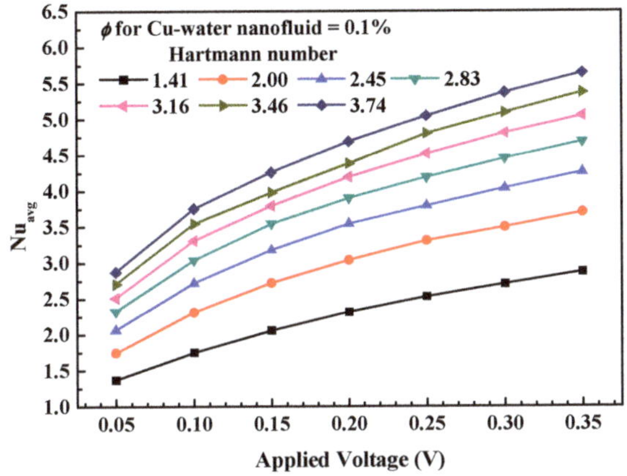

Figure 15. Variation of average Nusselt number with applied voltage.

3.4. *Influence of Various Nanofluids*

The thermal performance of the MHD pump is compared using various nanofluids. Three types of nanofluids including Cu-water, TiO_2-water and Al_2O_3-water are considered with a volume fraction of 0.1%. For performance comparison, the volume fraction of nanoparticles in nanofluids is kept constant. To evaluate the thermal performance of MHD pumps with various nanofluids, the heat transfer rate, efficiency and Nusselt number variation are considered.

Figure 16 shows variation of the heat removal rate for the varied Hartmann number. As the Hartmann number is increased, the heat removal rate increased. For example, as the Hartmann number increased from 1.41 to 3.74 at an applied voltage value of 0.35 V, the heat removal rate increased by 18.0% for Cu-water nanofluids. For the same applied voltage, higher heat removal rates are observed for Cu-water nanofluid as compared to TiO_2-water and Al_2O_3-water nanofluids. As previously noted, for a lower Hartmann number, the rate of change heat removal rate is large, whereas for higher Hartmann number, the rate of change of heat removal rate is small. The Cu-based nanofluid showed a better heat transfer rate owing to the high thermal conductivity of copper nanoparticles.

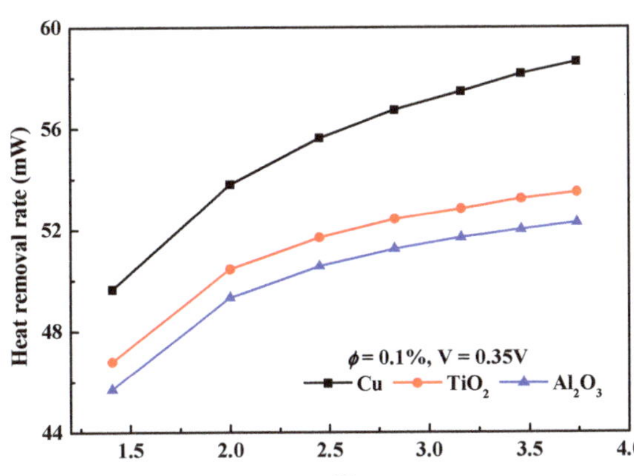

Figure 16. Variation of heat removal rate with various nanofluids at different Hartmann numbers.

Figure 17 shows variation of the efficiency for the varied Hartmann number. As the Hartmann number is increased, the efficiency decreased. For example, as the Hartmann number increased from 1.41 to 3.74 at applied voltage value of 0.35 V, efficiency decreased from 29.16% to 4.92% for Cu-water nanofluid. For the same applied voltage, higher efficiencies are observed for Cu-water nanofluid as compared to TiO_2-water and Al_2O_3-water nanofluids. For lower Hartmann number, the rate of change efficiency is large, whereas for higher Hartmann number, the rate of change of efficiency is small. This is because the dominance of magnetic force increased as the Hartmann number increased. The Cu-based nanofluid shows better efficiency owing to high thermal conductivity of copper nanoparticles.

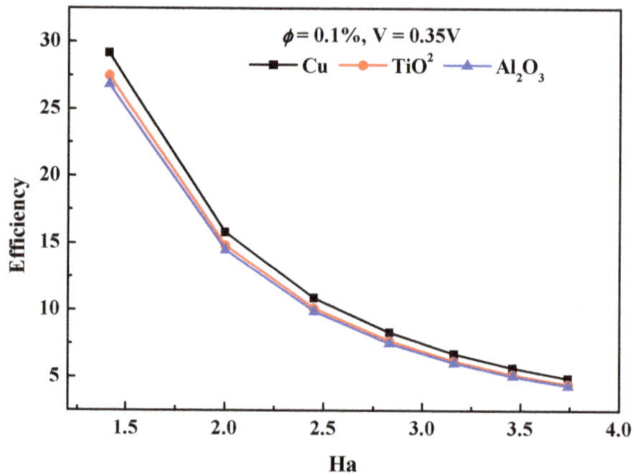

Figure 17. Variation of efficiency with various nanofluids at different Hartmann numbers.

Figure 18 shows variation of the average Nusselt number for the varied Hartmann number. As the Hartmann number is increased, the average Nusselt number increased. For example, as the Hartmann number increased from 1.41 to 3.74 at an applied voltage value of 0.35 V, the average Nusselt number increased by 96.5% for Cu-water nanofluid. For the same applied voltage, higher average Nusselt numbers are observed for Cu-water nanofluid as compared to TiO_2-water and Al_2O_3-water nanofluids. Interestingly, the Nusselt number for the TiO_2 based nanofluid and Al_2O_3 based nanofluid are found to be close. The Cu-based nanofluid showed a better average Nusselt number owing to the high thermal conductivity of copper nanoparticles.

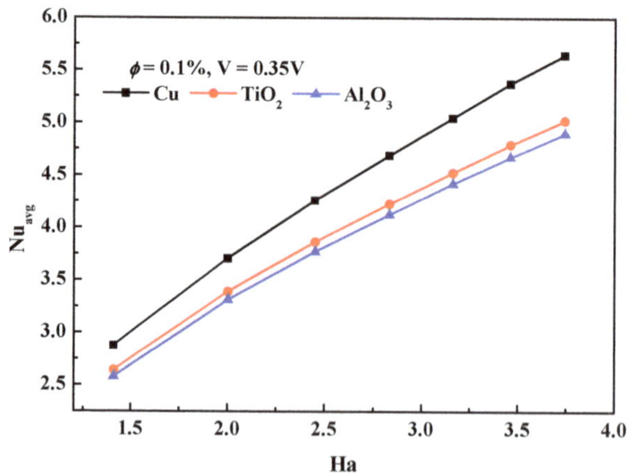

Figure 18. Variation of average Nusselt number with various nanofluids at different Hartmann numbers.

4. Conclusions

Magnetohydrodynamic pump-based microchannel cooling is proposed for cooling heat dissipating elements. The proposed magnetohydrodynamic pump has many advantages including vibration-free and noise-free applications. In the present study, the applied voltage and Hartmann number are varied to evaluate the effect on the MHD pump performance considering normal current density, magnetic flux density, volumetric Lorentz force, shear stress and pump flow velocity as evaluating parameters. The MHD pump-based microchannel cooling system performance with Cu-water nanofluid is evaluated considering the maximum temperature of the heat dissipating element, heat removal rate, efficiency, thermal field, flow field and Nusselt number for various applied voltages and Hartmann numbers. It is found that for a low Hartmann number, the Lorentz force increased with an increase in the applied voltage and Hartmann number. As the applied voltage increased from 0.05 V to 0.35 V at a Hartmann number of 1.41, the heat removal rate increased by 39.5%. The results revealed that for a low Hartmann number, the average Nusselt number increased with increase in the applied voltage and Hartmann number. As the applied voltage increased from 0.05 V to 0.35 V at a Hartmann number of 1.41, the average Nusselt number increased by 112.6%. In addition, the influence of various nanofluids including Cu-water, TiO_2-water and Al_2O_3-water nanofluids on heat transfer performance of MHD pump-based microchannels is evaluated. At the Hartmann number value of 3.74 and applied voltage value of 0.35 V, average Nusselt numbers are 12.3% and 15.1% higher for Cu-water nanofluid compared to TiO_2-water and Al_2O_3-water nanofluids, respectively. The MHD pump is more useful in cases where space and noise constraint are of particular interest. Especially in the microelectronics device cooling, the removal of heat is important and due to miniaturization, the MHD pump for cooling provides a promising option. The investigations provide an opportunity to further explore the application of MHD pumps in electronics cooling.

Author Contributions: Conceptualization, J.-H.S.; M.S.P. and M.-Y.L.; methodology, J.-H.S.; software, M.S.P.; validation, J.-H.S. and M.S.P.; Numerical investigation, J.-H.S. and M.S.P.; resources, M.-Y.L. and S.P.; data reduction, M.S.P. and S.P.; writing—original draft preparation, J.-H.S. and M.S.P.; writing—review and editing, M.-Y.L., and S.P.; visualization, M.S.P.; supervision, M.-Y.L.; project administration, M.-Y.L.; funding acquisition, M.-Y.L. All authors have read and agreed to the published version of the manuscript.

Funding: This research received no external funding.

Acknowledgments: This work was supported by the Dong-A University research fund.

Conflicts of Interest: The authors declare no conflict of interest.

Nomenclature

A	cross-sectional area (m^2)
$\vec{B}$	magnetic field vector (T)
B	magnitude of the magnetic field (T)
C_p	specific heat at constant pressure (J/kg-K)
D_h	hydraulic diameter (m)
$\vec{E}$	electric field vector (V/m)
$\vec{F}$	electromagnetic force (N)
h_{avg}	average heat transfer coefficient (W/m^2-K)
Ha	Hartmann number
$\vec{J}$	current density (A/m^2)
L	characteristic length (mm)
MHD	magnetohydrodynamic
Nu_{avg}	average Nusselt number
P	pressure (Pa)
Q	heat transfer rate (W)

T	temperature (°C/K)
t	time (s)
$\vec{V}$	velocity (m/s)
Greek symbols	
∇	gradient operator
α	thermal diffusivity (m^2/s)
σ	electrical conductivity (S/m)
ρ	density (kg/m^3)
ν	kinematic fluid viscosity (m^2/s)
μ	dynamic viscosity (Pa-s)
k	thermal conductivity (W/m-K)
ϕ	volume fraction (%)
Subscripts	
avg	average
bulk	bulk property
conv	convective heat transfer
f	fluid
in	inlet
LMTD	logarithmic mean temperature difference
n	nanoparticle
nf	nanofluid
out	outlet
wall	wall

References

1. Al-Habahbeh, O.M.; Al-Saqqa, M.; Safi, M.; Abo Khater, T. Review of magnetohydrodynamic pump applications. *Alexandria Eng. J.* **2016**, *55*, 1347–1358. [CrossRef]
2. Derakhshan, S.; Rezaee, M.; Sarrafha, H. A Molecular Dynamics Study of Description Models for Shear Viscosity in Nanochannels: Mixtures and Effect of Temperature. *Nanoscale Microscale Thermophys. Eng.* **2015**, *19*, 206–220. [CrossRef]
3. Derakhshan, S.; Adibi, I.; Sarrafha, H. Numerical study of electroosmotic micropump using Lattice Boltzmann method. Comput. *Fluids* **2015**, *114*, 232–241. [CrossRef]
4. Tay, F.E.H. *Literature Review for Micropumps;* Springer: Boston, MA, USA, 2002; pp. 3–24.
5. Zhang, X.; Jaluria, Y. Optimization of microchannel-based cooling systems. *Numer. Heat Transf. Part A Appl.* **2018**, *74*, 1053–1067. [CrossRef]
6. Patil, M.S.; Seo, J.H.; Panchal, S.; Jee, S.W.; Lee, M.Y. Investigation on thermal performance of water-cooled Li-ion pouch cell and pack at high discharge rate with U-turn type microchannel cold plate. *Int. J. Heat Mass Transf.* **2020**, *155*, 119728. [CrossRef]
7. Lemoff, A.V.; Lee, A.P. AC magnetohydrodynamic micropump. *Sens. Actuators B Chem.* **2000**, *63*, 178–185. [CrossRef]
8. Rivero, M.; Cuevas, S. Analysis of the slip condition in magnetohydrodynamic (MHD) micropumps. *Sens. Actuators B Chem.* **2012**, *166–167*, 884–892. [CrossRef]
9. Zhao, G.; Jian, Y.; Chang, L.; Buren, M. Magnetohydrodynamic flow of generalized Maxwell fluids in a rectangular micropump under an AC electric field. *J. Magn. Magn. Mater.* **2015**, *387*, 111–117. [CrossRef]
10. Yousofvand, R.; Derakhshan, S.; Ghasemi, K.; Siavashi, M. MHD transverse mixed convection and entropy generation study of electromagnetic pump including a nanofluid using 3D LBM simulation. *Int. J. Mech. Sci.* **2017**, *133*, 73–90. [CrossRef]
11. Moghaddam, S. Analytical solution of MHD micropump with circular channel. *Int. J. Appl. Electromagn. Mech.* **2012**. [CrossRef]
12. Miroshnichenko, I.V.; Sheremet, M.A.; Oztop, H.F.; Al-Salem, K. MHD natural convection in a partially open trapezoidal cavity filled with a nanofluid. *Int. J. Mech. Sci.* **2016**, *119*, 294–302. [CrossRef]

13. Kefayati, G.H.R. Simulation of double diffusive MHD (magnetohydrodynamic) natural convection and entropy generation in an open cavity filled with power-law fluids in the presence of Soret and Dufour effects (Part I: Study of fluid flow, heat and mass transfer). *Energy* **2016**, *107*, 889–916. [CrossRef]
14. Kefayati, G.H.R. Simulation of double diffusive MHD (magnetohydrodynamic) natural convection and entropy generation in an open cavity filled with power-law fluids in the presence of Soret and Dufour effects (part II: Entropy generation). *Energy* **2016**, *107*, 917–959. [CrossRef]
15. Shirvan, K.M.; Öztop, H.F.; Al-Salem, K. Mixed magnetohydrodynamic convection in a Cu-water-nanofluid-filled ventilated square cavity using the Taguchi method: A numerical investigation and optimization. *Eur. Phys. J. Plus* **2017**, *132*, 204. [CrossRef]
16. Kiyasatfar, M.; Pourmahmoud, N.; Golzan, M.; Mirzaee, I. Investigation of thermal behavior and fluid motion in direct current magnetohydrodynamic pumps. *Therm. Sci.* **2014**. [CrossRef]
17. Larimi, M.M.; Ghanaat, A.; Ramiar, A.; Ranjbar, A.A. Forced convection heat transfer in a channel under the influence of various non-uniform transverse magnetic field arrangements. *Int. J. Mech. Sci.* **2016**, *118*, 101–112. [CrossRef]
18. Kolsi, L.; Alrashed, A.A.A.A.; Al-Salem, K.; Oztop, H.F.; Borjini, M.N. Control of natural convection via inclined plate of CNT-water nanofluid in an open sided cubical enclosure under magnetic field. *Int. J. Heat Mass Transf.* **2017**, *111*, 1007–1018. [CrossRef]
19. Kefayati, G.H.R. FDLBM simulation of magnetic field effect on natural convection of non-Newtonian power-law fluids in a linearly heated cavity. *Powder Technol.* **2014**, *256*, 87–99. [CrossRef]
20. Kefayati, G.H.R. Simulation of magnetic field effect on non-Newtonian blood flow between two-square concentric duct annuli using FDLBM. *J. Taiwan Inst. Chem. Eng.* **2014**, *45*, 1184–1196. [CrossRef]
21. Kefayati, G.H.R. Simulation of vertical and horizontal magnetic fields effects on non-Newtonian power-law fluids in an internal flow using FDLBM. *Comput. Fluids* **2015**, *114*, 12–25. [CrossRef]
22. Sheikholeslami, M.; Ganji, D.D. Ferrohydrodynamic and magnetohydrodynamic effects on ferrofluid flow and convective heat transfer. *Energy* **2014**, *75*, 400–410. [CrossRef]
23. Sheikholeslami, M.; Gorji-Bandpy, M.; Ganji, D.D. Numerical investigation of MHD effects on Al_2O_3-water nanofluid flow and heat transfer in a semi-annulus enclosure using LBM. *Energy* **2013**, *60*, 501–510. [CrossRef]
24. Patil, M.S.; Bang, Y.M.; Seo, J.H.; Kim, D.W.; Cho, B.D.; Lee, M.Y. Experimental investigation of heat transfer characteristics of battery management system and electronic control unit of neighborhood electric vehicle. In *Lecture Notes in Electrical Engineering*; Springer: Cham, Switzerland, 2017; Volume 415, pp. 205–211. ISBN 9783319509037.
25. Patil, M.S.; Seo, J.; Panchal, S.; Lee, M. Numerical study on sensitivity analysis of factors influencing liquid cooling with double cold-plate for lithium-ion pouch cell. *Int. J. Energy Res.* **2020**, 1–27. [CrossRef]
26. Joye, D.D.; Bushinsky, J.P.; Saylor, P.E. Mixed Convection Heat Transfer at High Grashof Number in a Vertical Tube. *Ind. Eng. Chem. Res.* **1989**, *28*, 1899–1903. [CrossRef]
27. Patil, M.S.; Seo, J.H.; Bang, Y.M.; Lee, M.Y. Analysis of factors influencing on heat transfer characteristics of automobile LED headlamp. *Int. J. Control Autom.* **2016**, *9*, 263–272. [CrossRef]
28. Patil, M.S.; Seo, J.-H.; Bang, Y.-M.; Lee, M.-Y. Experimental Study on the Performance Characteristics of an Air-Cooled LED Cooling System for Headlamp of a Passenger Vehicle. *Adv. Sci. Technol. Lett.* **2016**, *120*, 120–128. [CrossRef]
29. Patil, M.S.; Seo, J.H.; Kim, C.M.; Lee, J.Y.; Lee, M.Y. Numerical study on magneto-acoustic thermal characteristics of micro-speaker for mobile phones. *Int. J. Heat Mass Transf.* **2021**, *164*, 120479. [CrossRef]
30. Khan, A.A.; Kim, K.Y. Evaluation of Various Channel Shapes of a Microchannel Heat Sink. *Int. J. Air-Cond. Refrig.* **2016**, *24*, 1650018. [CrossRef]
31. Saqib, M.; Shafie, S.; Khan, I.; Chu, Y.M.; Nisar, K.S. Symmetric MHD channel flow of nonlocal fractional model of BTF containing hybrid nanoparticles. *Symmetry* **2020**, *12*, 663. [CrossRef]
32. Permanasari, A.A.; Kuncara, S.; Puspitasari, P.; Sukarni, S.; Ginta, T.L.; Irdianto, W. Convective heat transfer characteristics of TiO_2-EG nanofluid as coolant fluid in heat exchanger. *AIP Conf. Proc.* **2019**, *2120*, 50014. [CrossRef]
33. Abu-Nada, E. Effects of variable viscosity and thermal conductivity of Al_2O_3-water nanofluid on heat transfer enhancement in natural convection. *Int. J. Heat Fluid Flow* **2009**, *30*, 679–690. [CrossRef]
34. Pak, B.C.; Cho, Y.I. Hydrodynamic and heat transfer study of dispersed fluids with submicron metallic oxide particles. *Exp. Heat Transf.* **1998**. [CrossRef]

35. Zhong, D.; Zhong, H.; Wen, T. Investigation on the thermal properties, heat transfer and flow performance of a highly self-dispersion TiO_2 nanofluid in a multiport mini channel. *Int. Commun. Heat Mass Transf.* **2020**. [CrossRef]
36. Batchelor, G.K. The effect of Brownian motion on the bulk stress in a suspension of spherical particles. *J. Fluid Mech.* **1977**. [CrossRef]
37. Vand, V. Theory of viscosity of concentrated suspensions. *Nature* **1945**, *155*, 364–365. [CrossRef]
38. Wang, X.; Xu, X.; Choi, S.U.S. Thermal conductivity of nanoparticle-fluid mixture. *J. Thermophys. Heat Transf.* **1999**. [CrossRef]
39. Duangthongsuk, W.; Wongwises, S. Measurement of temperature-dependent thermal conductivity and viscosity of TiO_2-water nanofluids. *Exp. Therm. Fluid Sci.* **2009**. [CrossRef]
40. Bobbo, S.; Fedele, L.; Benetti, A.; Colla, L.; Fabrizio, M.; Pagura, C.; Barison, S. Viscosity of water based SWCNH and TiO_2 nanofluids. *Exp. Therm. Fluid Sci.* **2012**. [CrossRef]
41. Timofeeva, E.V.; Gavrilov, A.N.; McCloskey, J.M.; Tolmachev, Y.V.; Sprunt, S.; Lopatina, L.M.; Selinger, J.V. Thermal conductivity and particle agglomeration in alumina nanofluids: Experiment and theory. *Phys. Rev. E Stat. Nonlinear Soft Matter Phys.* **2007**. [CrossRef]
42. Xuan, Y.; Roetzel, W. Conceptions for heat transfer correlation of nanofluids. *Int. J. Heat Mass Transf.* **2000**. [CrossRef]
43. Hwang, S.; Jeong, J.H. Performance Comparison of Modified Offset Strip Fins Using a CFD Analysis. *Int. J. Air-Conditioning Refrig.* **2016**, *24*, 1650015. [CrossRef]
44. Aoki, L.P.; Schulz, H.E.; Maunsell, M.G. An MHD Study of the Behavior of an Electrolyte Solution using3D Numerical Simulation and Experimental results. In Proceedings of the COMSOL Conference, Boston, MA, USA, 14 October 2013.
45. Kandev, N.; Kagan, V.; Daoud, A. Electromagnetic DC pump of liquid aluminium: Computer simulation and experimental study. *Fluid Dyn. Mater. Process.* **2010**. [CrossRef]
46. Zakeri, R.; Sabouri, M.; Maleki, A.; Abdelmalek, Z. Investigation of Magneto Hydro-Dynamics Effects on a Polymer Chain Transfer in Micro-Channel Using Dissipative Particle Dynamics Method. *Symmetry* **2020**, *12*, 397. [CrossRef]
47. Daoud, A.; Kandev, N. Magneto-hydrodynamic numerical study of DC electromagnetic pump for liquid metal. In Proceedings of the COMSOL Conference, Hannover, Germany, 5 November 2008.
48. Ekanayake, G.; Patil, M.S.; Seo, J.-H.; Lee, M.-Y. Numerical study of fin geometry on the heat transfer characteristics of 72 V ECU heatsink for an electric three-wheeler. *J. Mech. Sci. Technol.* **2019**, *33*, 1451–1462. [CrossRef]
49. Seo, J.H.; Patil, M.S.; Cho, C.P.; Lee, M.Y. Heat transfer characteristics of the integrated heating system for cabin and battery of an electric vehicle under cold weather conditions. *Int. J. Heat Mass Transf.* **2018**. [CrossRef]
50. Selimefendigil, F.; Öztop, H.F. Analysis of MHD mixed convection in a flexible walled and nanofluids filled lid-driven cavity with volumetric heat generation. *Int. J. Mech. Sci.* **2016**, *118*, 113–124. [CrossRef]

symmetry

MDPI

Article

Thermo-Hydraulic Performance Analysis on the Effects of Truncated Twisted Tape Inserts in a Tube Heat Exchanger

Mehdi Ghalambaz [1,2], Ramin Mashayekhi [3], Hossein Arasteh [4], Hafiz Muhammad Ali [5], Pouyan Talebizadehsardari [6,7,*] and Wahiba Yaïci [8,*]

1 Institute of Research and Development, Duy Tan University, Da Nang 550000, Vietnam; mehdighalambaz@duytan.edu.vn
2 Faculty of Electrical–Electronic Engineering, Duy Tan University, Da Nang 550000, Vietnam
3 Research & Development Team, Couette Limited, Altrincham WA14 2PX, UK; info@couette.co.uk
4 Department of Mechanical Engineering, Isfahan University of Technology, Isfahan 8415683111, Iran; h.arasteh@iut.ac.ir
5 Mechanical Engineering Department, King Fahd University of Petroleum and Minerals, Dhahran 31261, Saudi Arabia; h.m.ali@qmul.ac.uk
6 Metamaterials for Mechanical, Biomechanical and Multiphysical Applications Research Group, Ton Duc Thang University, Ho Chi Minh City 758307, Vietnam
7 Faculty of Applied Sciences, Ton Duc Thang University, Ho Chi Minh City 758307, Vietnam
8 CanmetENERGY Research Centre, Natural Resources Canada, 1 Haanel Drive, Ottawa, ON K1A 1M1, Canada
* Correspondence: ptsardari@tdtu.edu.vn (P.T.); wahiba.yaici@canada.ca (W.Y.); Tel.: +44-115-837-6859 (P.T.); +1-613-996-3734 (W.Y.)

Received: 10 September 2020; Accepted: 29 September 2020; Published: 9 October 2020

Abstract: This paper investigates the convective heat transfer in a heat exchanger equipped with twisted tape elements to examine effects of the twisted tape truncation percentage, pitch value, position and Reynolds number using 3D numerical simulation. A symmetric heat flux is applied around the tube as the studied heat exchanger. Based on the influences in both heat transfer enhancement and pressure drop, the performance evaluation criterion (PEC) is utilized. Inserting twisted tape elements and reducing the pitch value significantly augment the Nusselt number, friction coefficient and PEC number compared to the plain tube. For the best case with a Reynolds number of 1000, the average Nusselt number increases by almost 151%, which is the case of fully fitted twisted tape at a pitch value of L/4. Moreover, increasing the twisted tape truncation percentage reduces both heat transfer and pressure drop. Furthermore, the highest heat transfer rate is achieved when the truncated twisted tape is located at the entrance of the tube. Finally, it is concluded that for P = L, L/2, L/3 and L/4, the optimum cases from the viewpoint of energy conservation are twisted tapes with truncation percentages of 75, 50, 50 and 0%, in which the related PEC numbers at a Reynolds number of 1000 are almost equal to 1.08, 1.24, 1.4 and 1.76, respectively.

Keywords: truncated twisted tape; secondary flow; laminar convective heat transfer; performance evaluation criterion; CFD; heat exchanger

1. Introduction

Modifying heat transfer is a principal matter of concern for energy conservation and also advantageous from an economic viewpoint. Since heat exchangers are commonly used in almost all areas of industrial activities, including fuel cells [1], electronic device cooling [2], solar air collectors [3], aerospace engineering [4], flame stabilization [5], refrigeration [6], electric vehicles [7], natural gas liquefaction [8], air dehumidification [9] and so forth, increasing the thermal performance is of vital importance. To satisfy such a need, in the last two decades, researchers have proposed different passive techniques for heat transfer augmentation, including roughness elements [10], twisted tapes [11] and wires inserts [12]. Such equipment modifies heat transfer using a higher heat transfer surface, generating a swirl flow, as well as disturbing hydrodynamic and thermal boundary layers of the fluid inside the tube, which directly improves the heat transfer coefficient.

Among the proposed passive techniques in the literature, applying twisted tape inserts with different geometries in various systems, such as solar collectors [13,14] and solar water heating systems [15–17], has been the focus of attention among experts of the field. Due to large numbers of applications, embedding the twisted tape inserts in double-pipe heat exchangers has recently been studied by many researchers. Man et al. [18] experimentally investigated twisted tape inserts in an inner tube of a double pipe with clockwise and counterclockwise swirling directions. They showed that the heat exchanger thermal performance increases in the presence of twisted tape inserts, presenting a maximum value of 1.42 for the performance evaluation criterion (PEC) number. Lim et al. [19] performed a pitch investigation of twisted tape inserts in a double pipe and reported heat transfer characteristics for different Reynolds number values. Another evaluation of thermal efficiency of a double pipe equipped with twisted tapes using the ε-NTU (Effectiveness—Number of transferred unit) method was performed by Ravi Kumar et al. [20]. They found higher values for both the effectiveness and NTU of the heat exchanger when the twisted tapes were applied. Employing nanofluids [21], hybrid nanofluids [22,23], nano-encapsulated phase change suspensions [24–26] and boiling heat transfer [27,28] are also other promising approaches to improve heat transfer.

Besides, many papers focused on the geometrical aspects of twisted tapes, such as the effects of pitch and width ratios. Jaramillo et al. [29] assessed a parabolic trough collector equipped with twisted tape. They showed that the thermal performance of the collector increased as the twisted ratio reduced at low Reynolds numbers. Mwesigye et al. [30] studied a parabolic trough collector with wall-detached twisted tape. They indicated that higher values of twist ratio and lower values of width resulted in the enhancement of the optimal Reynolds number. Esfe et al. [31] inserted the twisted tapes in a tri-lobbed tube for Reynolds numbers of 5000 to 20,000. Their results revealed that increasing the tape ratio, which is the tape radial length to pipe diameter ratio, enhances the Nusselt number, friction coefficient and overall thermal performance.

Another technique for a higher heat transfer rate using twisted tape is to increase the flow mixing and secondary flow effects by changing the geometry of the twisted tape surface. Saylroy and Eiamsa-ard [32] examined the thermohydraulic performance of a tube equipped with a square-cut twisted tape. These researchers found a maximum PEC value of 1.32 for their captured models in which the thermal performance improved 1.32 times in comparison with the classically twisted tape inserts. In another study [33], they evaluated a multi-channel twisted tape and showed laminar convection heat transfer with twisted tapes embedded in the channel. He et al. [34] examined the thermal behavior of cross hollow, twisted tape inserts in a pipe. They showed PEC values of 0.87–0.98 for Reynolds numbers of 5600–18,000. Samruaisin et al. [35] investigated numerically and experimentally the free space ratio and twisted tape arrangement for regularly spaced quadruple twisted tape inserts in a tube, in a turbulent flow regime. In the range of their operational conditions, they presented a maximum PEC number equal to 1.27. Centrally perforated twisted tape elements in a tube were investigated by Ruengpayungsak et al. [36] for convective heat transfer enhancement under laminar and turbulent flow regimes. Maximum PEC values of 8.92 and 1.33 were reported for laminar and turbulent regimes, respectively. Hasanpour et al. [37] optimized a corrugated tube heat exchanger equipped with twisted tape elements with the purpose of heat transfer modification and pressure drop reduction. They showed that the V-cut twisted tape model causes maximum heat transfer, and the perforated twisted tape model leads to the smallest pressure drop.

Despite using classical twisted tape elements with a gap between the tape and walls, the use of overlapped twisted tapes due to the lower pressure drop penalty was also studied by some researchers. Hong et al. [38] carried out an empirical study of a spiral-grooved tube equipped with twin overlapped twisted tapes in a turbulent regime with Reynolds numbers of 8000 to 22,000. They showed an enhancement in both heat transfer and friction coefficient augmented by the overlapped twisted ratio. In another study [39], they employed overlapped multiple twisted tapes in a similar experimental study. They showed that entropy generation increases and decreases due to friction resistance and heat transfer, respectively, as the tape number changes and overlapped twisted ratio decreases. Eiamsa-Ard and Samravysin [40] carried out an empirical study to investigate the thermal performance of a tube heat exchanger equipped with overlapped quadruple twisted tape inserts compared to typical quadruple twisted tape elements with a Reynolds number ranging from 5000 to 20,000, under a turbulent regime. They reported a maximum value of 1.58 for the PEC number in the case of overlapped quadruple counter tapes in a cross arrangement at a Reynolds number of 5000. Later, overlapped dual twisted tapes, along with Al_2O_3 nanofluid, were studied by Rudrabhiramu et al. [41], with the objective of improving the thermal performance of a heat exchanger. They reported that using 1% nanofluid volume concentration and twisted tape with a twist ratio of two causes the best result.

Some researchers also benefited from compound techniques of heat transfer enhancement using nanofluid along with twisted tape inserts. Qi et al. [42] experimentally investigated the convective nanofluid heat transfer in a tube using rotating and static built-in twisted tape elements. They reported a 101.6% enhancement in heat transfer by using rotating twisted tape inserts along with the nanofluid. In another empirical study, Sunder et al. [43] examined the thermal performance of a solar water heater employing nanofluid and twisted tape inserts as the heat transfer enhancement techniques. They showed a 49.75% improvement using the best configuration of twisted tape.

The above literature review indicates that there are a substantial number of papers using twisted tape inserts in different ways and techniques to reach a model for heat transfer rate maximization and pressure drop minimization. However, reviewing the preceding papers reveals that the hydrothermal investigation of truncated twisted tapes in tubes has not been studied so far. Motivated by this research gap, this study examines the twisted tape truncation percentage and its position in the tube at different pitch values and Reynolds numbers. This paper provides guidelines for the novel usage of twisted tapes in tubes toward higher performance.

2. Problem Statement and Boundary Conditions

Figure 1 schematically shows the tube with twisted tape. The height of the proposed twisted tape (H) is 95% of the tube diameter (D = 20 mm) with a thickness (t) of 0.4 mm. Four different twisted tape pitches (P) are investigated, as shown in Figure 1a. To study the position of truncated twisted tape, three different locations, i.e., entrance, center and exit, are examined, as shown in Figure 1b. Note that the length of the tube is 400 mm. Water enters the tube at 300 K with uniform velocity, and a symmetrical constant heat flux of 5000 W/m^2 is applied on the walls, while the twisted tape walls are thermally insulated. A pressure outlet condition is also employed for the tube outlet [44–46].

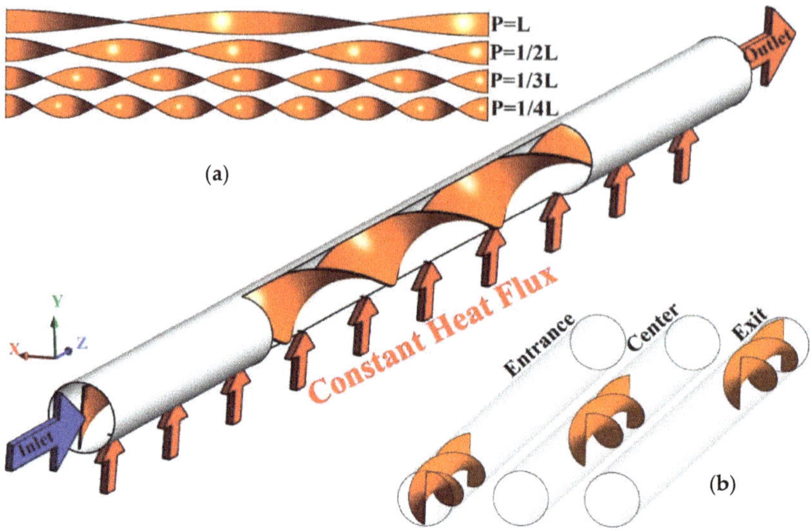

Figure 1. Schematic of the proposed system for (**a**) different twisted tape pitches and (**b**) different twisted tape positions.

3. Governing Equations

This research investigates a steady laminar flow of incompressible nanofluid, neglecting the effects of radiation and viscosity losses. The governing equations are defined as follows:

$$\nabla.\,(\rho\vec{V}) = 0 \tag{1}$$

$$\nabla.(\rho\vec{V}\vec{V}) = -\nabla p + \mu\nabla^2\vec{V} \tag{2}$$

$$\nabla.(\rho\vec{V}C_pT) = \nabla.(k\nabla T) \tag{3}$$

Different parameters calculated in this study, including the average and local heat transfer coefficient and Nusselt number, friction factor and PEC are defined as follows [36,37]:

$$D_h = \frac{4A}{P} \tag{4}$$

$$f = \frac{2\Delta P}{\rho V^2}\frac{D_h}{L} \tag{5}$$

$$h_x = \frac{q''}{T_w - T_b} \tag{6}$$

$$Nu_x = \frac{h_{xD_h}}{k} \tag{7}$$

$$Nu_{avg} = \frac{1}{l}\int_0^l Nu_x dx \tag{8}$$

$$PEC = \frac{Nu/Nu_0}{(f/f_0)^{1/3}} \tag{9}$$

4. Numerical Procedure

The commercial ANSYS Fluent computational fluid dynamics (CFD) code is employed to perform the simulation and solve the equations [47–49]. The velocity–pressure coupling is resolved using the coupled algorithm, and the convection terms are discretized using the second-order upwind scheme. The convergence criteria of 10^{-6} are also selected.

4.1. Grid Study

Figure 2 illustrates the computational mesh with finer mesh generated in the near-wall region due to velocity and temperature boundary layers and high gradients of variables. A fully structured mesh is created as shown using ANSYS meshing to have a high degree of quality and faster convergence in CFD FLUENT code, which also results in less computational time.

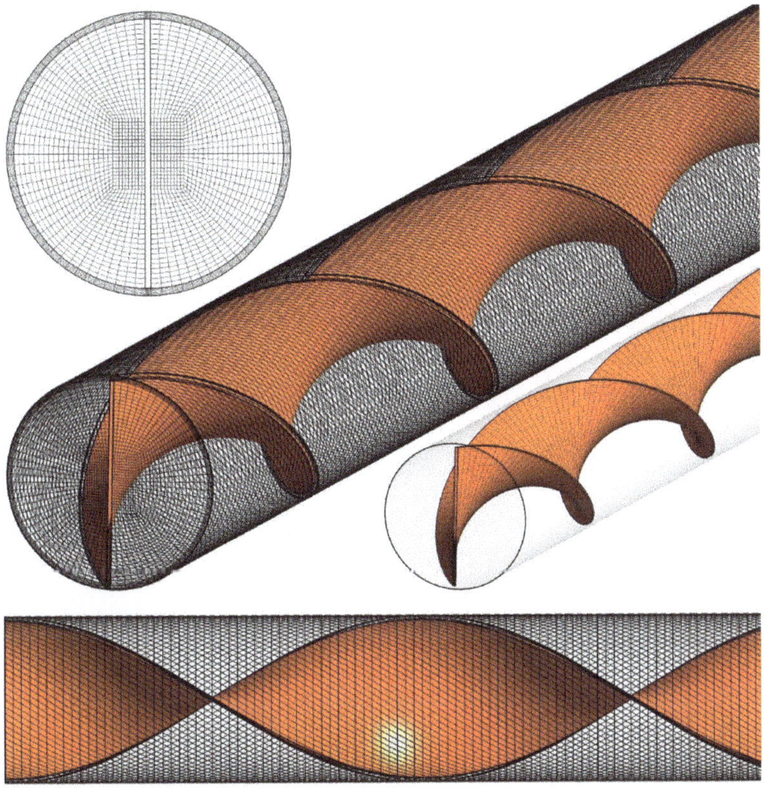

Figure 2. The meshing of the computational domain.

Different grid numbers are examined for grid independency analysis in the case of fully filled twisted tape with a pitch of L/4 and a Reynolds number of 1000. The average Nusselt number is selected as the selection criterion [50] presented in Table 1. As listed, case 3 is chosen for all the simulations, since using finer meshing leads to relative errors of less than one percent.

Table 1. Grid independence analysis.

Case	Number of Elements	Nusselt Number	Error (%)
1	750,000	41.61	-
2	1,250,000	39.12	5.98
3	1,750,000	38.04	2.74
4	2,250,000	37.85	0.52
5	3,000,000	37.7	0.39

4.2. Validation

To obtain a reliable result, the experimental results of Qi et al. [42] for the average Nusselt number for laminar fluid flow in a heat exchanger using stationary twisted tape are used. They experimentally examined pure water and nanofluid flow for different Reynolds numbers for a twisted tape length of 1600 mm and a pitch size of 100 mm with a width and thickness of 16 and 2 mm, respectively. Figure 3 displays the average Nusselt number for the cases of pure water and stationary twisted tapes for different Reynolds numbers. As shown, the results are in excellent agreement with the experimental data of Qi et al. [42], where the maximum difference is less than 2%.

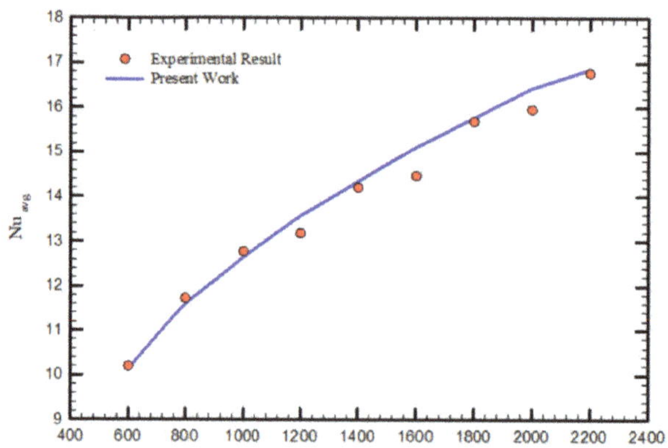

Figure 3. Validation of the present numerical results with the experimental data of Qi et al. [42].

5. Results and Discussion

Numerical simulations are performed at four pitch values (*P*) of L, L/2, L/3 and L/4, four Reynolds numbers (*Re*) of 250, 500, 750 and 1000, three twisted tape truncation (λ) percentages of 25, 50 and 75% and three positions of twisted tape at the entrance, center and exit of the tube, which are investigated in the following:

5.1. Effect of Twisted Tape Pitch

In the first step, the effect of twisted tape pitch on the hydrothermal characteristics of the tube is examined and analyzed. Figure 4 represents the local Nusselt number (*Nu*) throughout the tube length for the plain tube (PT) and four twisted tape pitch values at $Re = 250$ while the twisted tape is fully fitted in the tube (no truncation). Using the twisted tape and decreasing its pitch magnitude noticeably augments the local *Nu* along the tube length. The reason is that the twisted tape inserts create secondary flow as a result of flow swirling, which consequently improves the flow mixing, disturbs the thermal boundary layer and increases the heat transfer rate [51]. In other words, the twisted tape redirects the colder core fluid with a better cooling capacity to the heated walls of the tube where cooling is required.

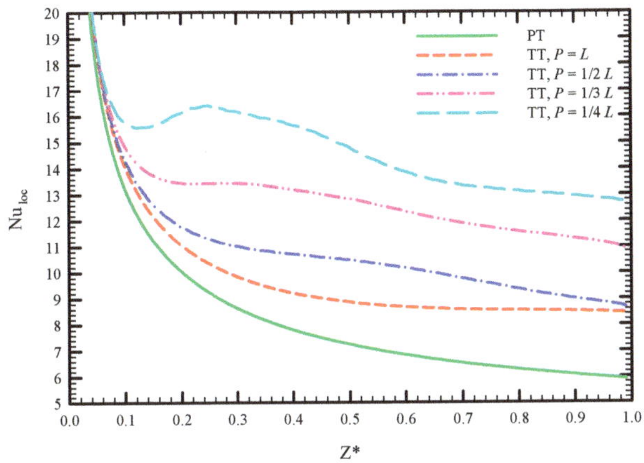

Figure 4. Local *Nu* along the tube length for plain tube (PT) and twisted tape inserts with P = L, L/2, L/3, L/4 at $Re = 250$.

The cause of the heat transfer enhancement in Figure 4 can be seen in Figure 5, in which the streamlines colored by velocity magnitude are illustrated for PT and four twisted tape pitch values. It is visible that as the twisted tape pitch value decreases, the flow path undergoes more changes. This is because more swirl flow fronts can be seen in lower pitch values with higher radial velocity, implying stronger secondary and mixing flow. As a result, it leads to a more effective redirection of core colder fluid towards the heated wall and, consequently, more heat is transferred between the fluid and heated wall.

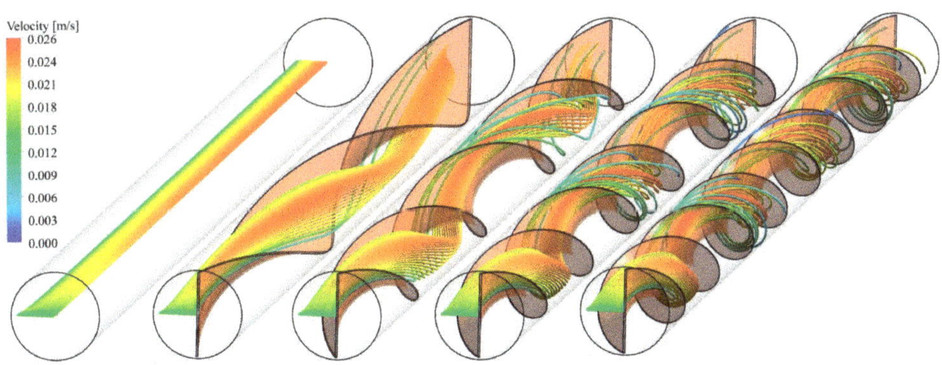

Figure 5. Streamlines colored by velocity magnitude for PT and twisted tape inserts with P = L, L/2, L/3, L/4 at $Re = 250$.

Figure 6 shows how decreasing the twisted tape pitch magnitude affects the cooling of the heated wall at $Re = 250$. The temperature distribution on the heated wall implies that the secondary and mixing flow intensity affect the heated wall. In PT, the uniform enhancement of temperature along the tube length is visible, showing the thermal boundary layer development without any disturbance. On the contrary, as the twisted tape is inserted in the tube, the change in temperature distribution on the heated wall is visible. The temperature value on the heated wall decreases along the tube length and strengthens as the pitch value reduces. In higher pitch values, some hotspot regions are visible on the heated wall temperature contour; however, these regions decline at lower pitch values, resulting in better cooling performance of the system.

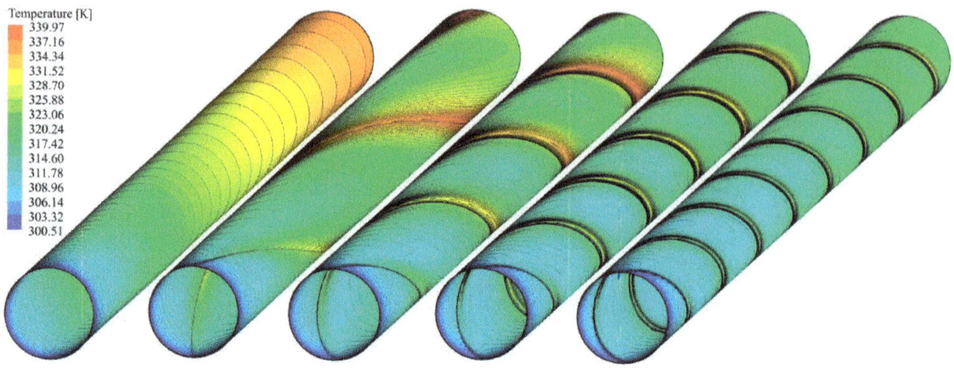

Figure 6. Temperature contours on the heated wall for PT and twisted tape inserts with P = L, L/2, L/3, L/4 at $Re = 250$.

To better understand the temperature distribution, four different cross-sections are defined along the tube length and are displayed in Figure 7. In the following, different parameters, such as temperature and velocity, are illustrated in them.

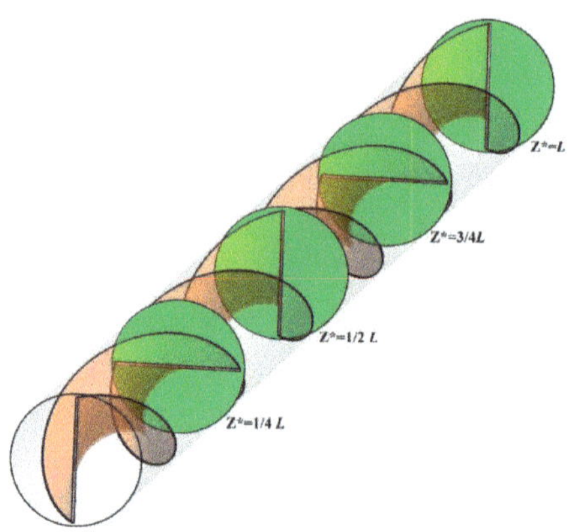

Figure 7. Generated cross-sectional surfaces throughout the tube length for post-processing purposes.

Figure 8 demonstrates the cross-sectional temperature contours on the surfaces shown in Figure 7 for PT and four twisted tape pitch values at $Re = 250$. The PT temperature contours show the normal development of the thermal boundary layer throughout the tube, leading to a great drop in heat transfer along the tube. On the other hand, the thermal boundary layer disturbance is intensified and gets thinner as the twisted tape is inserted in the tube, which is more effective for a lower pitch in heat transfer between the fluid and heated wall. Besides, fewer hotspot regions causing a reduction in heat transfer enhancement is visible at lower pitch values of the twisted tape, proving the cooling process improvement. Another point that can be noticed in this figure is that the presence of twisted tape and lowering its pitch value redirects the core colder fluid to the vicinity of the hot wall. The twisted tape causes more efficient heat dissipation from the wall and, consequently, more heat is transferred from the wall to the fluid. It should be noted that the contours are almost symmetrical, related to the center of the tube in all cases.

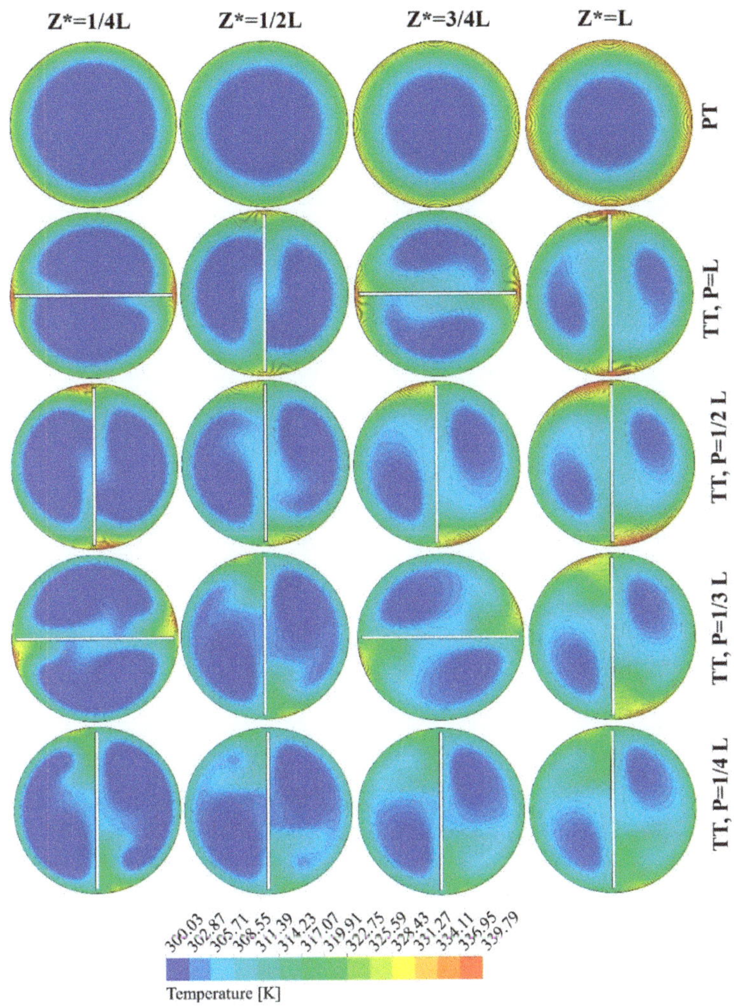

Figure 8. Cross-sectional temperature contours for PT and twisted tape inserts with P = L, L/2, L/3, L/4 at $Re = 250$.

Figure 9 displays the cross-sectional velocity contours on the surfaces shown in Figure 7 for PT and four twisted tape pitch values at $Re = 250$. The velocity contours imply the intensity of secondary and mixing flow in the presence of twisted tape inserts. In other words, the higher velocity of the fluid near the heated wall shows a stronger secondary flow, and, as a result, a higher fluid momentum near the wall and a better cooling process could be achieved. It is visible in this figure that inserting twisted tape with a pitch value of L results in a high-velocity region of the fluid, which is strengthened as the twisted tape pitch value decreases, resulting in the stronger secondary flow observed in Figure 5. It should be noted that the contours are almost symmetrical, related to the center of the tube in all cases; however, for the case with P = L, it is almost symmetrical related to the twisted tape plane.

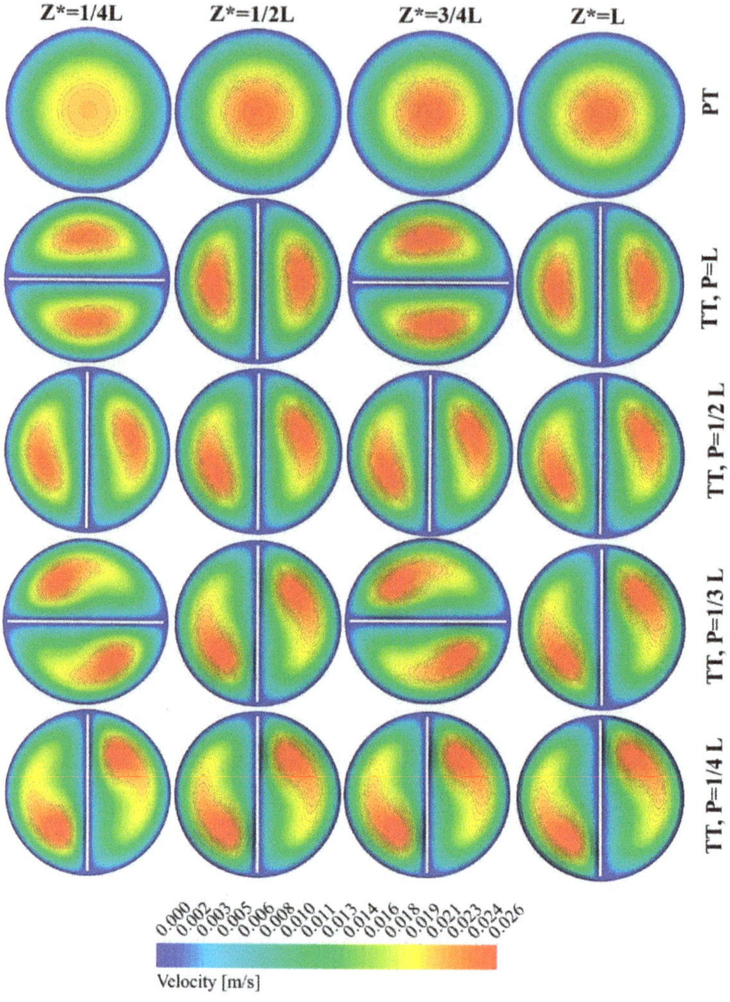

Figure 9. Cross-sectional velocity contours for PT and twisted tape inserts with P = L, L/2, L/3, L/4 at $Re = 250$.

To better quantify the heat transfer modification in the presence of twisted tape inserts with different pitch values, Figure 10 is provided for various *Re*. *Nu* increases as *Re* changes due to the more effective advection phenomenon and fluid momentum in higher fluid velocities. As seen in this figure, as an example, using twisted tape with pitch values of L, L/2, L/3 and L/4 increases the average *Nu* by about 26.87, 55.03, 86.59 and 151.42% at $Re = 1000$ compared with the PT, respectively.

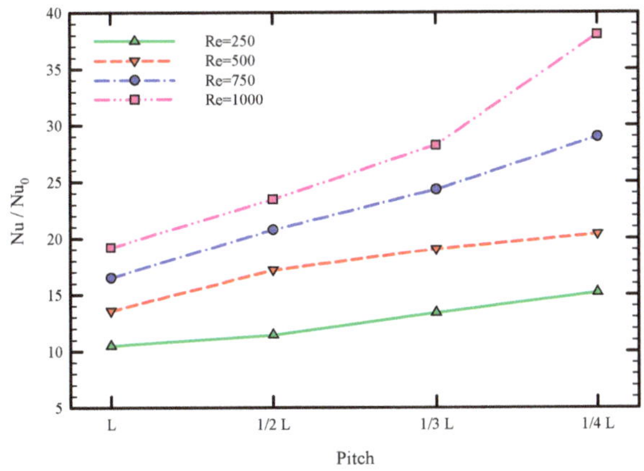

Figure 10. Variations of average *Nu* ratio with respect to PT at various pitch values and *Re*.

Figure 11 shows the variations of the friction coefficient ratio with respect to PT for different pitch values and *Re*. This figure shows that applying twisted tape enhances the friction coefficient ratio due to the added surface area and flow blockage [52]. Furthermore, higher values for the friction coefficient ratio are observed in lower pitch values due to the creation of the more intense secondary flow shown in Figure 5; however, when the twisted tape pitch is equal to L, its enhancement is insignificant due to the creation of very weak secondary flow in the tube. Moreover, this figure reveals that the friction coefficient ratio increases as *Re* changes because more intense swirling flow.

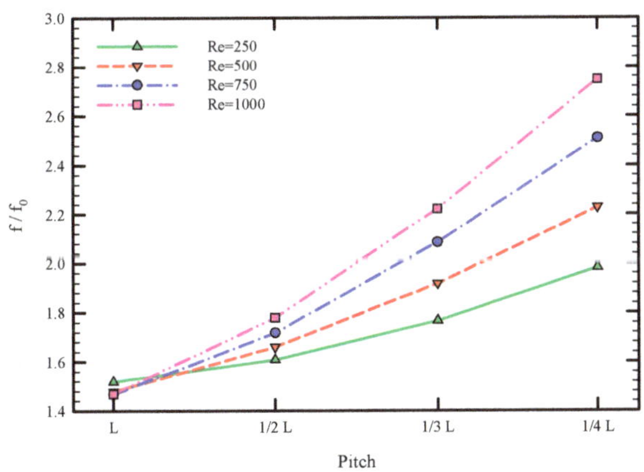

Figure 11. Variations of friction coefficient ratio with respect to PT at various pitch values and *Re*.

5.2. Effect of Truncated Twisted Tape Position and Percentage

So far, it has been shown that the twisted tape with a pitch of L/4 results in the best thermal performance of the system. Therefore, the following simulations are performed for this pitch value. To show the effect of twisted tape truncation percentage on local *Nu* throughout the tube length for different twisted tape positions, Figure 12 is provided at $Re = 250$. In Figure 12a, the twisted tape is embedded at the entrance of the tube with different values for λ. It is visible that as the flow enters the twisted tape at the tube inlet, the thermal boundary layer is disturbed and, due to the creation of secondary flow, the local *Nu* increases; however, as the flow passes the twisted tape, the thermal boundary layer starts to develop normally, and as a result, it tends to develop towards the local *Nu* curve of PT. Consequently, lower values of λ cause a higher heat transfer rate when the twisted tape is inserted at the tube entrance.

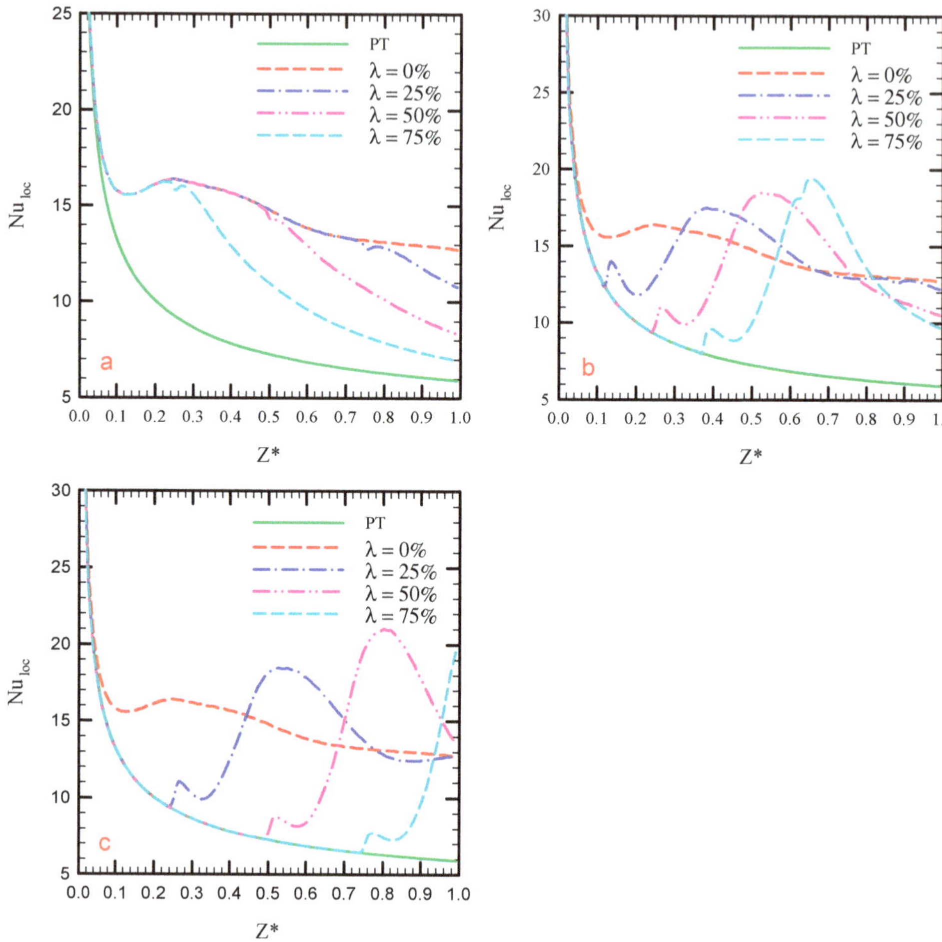

Figure 12. Local Nu along the tube length for different λ values at three twisted tape positions for (**a**) entrance, (**b**) center, (**c**) exit and P = L/4.

In Figure 12b, the twisted tape is embedded at the center of the tube with different values for λ. In this figure, the local Nu curve starts to grow as the fluid reaches the twisted tape due to the thermal boundary layer disturbance. It is visible that the local Nu at the truncated cases surpasses the fully fitted twisted tape, and a higher maximum value for Nu is visible for higher values of λ. As the flow passes the twisted tape, the thermal boundary layer starts to grow and becomes fully developed.

In Figure 12c, the twisted tape is embedded at the exit of the tube with different values for λ. This figure also shows the local Nu enhancement as the fluid enters the twisted tape as a result of thermal boundary layer disturbance and stronger mixing flow. In this case, after a sudden increase in Nu at the twisted tape entrance, the local Nu decreases, and its curve tends to reach the value of the fully fitted twisted tape.

Figure 13 displays the streamlines colored by velocity magnitude in different twisted tape truncation percentage values inserted at the tube entrance for $Re = 250$. In the truncated cases, the flow path swirls to the end of tube length, but the secondary flow intensity is reduced as the fluid passes the twisted tape. As a result, truncating the twisted tape results in a reduction in heat transfer rate but less material is used, causing fewer production expenses, and also the pressure loss penalty reduces due to less flow lockage and contact area between the fluid and solid.

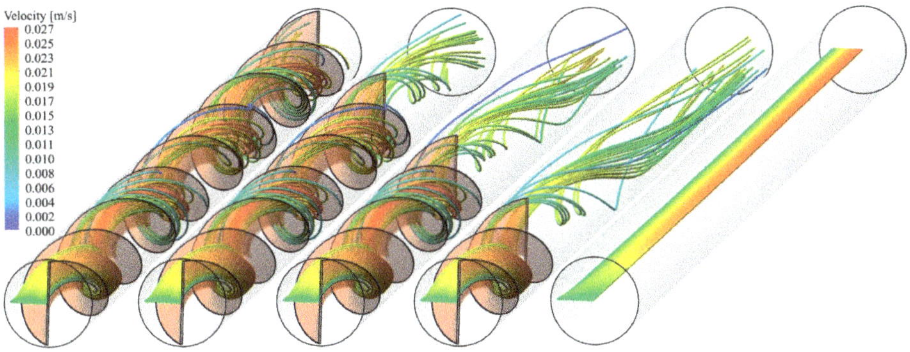

Figure 13. Streamline colored by velocity magnitude for different λ values at Re = 250 and P = L/4.

To show the effect of truncated twisted tape inserts and changes in the flow patch shown in Figure 13 on the temperature distribution of the heated wall, Figure 14 illustrates the temperature contours at the wall. It is visible that although there are some hotspots after the truncated twisted tape, the number of hotspots is still lower than for the PT, showing the effective cooling process even after the fluid passes the twisted tape. This figure proves the presence of secondary flow (not as intense as it is in the fully twisted tape insert) after the fluid passes the truncated twisted tape.

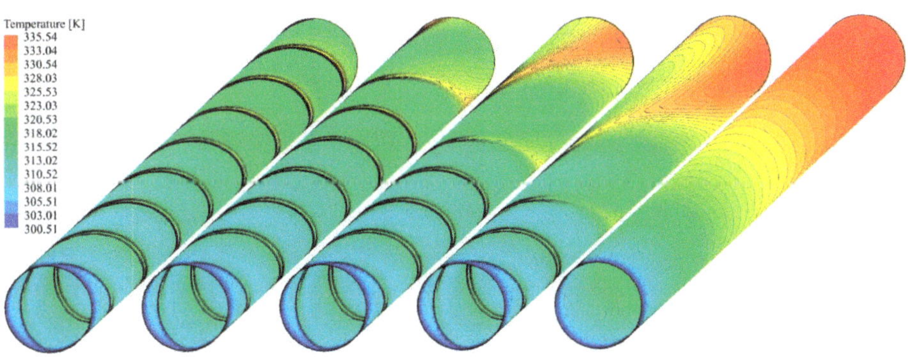

Figure 14. Temperature contours on the heated wall for different λ values at Re = 250 and P = L/4.

To clearly show the effect of twisted tape position on the local *Nu* throughout the tube length at different twisted tape truncation percentage values, Figure 15 illustrates the local *Nu* for three different twisted tape truncation percentage values of 25, 50 and 75% at $Re = 250$ for different positions, as shown in Figure 1b. This figure shows that when the twisted tape is at the tube entrance, no sudden increase in *Nu* is visible; in contrast, there is a maximum value for *Nu* along the tube length when the twisted tape is inserted at the center and exit of the tube. For all values of λ, the highest maximum *Nu* throughout the tube length corresponds to the layouts where twisted tape is inserted at the tube exit, which is higher as λ changes.

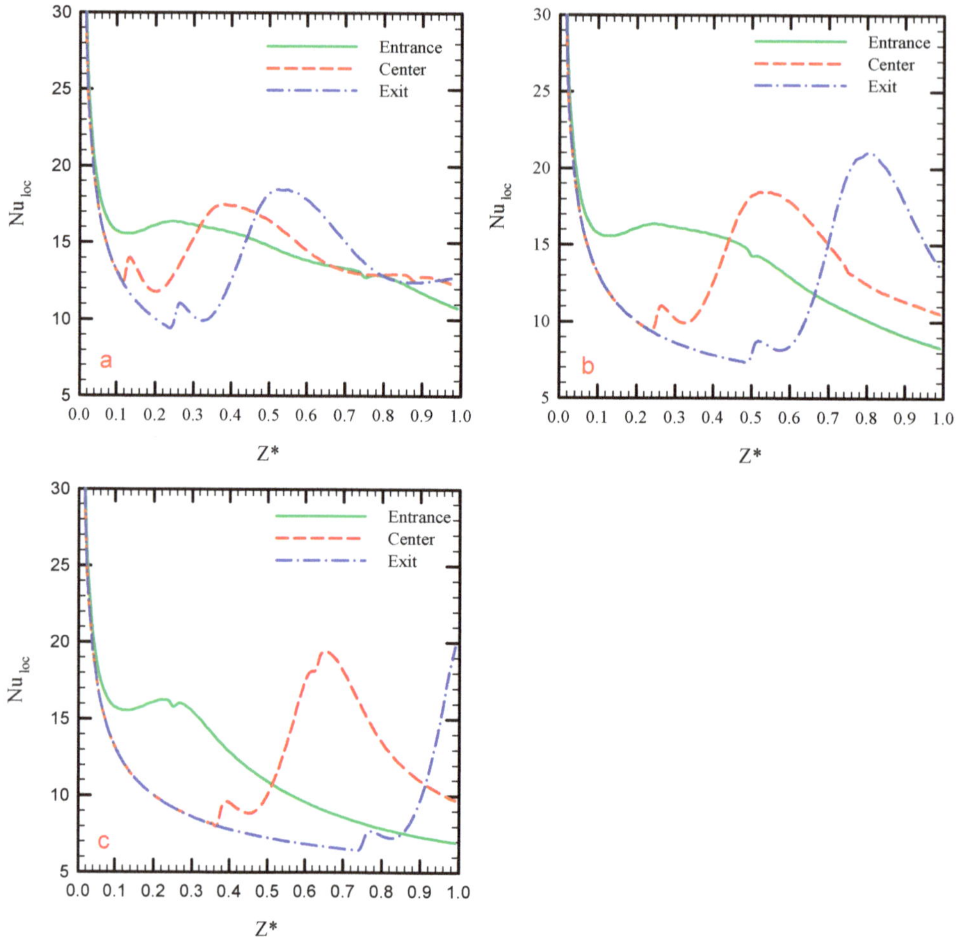

Figure 15. Local *Nu* along the tube length for different twisted tape positions at three λ values of (**a**) 25%, (**b**) 50%, (**c**) 75% and P = L/4.

Figure 16 illustrates the streamlines colored by velocity magnitude at different positions of the twisted tape in the tube at a twisted tape truncation percentage of 75% and $Re = 250$. This figure shows that when the twisted tape is embedded at the tube entrance, the whole flow patch is affected. When the twisted tape is at the tube center, half of the tube length is affected, and for the case in which twisted tape is at the tube exit, the flow path before reaching the twisted tape is similar to that of PT.

The effect of the position of the truncated twisted tape inserts on the temperature distribution of the heated wall is depicted in Figure 17. The same temperature distribution before the flow reaches the twisted tape as that of PT for the heated wall temperature distribution is also visible in this figure. Moreover, flow mixing as the fluid passes the twisted tape and the strong secondary flow in the twisted tape regions are observable in this figure.

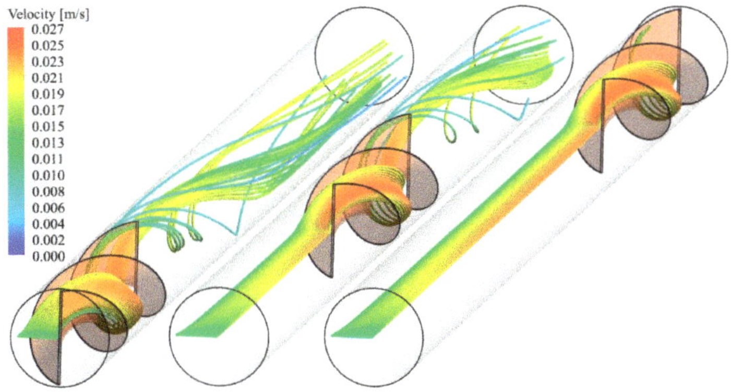

Figure 16. Streamline colored by velocity magnitude for different twisted tape positions at Re = 250, P = L/4 and λ = 0.75.

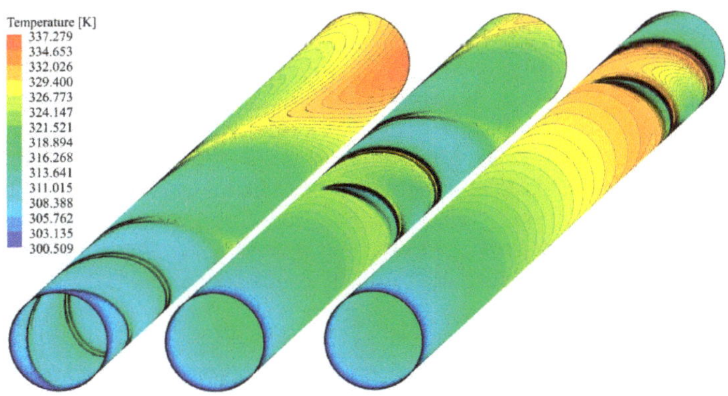

Figure 17. Temperature contours on the heated wall for different twisted tape positions at Re = 250, P = L/4 and λ = 0.75.

To quantify the heat transfer rate for different truncation values and positions of the twisted tape, Figure 18 is provided to show the variations of the *Nu* ratio compared with PT at various λ and Re values, different twisted tape positions and a pitch of L/4. For all twisted tape positions and *Re*, higher values of λ result in lower *Nu* and heat transfer due to the fact that the secondary and mixing flows in a tube fully fitted with twisted tape are much stronger than for a tube equipped with a truncated twisted tape, as shown in Figure 14. As seen in this figure, using twisted tape with a pitch value of L/4 at the entrance of the tube and λ values of 0, 25, 50 and 75% increase the average *Nu* by about 71.26, 68.50, 57.59 and 37.34% at $Re = 250$ and 151.42, 133.99, 109.52 and 71.43% at $Re = 1000$ in comparison with the PT, respectively.

Considering the position of the twisted tape, for all values of λ at Re of 1000, when the truncated twisted tape is placed at the tube entrance, a higher *Nu* is obtained, followed by the cases with twisted tape at the tube center, and the lowest *Nu* correspond to the cases with twisted tape at the tube exit. This is due to the fact that as the fluid passes the twisted tape, the flow is still affected by the twisted tape and swirl flow is visible to the end of the tube, as shown in Figure 16. Thus, for the case of twisted tape inserts in the tube entrance, more of the tube length experiences swirl flow, and as a result, the cooling process improves.

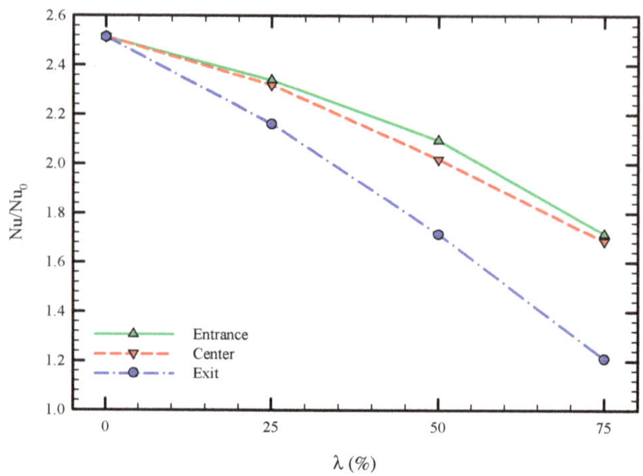

Figure 18. Variations of average *Nu* ratio with respect to PT at different twisted tape positions and λ values for Re = 1000.

Figure 19 represents the variations of friction coefficient ratio with respect to PT at various λ for *Re* of 1000, different twisted tape positions and a pitch of L/4. It is visible that for all twisted tape positions, as λ increases, the friction coefficient ratio is reduced due to the lower level of solid material and, as a result, less flow blockage and smaller contact area between the fluid flow and solid. Furthermore, for all values of λ, there are almost similar values for the friction coefficient ratio for the cases where twisted tape is placed at the tube entrance and center; however, lower values for the friction coefficient ratio are observable when the twisted tape is embedded at the tube exit. The cause of this scenario can be attributed to the fact that when the twisted tape is at the tube exit, no swirl flow is generated before the fluid reaches the twisted tape.

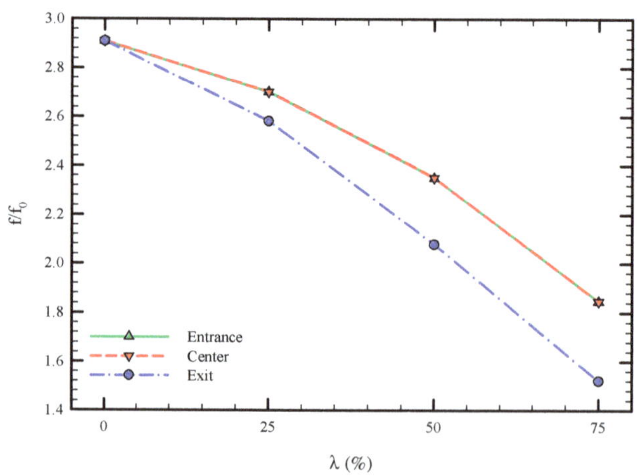

Figure 19. Variations of friction coefficient ratio with respect to PT at different twisted tape positions and λ values for Re = 1000.

As discussed above, the application of twisted tape inserts in the captured cases increases both the heat transfer rate as a desirable outcome and the friction coefficient as an undesirable result. Therefore, there is an interplay between the benefits of using twisted tape in heat transfer enhancement and their side effects in forcing more pumping power into the system. To analyze this issue, the dimensionless PEC number introduced in Equation (9) is discussed here. In fact, this number is used to evaluate the practical use of any modified heat transfer technique from the viewpoint of energy-saving potential. Generally, higher values of PEC imply superior energy saving. Figure 20 illustrates the PEC parameter for all cases investigated in this study for Re of 1000. It is visible that decreasing the twisted tape pitch value from Figure 20a–d enhances the PEC number, showing the fact that applying twisted tape with a lower pitch value is efficient from the viewpoint of both heat transfer enhancement and energy saving. It is also visible that as the twisted tape pitch increases, the sensitivity of PEC to the λ value is reduced.

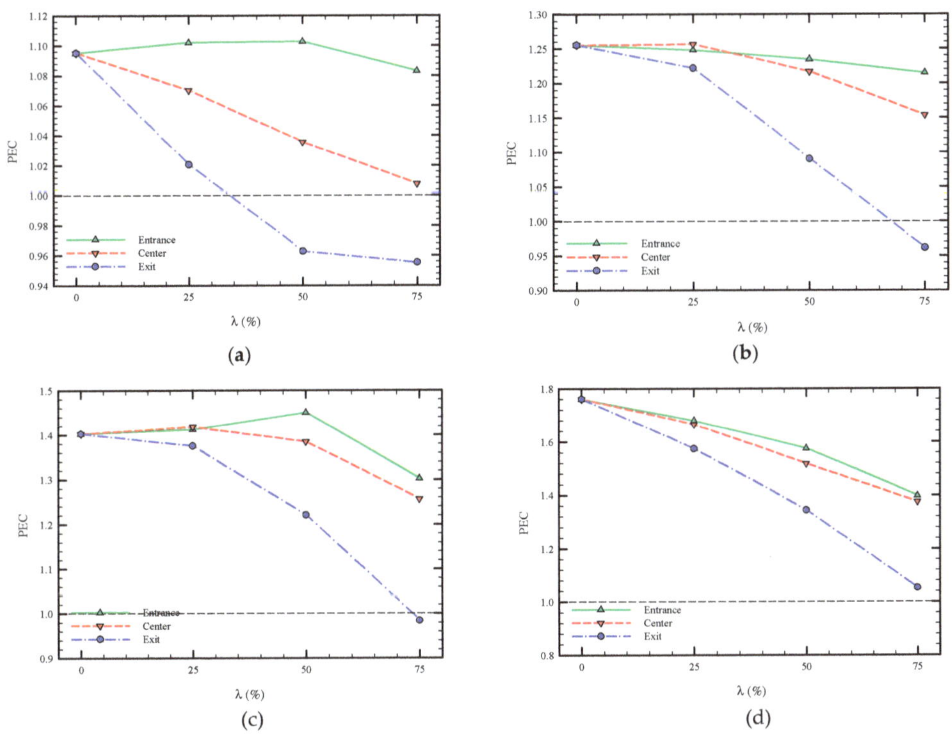

Figure 20. Variations of PEC number at different twisted tape positions, λ and P values. (**a**) P = L, (**b**) P = L/2, (**c**) P = L/3 and (**d**) P = L/4 for Re = 1000.

Furthermore, for all λ values, placing the twisted tape at the tube entrance leads to higher PEC magnitudes. Therefore, for P = L, L/2, L/3 and L/4, the optimum cases from the viewpoint of energy saving are twisted tapes with λ = 75, 50, 50 and 0%, for which the related PEC numbers at $Re = 1000$ are almost equal to 1.08, 1.24, 1.4 and 1.76, respectively. In addition, PEC numbers in all cases are tabulated in Appendix A of this paper.

6. Conclusions

This paper investigated the laminar convection heat transfer in a tube equipped with twisted tape inserts. A parametric study was conducted to evaluate the impact of key design variables, including the twisted tape truncation percentage, pitch value, position in the tube and Reynolds number on the fluid flow and heat transfer characteristics of such a configuration using symmetric heat flux around the tube. To determine the trade-off between the heat transfer enhancement and the pressure drop penalty, the PEC number was calculated. The obtained results indicated that using the twisted tape and reducing its pitch value increases the Nusselt number, friction coefficient and PEC number due to the generation of efficient secondary and mixing flow. The average Nusselt number increased by about 151.42 for a Reynolds number of 1000 in the case of fully fitted twisted tape at a pitch value of L/4. It was also found that increasing the twisted tape truncation percentage reduced both heat transfer and pressure drop in comparison with the fully fitted twisted tape case. Moreover, the best position for the truncated twisted tapes was at the tube entrance to reach the highest thermal performance. Ultimately, it was concluded that for P = L, L/2, L/3 and L/4, the optimum cases from the viewpoint of energy saving are twisted tapes with λ = 75, 50, 50 and 0%, for which the related PEC numbers at a Reynolds number of 1000 are almost equal to 1.08, 1.24, 1.4 and 1.76, respectively. The finding of this research provides a framework for researchers working in this area toward higher performance based on energy saving according to both heat transfer enhancement and pressure drop reduction.

Author Contributions: R.M. and P.T. performed and designed the numerical simulations; M.G., H.A., R.M., H.M.A. and W.Y. analyzed the data and performed the discussion; M.G., H.A., R.M., P.T., H.M.A. and W.Y. wrote the paper. All authors have read and agreed to the published version of the manuscript.

Funding: The APC was funded by Dr Wahiba Yaïci from CanmetENERGY Research Centre, Natural Resources Canada.

Conflicts of Interest: The authors declare no conflict of interest.

Appendix A

Table A1. PEC in all cases for twisted tape at the entrance.

λ (%) \ Re	250	500	750	1000	
0	1.011295	1.021043	1.059828	1.095052	P = L
25	1.01388	1.017496	1.045332	1.070205	
50	1.004123	1.006426	1.02249	1.035571	
75	0.995671	0.99766	1.003663	1.007723	
0	1.082103	1.24311	1.265074	1.255414	P = L/2
25	1.089229	1.232876	1.263873	1.256518	
50	1.092546	1.194338	1.220967	1.216784	
75	1.074308	1.131182	1.149658	1.153739	
0	1.227265	1.311893	1.387978	1.402961	P = L/3
25	1.227311	1.319265	1.401682	1.418449	
50	1.210159	1.302567	1.334738	1.384967	
75	1.199981	1.258698	1.265143	1.25622	
0	1.337716	1.336512	1.555087	1.761103	P = L/4
25	1.338476	1.40221	1.515203	1.665051	
50	1.315334	1.37707	1.476264	1.517847	
75	1.29284	1.345982	1.368546	1.374241	

Table A2. PEC in all cases for twisted tape at the center.

λ (%) \ Re	250	500	750	1000	
0	1.011295	1.021043	1.059828	1.095052	P = L
25	1.030648	1.043332	1.079246	1.102066	
50	1.033091	1.058554	1.08903	1.102768	
75	1.018142	1.046153	1.071832	1.083035	
0	1.082103	1.24311	1.265074	1.255414	P = L/2
25	1.109637	1.255035	1.268496	1.248454	
50	1.130221	1.236448	1.245266	1.234765	
75	1.098127	1.18203	1.205088	1.215016	
0	1.227265	1.311893	1.387978	1.402961	P = L/3
25	1.257101	1.318361	1.391219	1.412675	
50	1.246763	1.292509	1.340597	1.450322	
75	1.18302	1.256449	1.293239	1.302188	
0	1.337716	1.336512	1.555087	1.761103	P = L/4
25	1.357279	1.340645	1.49818	1.680165	
50	1.332735	1.385675	1.507737	1.575765	
75	1.237176	1.333577	1.375402	1.397482	

Table A3. PEC in all cases for twisted tape at the exit.

λ (%) \ Re	250	500	750	1000	
0	1.011295	1.021043	1.059828	1.095052	P = L
25	0.984648	0.983887	1.002881	1.02074	
50	0.95755	0.953732	0.958557	0.962848	
75	0.956485	0.954873	0.955329	0.955342	
0	1.082103	1.24311	1.265074	1.255414	P = L/2
25	1.06148	1.185216	1.225127	1.222088	
50	1.005875	1.052004	1.083942	1.090425	
75	0.959385	0.95912	0.960504	0.961318	
0	1.227265	1.311893	1.387978	1.402961	P = L/3
25	1.18397	1.29666	1.343979	1.375853	
50	1.115867	1.214709	1.214682	1.220923	
75	0.971801	0.980115	0.98335	0.98392	
0	1.337716	1.336512	1.555087	1.761103	P = L/4
25	1.294428	1.382539	1.469811	1.574971	
50	1.252857	1.297774	1.344916	1.343726	
75	1.021773	1.049235	1.050204	1.052115	

References

1. Hosseini, M.; Afrouzi, H.H.; Arasteh, H.; Toghraie, D. Energy analysis of a proton exchange membrane fuel cell (PEMFC) with an open-ended anode using agglomerate model: A CFD study. *Energy* **2019**, *188*, 116090. [CrossRef]
2. Purusothaman, A. Investigation of natural convection heat transfer performance of the QFN-PCB electronic module by using nanofluid for power electronics cooling applications. *Adv. Powder Technol.* **2018**, *29*, 996–1004. [CrossRef]
3. Dabiri, S.; Khodabandeh, E.; Poorfar, A.K.; Mashayekhi, R.; Toghraie, D.; Zade, S.A.A. Parametric investigation of thermal characteristic in trapezoidal cavity receiver for a linear Fresnel solar collector concentrator. *Energy* **2018**, *153*, 17–26. [CrossRef]

4. Sakanova, A.; Tong, C.F.; Tseng, K.J.; Simanjorang, R.; Gupta, A.K. Weight consideration of liquid metal cooling technology for power electronics converter in future aircraft. In Proceedings of the 2016 IEEE 2nd Annual Southern Power Electronics Conference (SPEC), Auckland, New Zealand, 5–8 December 2017; pp. 1–5.
5. Ravi, R.; Pachamuthu, S.; Kasinathan, P. Computational and experimental investigation on effective utilization of waste heat from diesel engine exhaust using a fin protracted heat exchanger. *Energy* **2020**, *200*, 117489. [CrossRef]
6. Kadam, S.T.; Gkouletsos, D.; Hassan, I.; Rahman, M.A.; Kyriakides, A.-S.; Papadopoulos, A.I.; Seferlis, P. Investigation of binary, ternary and quaternary mixtures across solution heat exchanger used in absorption refrigeration and process modifications to improve cycle performance. *Energy* **2020**, *198*, 117254. [CrossRef]
7. Jung, J.; Jeon, Y.; Cho, W.; Kim, Y. Effects of injection-port angle and internal heat exchanger length in vapor injection heat pumps for electric vehicles. *Energy* **2020**, *193*. [CrossRef]
8. Wu, J.; Ju, Y. Design and optimization of natural gas liquefaction process using brazed plate heat exchangers based on the modified single mixed refrigerant process. *Energy* **2019**, *186*, 115819. [CrossRef]
9. Chai, S.; Sun, X.; Zhao, Y.; Dai, Y. Experimental investigation on a fresh air dehumidification system using heat pump with desiccant coated heat exchanger. *Energy* **2019**, *171*, 306–314. [CrossRef]
10. Toghraie, D.; Mashayekhi, R.; Niknejadi, M.; Arasteh, H. Hydrothermal performance analysis of various surface roughness configurations in trapezoidal microchannels at slip flow regime. *Chin. J. Chem. Eng.* **2020**, *28*, 1522–1532. [CrossRef]
11. Ghalambaz, M.; Arasteh, H.; Mashayekhi, R.; Keshmiri, A.; Talebizadehsardari, P.; Yaïci, W. Investigation of Overlapped Twisted Tapes Inserted in a Double-Pipe Heat Exchanger Using Two-Phase Nanofluid. *Nanomaterials* **2020**, *10*, 1656. [CrossRef] [PubMed]
12. Hamid, K.A.; Azmi, W.; Mamat, R.; Sharma, K. Heat transfer performance of TiO2–SiO2 nanofluids in a tube with wire coil inserts. *Appl. Therm. Eng.* **2019**, *152*, 275–286. [CrossRef]
13. Song, X.; Dong, G.; Gao, F.; Diao, X.; Zheng, L.; Zhou, F. A numerical study of parabolic trough receiver with nonuniform heat flux and helical screw-tape inserts. *Energy* **2014**, *77*, 771–782. [CrossRef]
14. Murugan, M.; Vijayan, R.; Saravanan, A.; Jaisankar, S. Performance enhancement of centrally finned twist inserted solar collector using corrugated booster reflectors. *Energy* **2019**, *168*, 858–869. [CrossRef]
15. Jaisankar, S.; Radhakrishnan, T.; Sheeba, K. Studies on heat transfer and friction factor characteristics of thermosyphon solar water heating system with helical twisted tapes. *Energy* **2009**, *34*, 1054–1064. [CrossRef]
16. Ananth, J.; Jaisankar, S. Investigation on heat transfer and friction factor characteristics of thermosiphon solar water heating system with left-right twist regularly spaced with rod and spacer. *Energy* **2014**, *65*, 357–363. [CrossRef]
17. Saravanan, A.; Senthilkumaar, J.; Jaisankar, S. Experimental studies on heat transfer and friction factor characteristics of twist inserted V-trough thermosyphon solar water heating system. *Energy* **2016**, *112*, 642–654. [CrossRef]
18. Man, C.; Lv, X.; Hu, J.; Sun, P.; Tang, Y. Experimental study on effect of heat transfer enhancement for single-phase forced convective flow with twisted tape inserts. *Int. J. Heat Mass Transf.* **2017**, *106*, 877–883. [CrossRef]
19. Lim, K.Y.; Hung, Y.M.; Tan, B.T. Performance evaluation of twisted-tape insert induced swirl flow in a laminar thermally developing heat exchanger. *Appl. Therm. Eng.* **2017**, *121*, 652–661. [CrossRef]
20. Kumar, N.R.; Bhramara, P.; Kirubeil, A.; Sundar, L.S.; Singh, M.K.; Sousa, A. Effect of twisted tape inserts on heat transfer, friction factor of Fe3O4 nanofluids flow in a double pipe U-bend heat exchanger. *Int. Commun. Heat Mass Transf.* **2018**, *95*, 53–62. [CrossRef]
21. Ho, C.; Chiou, Y.-H.; Yan, W.-M.; Ghalambaz, M. Transient cooling characteristics of Al2O3-water nanofluid flow in a microchannel subject to a sudden-pulsed heat flux. *Int. J. Mech. Sci.* **2019**, *151*, 95–105. [CrossRef]
22. Ghalambaz, M.; Doostani, A.; Izadpanahi, E.; Chamkha, A.J. Conjugate natural convection flow of Ag–MgO/water hybrid nanofluid in a square cavity. *J. Therm. Anal. Calorim.* **2019**, *139*, 2321–2336. [CrossRef]
23. Mehryan, S.; Ghalambaz, M.; Chamkha, A.J.; Izadi, M. Numerical study on natural convection of Ag–MgO hybrid/water nanofluid inside a porous enclosure: A local thermal non-equilibrium model. *Powder Technol.* **2020**, *367*, 443–455. [CrossRef]

24. Ho, C.; Liu, Y.-C.; Ghalambaz, M.; Yan, W. Forced convection heat transfer of Nano-Encapsulated Phase Change Material (NEPCM) suspension in a mini-channel heatsink. *Int. J. Heat Mass Transf.* **2020**, *155*. [CrossRef]
25. Ghalambaz, M.; Chamkha, A.J.; Wen, D. Natural convective flow and heat transfer of Nano-Encapsulated Phase Change Materials (NEPCMs) in a cavity. *Int. J. Heat Mass Transf.* **2019**, *138*, 738–749. [CrossRef]
26. Ghalambaz, M.; Groşan, T.; Pop, I. Mixed convection boundary layer flow and heat transfer over a vertical plate embedded in a porous medium filled with a suspension of nano-encapsulated phase change materials. *J. Mol. Liq.* **2019**, *293*. [CrossRef]
27. Chien, L.-H.; Cheng, Y.-T.; Lai, Y.-L.; Yan, W.-M.; Ghalambaz, M. Experimental and numerical study on convective boiling in a staggered array of micro pin-fin microgap. *Int. J. Heat Mass Transf.* **2020**, *149*, 119203. [CrossRef]
28. Chien, L.-H.; Liao, W.-R.; Ghalambaz, M.; Yan, W.-M. Experimental study on convective boiling flow and heat transfer in a microgap enhanced with a staggered arrangement of nucleated micro-pin-fins. *Int. J. Heat Mass Transf.* **2019**, *144*, 118653. [CrossRef]
29. Jaramillo, O.A.; Borunda, M.; Velazquez-Lucho, K.; Robles, M. Parabolic trough solar collector for low enthalpy processes: An analysis of the efficiency enhancement by using twisted tape inserts. *Renew. Energy* **2016**, *93*, 125–141. [CrossRef]
30. Mwesigye, A.; Bello-Ochende, T.; Meyer, J.P. Heat transfer and entropy generation in a parabolic trough receiver with wall-detached twisted tape inserts. *Int. J. Therm. Sci.* **2016**, *99*, 238–257. [CrossRef]
31. Esfe, M.H.; Mazaheri, H.; Mirzaei, S.S.; Kashi, E.; Kazemi, M.; Afrand, M. Effects of twisted tapes on thermal performance of tri-lobed tube: An applicable numerical study. *Appl. Therm. Eng.* **2018**, *144*, 512–521. [CrossRef]
32. Saysroy, A.; Eiamsa-Ard, S. Periodically fully-developed heat and fluid flow behaviors in a turbulent tube flow with square-cut twisted tape inserts. *Appl. Therm. Eng.* **2017**, *112*, 895–910. [CrossRef]
33. Saysroy, A.; Eiamsa-Ard, S. Enhancing convective heat transfer in laminar and turbulent flow regions using multi-channel twisted tape inserts. *Int. J. Therm. Sci.* **2017**, *121*, 55–74. [CrossRef]
34. He, Y.; Liu, L.; Li, P.; Ma, L. Experimental study on heat transfer enhancement characteristics of tube with cross hollow twisted tape inserts. *Appl. Therm. Eng.* **2018**, *131*, 743–749. [CrossRef]
35. Samruaisin, P.; Changcharoen, W.; Thianpong, C.; Chuwattanakul, V.; Pimsarn, M.; Eiamsa-Ard, S. Influence of regularly spaced quadruple twisted tape elements on thermal enhancement characteristics. *Chem. Eng. Process. Process. Intensif.* **2018**, *128*, 114–123. [CrossRef]
36. Ruengpayungsak, K.; Saysroy, A.; Wongcharee, K.; Eiamsa-Ard, S. Thermohydraulic performance evaluation of heat exchangers equipped with centrally perforated twisted tape: Laminar and turbulent flows. *J. Therm. Sci. Technol.* **2019**, *14*, JTST0002. [CrossRef]
37. Hasanpour, A.; Farhadi, M.; Sedighi, K. Intensification of heat exchangers performance by modified and optimized twisted tapes. *Chem. Eng. Process. Process. Intensif.* **2017**, *120*, 276–285. [CrossRef]
38. Hong, Y.; Du, J.; Wang, S. Experimental heat transfer and flow characteristics in a spiral grooved tube with overlapped large/small twin twisted tapes. *Int. J. Heat Mass Transf.* **2017**, *106*, 1178–1190. [CrossRef]
39. Hong, Y.; Du, J.; Wang, S. Turbulent thermal, fluid flow and thermodynamic characteristics in a plain tube fitted with overlapped multiple twisted tapes. *Int. J. Heat Mass Transf.* **2017**, *115*, 551–565. [CrossRef]
40. Eiamsa-Ard, S.; Samravysin, P.; Eiamsa-Ard, S.; Samruaysin, P. Characterization of Heat Transfer by Overlapped-Quadruple Counter Tapes. *J. Heat Transf.* **2018**, *140*, 114501. [CrossRef]
41. Rudrabhiramu, R.; Kumar, K.H.; Rao, K.M. Heat Transfer Enhancement Using Overlapped Dual Twisted Tape Inserts with Nanofluids. In *Advances in Automotive Technologies*; Springer Science and Business Media LLC.: Berlin/Heidelberg, Germany, 2020; pp. 123–130.
42. Qi, C.; Wang, G.; Yan, Y.; Mei, S.; Luo, T. Effect of rotating twisted tape on thermo-hydraulic performances of nanofluids in heat-exchanger systems. *Energy Convers. Manag.* **2018**, *166*, 744–757. [CrossRef]
43. Sundar, L.S.; Singh, M.K.; Punnaiah, V.; Sousa, A.C.M. Experimental investigation of Al2O3/water nanofluids on the effectiveness of solar flat-plate collectors with and without twisted tape inserts. *Renew. Energy* **2018**, *119*, 820–833. [CrossRef]
44. Tian, Z.; Abdollahi, A.; Shariati, M.; Amindoust, A.; Arasteh, H.; Karimipour, A.; Goodarzi, M.; Bach, Q.-V. Turbulent flows in a spiral double-pipe heat exchanger. *Int. J. Numer. Methods Heat Fluid Flow* **2019**, *30*, 39–53. [CrossRef]

45. Toghraie, D.; Mashayekhi, R.; Arasteh, H.; Sheykhi, S.; Niknejadi, M.; Chamkha, A.J. Two-phase investigation of water-Al2O3 nanofluid in a micro concentric annulus under non-uniform heat flux boundary conditions. *Int. J. Numer. Methods Heat Fluid Flow* **2019**, *30*, 1795–1814. [CrossRef]
46. Arasteh, H.; Salimpour, M.; Tavakoli, M.R. Optimal distribution of metal foam inserts in a double-pipe heat exchanger. *Int. J. Numer. Methods Heat Fluid Flow* **2019**, *29*, 1322–3142. [CrossRef]
47. Wei, L.; Arasteh, H.; Abdollahi, A.; Parsian, A.; Taghipour, A.; Mashayekhi, R.; Tlili, I. Locally weighted moving regression: A non-parametric method for modeling nanofluid features of dynamic viscosity. *Phys. A Stat. Mech. Appl.* **2020**, *550*, 124124. [CrossRef]
48. Rezaei, O.; Akbari, O.A.; Marzban, A.; Toghraie, D.; Pourfattah, F.; Mashayekhi, R. The numerical investigation of heat transfer and pressure drop of turbulent flow in a triangular microchannel. *Phys. E Low Dimens. Syst. Nanostruct.* **2017**, *93*, 179–189. [CrossRef]
49. Khodabandeh, E.; Toghraie, D.; Chamkha, A.J.; Mashayekhi, R.; Akbari, O.A.; Rozati, S.A. Energy saving with using of elliptic pillows in turbulent flow of two-phase water-silver nanofluid in a spiral heat exchanger. *Int. J. Numer. Methods Heat Fluid Flow* **2019**, *30*, 2025–2049. [CrossRef]
50. Ghaneifar, M.; Arasteh, H.; Mashayekhi, R.; Rahbari, A.; Mahani, R.B.; Talebizadehsardari, P. Thermohydraulic analysis of hybrid nanofluid in a multilayered copper foam heat sink employing local thermal non-equilibrium condition: Optimization of layers thickness. *Appl. Therm. Eng.* **2020**, *181*, 115961. [CrossRef]
51. Hong, S.W.; Bergles, A.E. Augmentation of Laminar Flow Heat Transfer in Tubes by Means of Twisted-Tape Inserts. *J. Heat Transf.* **1976**, *98*, 251–256. [CrossRef]
52. Sarviya, R.; Fuskele, V. Heat Transfer and Pressure Drop in a Circular Tube Fitted with Twisted Tape Insert Having Continuous Cut Edges. *J. Energy Storage* **2018**, *19*, 10–14. [CrossRef]

symmetry

MDPI

Article

A Study on Performance Characteristics of a Heat Pump System with High-Pressure Side Chiller for Light-Duty Commercial Electric Vehicles

Moo-Yeon Lee [1], Kunal Sandip Garud [1], Han-Byeol Jeon [2] and Ho-Seong Lee [2,*]

1 Department of Mechanical Engineering, Dong-A University, 37 Nakdong-Daero 550, Saha-gu, Busan 49315, Korea; mylee@dau.ac.kr (M.-Y.L.); kunalgarud1@gmail.com (K.S.G.)

2 Thermal Management R&D Center, KATECH, 303 Pungse-ro, Pungse-Myun, Cheonan 31214, Korea; jeonhb@katech.re.kr

* Correspondence: leehs@katech.re.kr; Tel.: +82-41-559-3094 or +82-10-4653-4216

Received: 15 June 2020; Accepted: 25 July 2020; Published: 27 July 2020

Abstract: One of barriers for the present heat pump system's application in an electric vehicle was decreased performance under cold ambient conditions due to the lack of evaporating heat source. In order to improve the heat pump's performance, a high-pressure side chiller was additionally installed, and the tested heat pump system was modified with respect to refrigerant flow direction along with operating modes. In the present work, the performance characteristics of the heat pump system with a high-pressure side chiller for light-duty commercial electric vehicles were studied experimentally under hot and cold ambient conditions, reflecting real road driving. The high-pressure side chiller was located after the electric compressor so that the highest refrigerant temperature transferred the heat to the coolant. The controlled coolant with discharged refrigerant from the electric compressor was used to heat up the cabin, transferring heat to the inlet air like the internal combustion engine vehicle's heating system, except with unused engine waste heat. In the cooling mode, for the exterior air temperature of 35 °C and interior air temperature of 25 °C, cooling performance along with the compressor speed showed that the system efficiency decreased by 16.4% on average, the cooling capacity increased by 8.0% on average and the compressor work increased by 27% on average. In heating mode, at the exterior and interior air temperature of −6.7 °C, compressor speed and coolant temperature variation with steady conditions were tested with respect to heating performance. In transient mode, to increase coolant temperature with a closed loop from −6.7 °C, tested system characteristics were studied along the compressor speed with respect to heating up the cabin. As the inlet air of the HVAC was maintained at −6.7 °C, even though the heat-up rate of the cabin room was a little slow, the cabin temperature reached 20 °C within 50 min and the temperature difference with the ambient air attained 28.7 °C.

Keywords: cooling and heating mode; heat pump; high-pressure chiller; light-duty commercial electric vehicle; system performance

1. Introduction

The development of efficient and ecofriendly engines which emit less waste heat is the focus of recent research to overcome the environmental issues of global warming and CO_2 emission [1]. Therefore, in the last decade, the research trend is shifting towards the development of efficient electric vehicles [2]. The challenge associated with full commercialization of electric vehicles is their lower driving range due to high power consumption by the thermal management system. The gasoline engines are using waste heat for cabin heating during cold climatic conditions but in the case of electric vehicles, cabin heating energy is consumed from the battery, which reduces the driving range of the vehicles more in cold climatic conditions [3]. Currently, gasoline vehicles and electric vehicles are using positive temperature coefficient (PTC) heaters widely for heating purposes despite its higher cost above 2 kW and higher power consumption. The driving range of electric vehicles with PTC heaters has reduced up to 24% more than that of electric vehicles without PTC heaters [4]. Heat pumps are the best and efficient alternative for the PTC heater because the second law of thermodynamics states that the coefficient of performance of the heat pump is above 1.0. However, the heating performance of a heat pump system decreases drastically in cold climatic conditions for electric vehicles [5]. To overcome this limitation and develop an efficient heat pump system for cold regions, several studies of an improved model of heat pump system for electric vehicles are presented.

Zhang et al. have proposed an R134a-based economized vapor injection heat pump system for electric vehicles to eliminate the key issues of drainage associated with external heat exchangers and the decrease in heating performance during cold weather conditions. The proposed heat pump system showed a 57.7% improvement in heating capacity, 2097 W of maximum capacity and coefficient of performance of 1.25 under the ambient temperature of −20 °C [5]. Qin et al. have developed refrigerant injection air-source heat pump for electric vehicle to overcome the poor heating performance, high battery power consumption and operational safety of traditional air-source heat pump systems in cold climate conditions. The electric scroll compressor of the refrigerant injection air-source heat pump was provided with injection portholes and the effect of the size and shape of these injection portholes was analyzed with respect to the system performance. The larger injection portholes showed enhancement in heating capacity and the refrigerant injection air-source heat pump presented 28.6% higher heating performance compared with the traditional air-source heat pump [6,7]. To develop the efficient heat pump system with less battery power consumption and extended driving range for electric vehicles in cold weather conditions, Qin et al. have proposed air-source heat pumps with refrigerant injection. Under the ambient temperature of −20 °C, the heating performance of the heat pump was tested for various in-car inlet temperatures and different fresh-air ratios. In addition, the effects of refrigerant injection and dryness were also tested on the heating performance of the heat pump. The proposed heat pump showed 31% improved heating capacity compared with the traditional heat pump above a −10 °C in-car inlet temperature [8]. Qin et al. have proved that the air-source heat pump is a potential candidate to heat up the cabin of electric vehicle in cold weather conditions. The experimental study on the proposed heat pump showed coefficient of performance above 1.7 at the ambient temperature of −20 °C [9]. Zhou et al. have developed a heat pump system with defrost technology for the thermal management of electric vehicles under low temperature and high humidity ambient conditions. With the proposed heat pump system with defrost technology, the defrosting time of external heat exchanger could be controlled within 100 s at the ambient conditions of temperature of −20 °C and relative humidity of 80% [10]. Ahn et al. have investigated the heating performance of a dual source heat pump which used a combination of air and waste heat as the heat sources in electrical vehicles. At lower ambient temperatures, the proposed dual heat source heat pump showed better heating performance compared with a single heat source heat pump which used air or waste heat as heat source [11]. Jung et al. have developed a simulation model to analyze the coefficient of performance and isentropic efficiency of an R134a heat pump with vapor injection. At an ambient temperature of −10 °C, the heat pump with single injection port and dual injection port has shown the coefficient of performance higher by 7.5% and 9.8%, respectively, and the isentropic efficiency higher by 11.2% and 22.9%, respectively, compared with the conventional heat pump [12]. Zhang et al. have proposed a desiccant integrated heat pump system to reduce the heat demand inside the cabin of electric vehicles. At an ambient temperature of −20 °C, the proposed heat pump system has shown reductions of 42% and 38% in cabin heat load and compressor electric power, respectively, compared with the traditional heat pump system [13]. Choi et al. have designed heat pump system with vapor injection to evaluate the maximum heating capacity and coefficient of performance at lower ambient temperatures under the influence of various injection positions and different intermediate pressures [14]. Lee et al. have experimentally analyzed the heating and cooling performances of hybrid heat pump which uses waste heat of electric devices for heating and air source for cooling of electric bus. The proposed hybrid heat pump shows a cooling capacity of 23 kW and heating coefficient of performance of 2.4 [15]. Kwon et al. have presented experimental and numerical studies on the heating performance of a heat pump system with vapor injection for electric vehicles in cold weather conditions. Compared with the Joule heating system for electric vehicles, vapor injection heat pump systems enabled the extended driving range with less battery power consumption and improved heating performance in cold weather conditions [16]. Liu et al. have proposed a propane-based heat pump system to improve the heating performance for electric vehicles in cold ambient conditions. The influences of compressor speed, outside ambient temperature, inside circulated air percentage, inside air volume flow rate and outside air velocity were experimentally investigated with respect to the performance of propane-based heat pump system. The propane-based heat pump system shows effective performance above the ambient temperature of −10 °C [17]. Li et al. have developed an R1234yf-based heat pump system and compared its performance with an R134a-based heat pump system for an electric vehicle under cold ambient conditions. The comparison of both heat pump systems was conducted for outside temperature, outside air velocity, inside temperature, inside air mass flow rate, compressor speed, charge, inner condenser width and economized vapor injection [18]. Ahn et al. have analyzed the heating performance and coefficient of performance of dehumidifying heat pump integrated with additional heat source for electric vehicles with less occupancy [19] Bellocchi et al. have proposed a heat pump integrated with regenerative heat exchanger for electric vehicle HVAC which decreases the power consumption by 17–52% and driving range reduction up to 6% [20]. Lee et al. have experimentally investigated air-source heat pump system for electric vehicles to evaluate the steady state performances of heating capacity and coefficient of performance and transient temperature for cabin heating performance. The proposed air-source heat pump system presents heating capacity of 3.10 kW and coefficient of performance of 3.26 at the ambient temperature of −10 °C [21]. Li et al. have developed a hybrid model of an air-cycle heat pump system with turbocharger, blower and regenerated heat exchanger for electric vehicles. The performance of air-cycle heat pumps was numerically examined for three different positions of blower and results show that the blower installed before the compressor achieves higher coefficient of performance and heating capacity. Under the same operating conditions of electric vehicles, the air-cycle heat pump system with turbocharger power 23% compared with the positive temperature coefficient (PTC) system [22]. Cho et al. have analyzed the heating performance of an R-134a based heat pump system which uses waste heat of electrical devices for an electric bus. The behaviors of compressor work, heating capacity and coefficient of performance are investigated for the outdoor temperature and volume flow rate [23]. Lee et al. have developed R744 based electric air conditioning system for fuel cell electric vehicles which showed superior performance characteristics compared with the conventional R-134a-based air conditioning system. The developed system presented cooling capacity of 6.4 kW and coefficient of performance of 2.5 [24]. Lee et al. have proposed R744-based heat pump system with stack coolant heat source to reduce the power consumption and improve the driving range of fuel cell electric vehicles in cold climatic conditions. The proposed system showed heating capacity of 5 kW for coolant flow rate of 5 L/min and ambient temperature of −20 °C [25]. Shi et al. have presented the experimental study to analyze the performance of economized vapor injection heat pump system with R32 refrigerant. The effect of injection pressure on heat capacity and power consumption of system was analyzed. The economized vapor injection heat pump system with R32 refrigerant shows better performance than that with R410A refrigerant [26].

From the conducted literature review, the issues associated with the efficient performance of heat pump system of electric vehicle need to be addressed in future for full commercialization of electric vehicles. To improve the performance of heat pump systems of light-duty commercial vehicles, an additional high-pressure side chiller is installed at the discharge end of the electric compressor. The main objective of the present work is to experimentally investigate the cooling and heating performance characteristics of the heat pump system with a high-pressure side chiller for light-duty commercial electric vehicle under the real road driving conditions.

2. Experimental Method

The experimental set-up of the heat pump system with a high-pressure side chiller is shown in Figure 1. In order to get the heat from the highest temperature condition and give it to the coolant, the developed chiller was installed at the discharge side of the compressor, hence it is called the high-pressure side chiller. The experimental set-up mainly comprised of electric compressor, condenser, evaporator, expansion device, chiller and heater core. The specifications of the primary components of the experimental set-up is presented in Table 1. The experimental set-up had three fluid circulation loops, namely the refrigerant circulation loop, coolant circulation loop and air circulation loop. To control the ambient conditions for the experiments, the experimental set-up was housed inside a psychometric calorimeter which consisted of a cooling coil, heating coil and humidifier and was controlled using a PID controller. Using this facility, the experimental test conditions were set to a temperature of 120 °C and pressure of 2500 kPa. The scroll-type electric compressor was driven using a compressor driver at a specific frequency and current. The power input measured by a power meter and current were used to evaluate the compressor work. The refrigerant loop comprised an electric compressor, chiller, condenser, expansion valve and evaporator, while the coolant loop consisted of a chiller, water pump and receiver tank. The chiller was the additional component in the heat pump system and the common component in the refrigerant loop and coolant loop which was located next to the electric compressor. The high-pressure and high-temperature refrigerant discharges from the electric compressor transferred the heat to the coolant in the chiller. The coolant, which carried the heat from the refrigerant, circulated in the coolant loop and transferred the heat to the air in heater core. The heated air in the heater core passed through the cabin of the electric vehicle for the heating purposes. To measure the flow rates of the refrigerant and coolant, a mass flow meter and coolant flow rate meter were installed in the refrigerant loop and coolant loop, respectively. The refrigerant loop consisted of auxiliary components, namely, an accumulator, electronic expansion valve with driver and three-way valve. The temperatures and pressures of the refrigerant and coolant were measured using temperature and pressure sensors which were installed at the inlet and outlet of the electric compressor, chiller, condenser, evaporator, heater core, electronic expansion valve and accumulator. The measured temperature and pressure data were monitored in a data logger and recorded in a computer. The range and accuracy of each instrument is presented in Table 2.

The experiments were performed in two modes, namely, cooling mode and heating mode. In cooling mode, the inlet air conditions to HVAC were set to temperature of 25 °C, relative humidity of 60% and flow rate of 450 m^3/h, the compressor speed was varied from 4000 rpm to 6000 rpm, the coolant inlet temperature was varied from 35 °C to 55 °C and the coolant volume flow rate was varied from 10 L/min to 20 L/min to evaluate the performance of heat pump system with a high-pressure side chiller. In the heating mode, the ambient air temperature and air temperature were set to −6.7 °C, air volume was fixed to 300 m^3/h, coolant flow rate was set to 10 L/min, air velocity was varied from 3 m/s to 5 m/s, compressor speed was varied from 2000 rpm to 6000 rpm and the coolant temperature was varied from −6.7 °C to 50 °C to evaluate the performance of the heat pump system with a high-pressure side chiller.

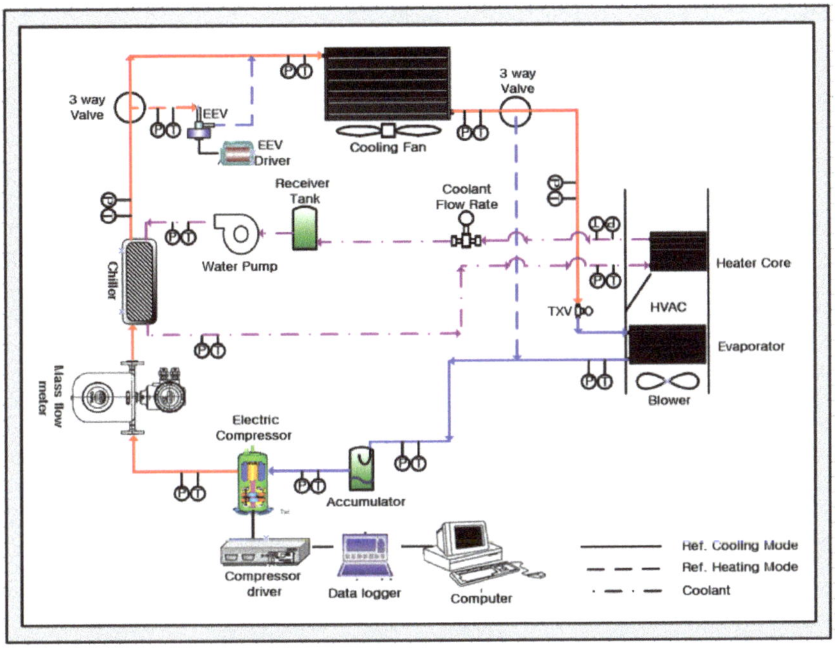

Figure 1. Experimental set-up of the heat pump system with a high-pressure side chiller.

Table 1. The specifications of primary components of the experimental set-up.

Component	Specification
Chiller	Plate-type inner-fin brazed-aluminum heat exchanger 150.3 W × 82.7 H × 57.5 D
Condenser (Size, mm)	PF(parallel flow)-type louvered-fin brazed-aluminum heat exchanger 568 W × 382 H × 20 D
Evaporator (Size, mm)	Laminated-type louvered-fin brazed-aluminum heat exchanger 216 W × 200 H × 48 D
Heater core (Size, mm)	PF(parallel-flow)-type louvered-fin brazed-aluminum heat exchanger 185 W × 165 H × 30 D
Compressor (Displacement, cm^3)	Scroll type (36.0)
Expansion device	Electric expansion valve (EEV)
Water pump (Coolant)	Electric-driven water [ump] (ethylene glycol:water = 50:50)

Table 2. Range and accuracy of each instrument.

Instrument	Range	Accuracy
Thermocouples (T-type)	−25~100 °C	±0.1 °C
Pressure gage (Sensors, PI3H)	Max. 2500 kPa	±0.1%
Different pressure transducer (Sensors, EJX110)	0~150 mmAq	±0.15%
Mass flow rate (Coriolis type)	Max. 600 kg/h	±0.2%
Volume flow rate (Electromagnetic type)	0.1~10 m/s	±0.5%
Humidity/Temperature Module (EE99-1)	0~100% RH	±2.0%

3. Uncertainty Analysis and Data Reduction

3.1. Uncertainty Analysis

The measured values of various parameters deviated from their actual values. The deviation between the measured value and the actual value of any parameter is termed as uncertainty associated with that parameter [27]. The accuracies of measuring instruments and errors during the measurements are the primary reasons for the uncertainties in various measuring and manipulating parameters [28]. The uncertainty analysis was conducted to assure the accuracy and reliability of the experimental data associated to the heat pump system with a high-pressure side chiller. The temperature, pressure, humidity, flow rate, air velocity and compressor speed were the measuring parameters during the experiments on the heat pump system with a high-pressure side chiller. The accuracies of the measuring instruments for various measuring parameters were the uncertainties associated with them, whereas the cooling capacity, coefficient of performance, chiller heat transfer rate, compressor power consumption, system efficiency and heater core performance were the performance (manipulating) parameters which were deducted from the experimental data of measuring parameters. The uncertainties in the performance parameters were evaluated using the concept of linearized fractional approximation which was proposed by Holman and Gajda [27,28]. Using this concept, the uncertainties in various performance (dependent) parameters due to uncertainties of various measuring (independent) parameters are evaluated by Equation (1) [29,30].

$$U_R = [(\frac{\partial R}{\partial X_1}U_1)^2 + (\frac{\partial R}{\partial X_2}U_2)^2 + (\frac{\partial R}{\partial X_3}U_3)^2 \ldots + (\frac{\partial R}{\partial X_n}U_n)^2]^{\frac{1}{2}} \tag{1}$$

where, X_1, X_2, X_3, … X_n are the independent parameters, R is the dependent parameter, U_1, U_2, U_3, … U_n are the uncertainties associated with independent parameters and U_R is the total uncertainty associated with dependent parameter.

The uncertainties associated with measuring parameters, temperature, pressure, humidity, coolant flow rate and air flow rate were 2.2%, 0.2%, 0.2%, 0.7% and 2.0%, respectively. Whereas, the uncertainties associated with performance (manipulating) parameters, cooling capacity, coefficient of performance, chiller heat transfer rate, compressor power consumption and heater core performance were 3.02%, 3.06%, 0.60%, 0.38% and 5.08%, respectively.

3.2. Data Reduction

Based on the measured data, the refrigerant-side heat transfer rate of the chiller was calculated by Equation (2) using the refrigerant enthalpy method (ASHRAE Standard 116, 1983). The air-side heat transfer rate is calculated by Equation (3) using both the air flow rate and enthalpy difference. The compressor power consumption was calculated using voltage and current values as expressed by Equation (4). The coefficient of performance (COP) of the tested system was calculated by Equation (5) using the compressor power consumed by compressor work. The heat transfer rate of coolant in the chiller was calculated based on its volume flow rate, density, temperature gain and specific heat capacity of coolant, as presented by Equation (6). The pressure ratio in cooling and heating modes was calculated using Equation (7) as the ratio of high pressure to low pressure.

$$\dot{Q}_{ref} = \dot{m}_{ref}\Delta h_{ref} \tag{2}$$

$$\dot{Q}_{air} = \dot{m}_{air}\Delta h_{air} \tag{3}$$

$$W_{comp} = VI \tag{4}$$

$$\mathrm{COP} = \frac{\dot{Q}_{air}}{W_{comp}} \tag{5}$$

$$\dot{Q}_{coolant} = \dot{V}_{coolant} \cdot \rho_{coolant} \cdot C_{p,coolant} \left(T_{coolant.in} - T_{coolant,out} \right) \tag{6}$$

$$\text{Pressure ratio} = \frac{P_{high}}{P_{low}} \tag{7}$$

where, $\dot{Q}_{ref}$ is the refrigerant side heat transfer rate of chiller (kW), $\dot{m}_{ref}$ is mass flow rate of refrigerant (kg/s), Δh_{ref} is the enthalpy difference for refrigerant (kJ/kg), $\dot{Q}_{air}$ is the air side heat transfer rate (kW), $\dot{m}_{air}$ is the mass flow rate of air (kg/s), Δh_{air} is the enthalpy difference for air (kJ/kg), W_{comp} is the compressor work (kW), V is voltage (v), I is current (A), COP is the coefficient of performance of the tested system, $\dot{Q}_{coolant}$ is coolant heat transfer rate in chiller, $\dot{V}_{coolant}$ is volume flow rate of coolant (L/min), $\rho_{coolant}$ is density of coolant (kg/m^3), $C_{p,coolant}$ is specific heat capacity of coolant (kJ/kg·K), $T_{coolant.in}$ is coolant inlet temperature (°C), $T_{coolant.out}$ is coolant outlet temperature (°C), and P_{high} and P_{low} are high pressure (MPa) and low pressure (MPa).

4. Results and Discussion

The experimental behaviors of cooling and heating performance characteristics of the heat pump system with a high-pressure side chiller under various realistic conditions are elaborated in this section.

4.1. Performance of Heat Pump System with a High-Pressure Side Chiller in Cooling Mode

This section explains the performance characteristics of the heat pump system with a high-pressure side chiller, namely, cooling capacity, compressor work, coefficient of performance and pressure characteristics for various compressor speeds, coolant inlet temperatures and coolant volume flow rates in cooling mode.

4.1.1. Cooling Capacity, Compressor Work and COP at Various Compressor Speeds

The effect of compressor speed on cooling capacity, compressor work and coefficient of performance (COP) of the heat pump system with a high-pressure side chiller during the cooling mode is shown in Figure 2a. The variation of cooling capacity, compressor work and coefficient of performance with compressor speed were experimentally recorded at HVAC inlet air conditions of temperature of 25 °C, relative humidity of 60% and flow rate of 450 m^3/h, air velocity of 3 m/s and coolant inlet conditions of 35 °C and volume flow rate of 10 L/min. As presented in Figure 2a, the cooling capacity increased from 3.8 kW to 4.7 kW with the rise in the compressor speed from 4000 rpm to 6000 rpm. The cooling capacity increased with compressor speed due to the increase in refrigerant flow rate [31]. The cooling capacity increased from 3.8 kW to 4.4 kW with an enhancement of 15.8% for the increase in the compressor speed from 4000 rpm to 5000 rpm. However, the cooling capacity increased from 4.4 kW to 4.7 kW with an enhancement of 6.82% for the increase in the compressor speed from 5000 rpm to 6000 rpm. Beyond the compressor speed of 5000 rpm, the increase in the cooling capacity reduced due to the rapid rise in the pressure ratio [31]. The rise in the pressure ratio affected the cycle capacity by increasing the degree of superheat associated with vapor refrigerant discharges from compressor and reducing the degree of subcooling of liquid refrigerant discharges from the condenser. In addition, the pressure ratio affected the compressor work. As presented in Figure 2a, the compressor work also increased with the increase in compressor speed. With the increase in the compressor speed from 4000 rpm to 5000 rpm, the compressor work increased from 1.07 kW to 1.37 kW with an enhancement of 28.0%. The compressor work increased from 1.37 kW to 1.71 kW with an enhancement of 24.8% for the increase in the compressor speed from 5000 rpm to 6000 rpm. Unlike cooling capacity, the percentage enhancement in compressor work was not reduced drastically with the rise in compressor speed. The variations of cooling capacity and compressor work with various compressor speeds affected the coefficient of performance. The coefficient of performance is the ratio of cooling capacity to compressor work. With the increase in the compressor speed, the percentage enhancement in the cooling capacity decreased drastically but the percentage enhancement in compressor work was not decreased drastically. Therefore, the coefficient of performance decreased with the increase in the compressor speed as shown in Figure 2a. With the increase in the compressor speed from 4000 rpm to 5000 rpm, the coefficient of performance decreased from 3.54 to 3.21 with a reduction of 9.32%, whereas, the coefficient of performance decreased from 3.21 to 2.75 with a reduction of 14.3% for the increase in the compressor speed from 5000 rpm to 6000 rpm. The percentage reduction in the coefficient of performance increased beyond a compressor speed of 5000 rpm due to characteristic variations of cooling capacity and compressor work with compressor speed, as presented in Figure 2a. The working of the heat pump system with a high-pressure side chiller for various compressor speeds is shown on a P-h diagram in Figure 2b. With respect to the P-h diagram for the refrigeration cycle, as compressor speed increased, low-side pressure decreased to under 2 bar. However, high-side pressure was stable due to similar coolant conditions to be supplied. The portion of heat transfer rate between a refrigerant and a coolant at the developed chiller showed about one-third of the condensing heat capacity along with compressor speed.

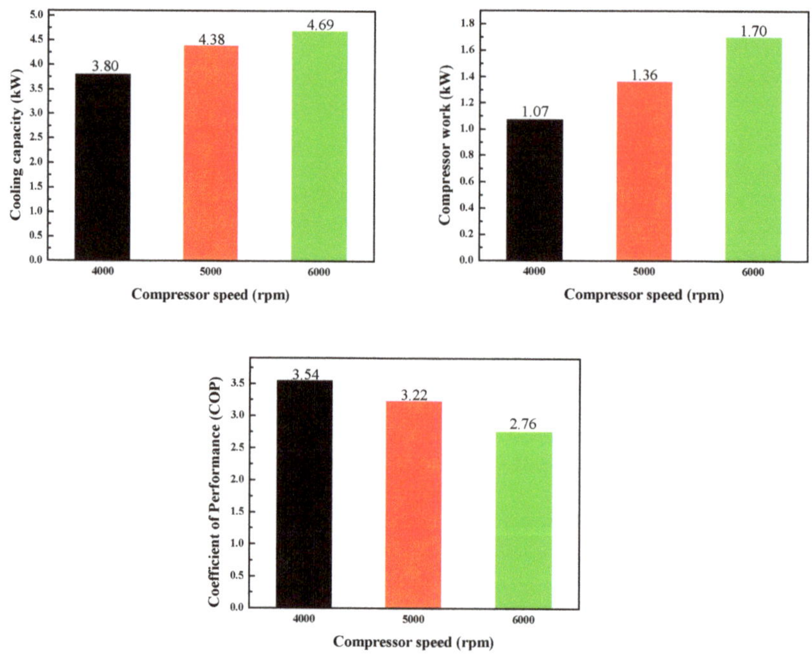

(a)

Figure 2. *Cont.*

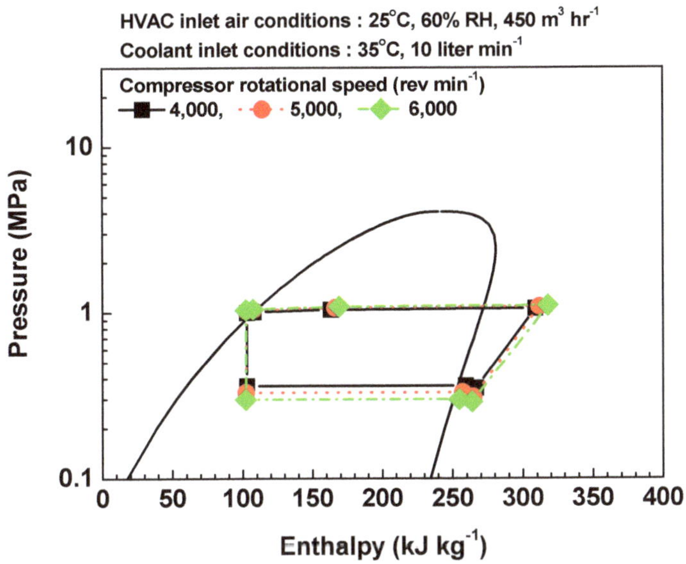

(**b**)

Figure 2. Effect of compressor speed on (**a**) cooling capacity, compressor work and coefficient of performance (COP) and (**b**) P-h diagram of heat pump system with chiller during the cooling mode.

4.1.2. Cooling Capacity, Compressor Work and COP at Various Coolant Inlet Temperatures

The variation of cooling capacity, compressor work and coefficient of performance of heat pump system with a high-pressure side chiller for various coolant inlet temperatures during the cooling mode is presented in Figure 3a. The behaviors of cooling capacity, compressor work and coefficient of performance for various coolant inlet temperatures were experimentally recorded at the HVAC inlet air conditions of temperature of 25 °C, relative humidity of 60% and flow rate of 450 m^3/h, air velocity of 3 m/s, coolant volume flow rate of 10 L/min and compressor speed of 5000 rpm. The coolant inlet temperature had very little effect on the cooling capacity. As shown in Figure 3a, with the increase in the coolant inlet temperature from 35 °C to 45 °C, the cooling capacity remained constant at 4.15 kW, whereas, with an increase in the coolant inlet temperature from 45 °C to 55 °C, the cooling capacity decreased from 4.15 kW to 4.06 kW, a reduction of 2.2%. With the increase in the coolant inlet temperature, the compressor work increased because of the increase in the high-pressure side of the compressor. The compressor work increased from 1.36 kW to 1.58 kW with an enhancement of 16.2% for the increase in the coolant inlet temperature from 35 °C to 45 °C. However, with the increase in the coolant inlet temperature from 45 °C to 55 °C, the compressor work increased from 1.58 kW to 1.67 kW, an enhancement of 5.70%. With an increase in the coolant inlet temperature, the percentage enhancement in the compressor decreased as shown in Figure 3a. The coefficient of performance varied with coolant inlet temperature due to variations in cooling capacity and compressor work with coolant inlet temperature. As shown in Figure 3a, the coefficient of performance decreased with the increase in the coolant inlet temperature because the cooling capacity was not affected much, and compressor work increased with coolant inlet temperature. Because of the characteristic variation of cooling capacity and compressor work presented in Figure 3a, the coefficient of performance decreased from 3.06 to 2.64 with a reduction of 13.7% for the increase in the coolant inlet temperature from 35 °C to 45 °C, while it decreased from 2.64 to 2.43, a reduction of 7.95%, for the increase in the coolant inlet temperature from 45 °C to 55 °C. The refrigeration cycle of the heat pump system with a high-pressure side chiller for various coolant inlet temperatures on the P-h diagram is shown in Figure 3b. High-side pressure increased with coolant inlet temperature in order to have certain temperature difference. The proportion of the heat transfer rate in the developed chiller with the increase in coolant temperature varied from one quarter to three-quarters of the total condensing heat capacity due to heat transfer potential, such as temperature difference among fluids.

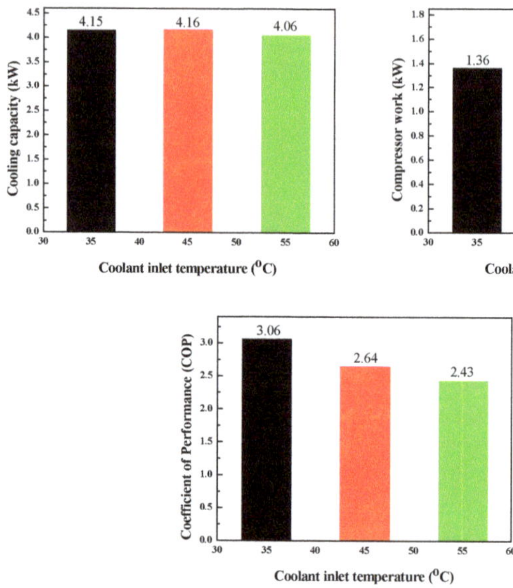

(**a**)

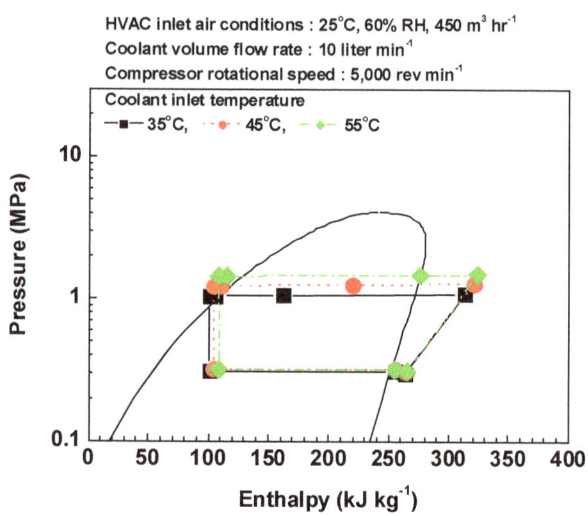

(**b**)

Figure 3. Effect of coolant inlet temperature on (**a**) cooling capacity, compressor work and coefficient of performance (COP) and (**b**) P-h diagram of heat pump system with chiller during the cooling mode.

4.1.3. Cooling Capacity, Compressor Work and COP at Various Coolant Volume Flow Rates

The behaviors of cooling capacity, compressor work and coefficient of performance of heat pump system with a high-pressure side chiller for various coolant volume flow rates during the cooling mode is presented in Figure 4a. The variations of cooling capacity, compressor work and coefficient of performance with coolant volume flow rate were experimentally recorded for HVAC inlet air conditions of a temperature of 25 °C, relative humidity of 60%, flow rate of 450 m^3/h, air velocity of 3 m/s, coolant inlet conditions of 35 °C and compressor speed of 5000 rpm. The coolant volume flow rate had no significant effect on the cooling capacity as shown in Figure 4a. With increase in the coolant volume flow rate from 10 L/min to 15 L/min, the cooling capacity reduced by 1.47% from 4.09 W to 4.03 W, whereas the cooling capacity remained constant at 4.03 W when the coolant volume flow rate increased from 15 L/min to 20 L/min. The compressor work decreased with the increase in the coolant volume flow rate due to the fall in the high-pressure side of the compressor. However, the decrease in the compressor work with coolant volume flow rate was not significant as shown in Figure 4a. The compressor work decreased from 1.33 kW to 1.3 kW with a reduction of 2.26% for the increase in coolant volume flow rate from 10 L/min to 15 L/min. With the increase in the coolant volume flow rate from 15 L/min to 20 L/min, the compressor work decreased by 1.54% from 1.3 kW to 1.28 kW. Due to variations in the cooling capacity and compressor work with coolant volume flow rate, the coefficient of performance also showed variation with coolant volume flow rate. As shown in Figure 4a, the effect of coolant volume flow rate on coefficient of performance was less because of less effect of coolant volume flow rate on cooling capacity and compressor work. Due to characteristic behaviors of cooling capacity and compressor work with coolant volume flow rate, as presented in Figure 4a, the coefficient of performance increased from 3.08 to 3.1 with an enhancement of 0.65% for the increase in the coolant volume flow rate from 10 L/min to 15 L/min, whereas, when coolant volume flow rate increased from 15 L/min to 20 L/min, the coefficient of performance increased by 1.29% from 3.1 to 3.14. Figure 4b presents the P-h diagram for the refrigeration cycle of the heat pump system with a high-pressure side chiller with various coolant volume flow rates. The working of this refrigeration cycle on the P-h diagram for various coolant volume flow rates was similar to that explained for the refrigeration cycle on the P-h diagram for various compressor speeds.

4.1.4. Pressure Characteristics

The pressure ratio between low and high pressure of refrigeration system affected compressor power consumption. The behaviors of pressure ratio of tested system with high-pressure side chiller for various operating conditions during the cooling mode is presented in Figure 5. The variations of pressure ratio with coolant operating conditions and compressor speed were experimentally recorded for HVAC inlet air conditions of temperature of 25 °C, relative humidity of 60%, flow rate of 450 m^3/h, and air velocity of 3 m/s. At coolant inlet conditions of 35 °C and 10 L/min, the compressor speed varied from 3000 rpm to 5000 rpm and pressure ratio was analyzed. With the increase in the compressor speed from 3000 rpm to 5000 rpm, the low pressure decreased; however, the high pressure remained constant in the range of 1100 kPa because the coolant and ambient conditions remained constant. As an effect of this the pressure ratio which is defined as the ratio of high pressure to low pressure, increased with the increase in the compressor speed. Pressure ratio increased by 26.8% from 3.05 to 3.87 with the increase in the compressor speed from 3000 rpm to 5000 rpm, as shown in Figure 5a. This was the reason behind the increase in the compressor power consumption as depicted Figure 2a. In the case when the coolant temperature increased from 35 °C to 55 °C, the low pressure remained constant, whereas the high pressure increased up to 1500 kPa. Therefore, the pressure ratio increased with the increase in the coolant temperature. The pressure ratio increased by 35.0% from 3.6 to 4.87 with the increase in the coolant temperature from 35 °C to 55 °C, as shown in Figure 5b. The effect of coolant temperature on the pressure ratio caused an increase in the compressor power consumption with an increase in the coolant temperature as shown in Figure 3a. However, the pressure ratio affected little with the variation of coolant volume flow rate from 10 L/min to 20 L/min because of the same operating conditions of coolant temperature and compressor speed.

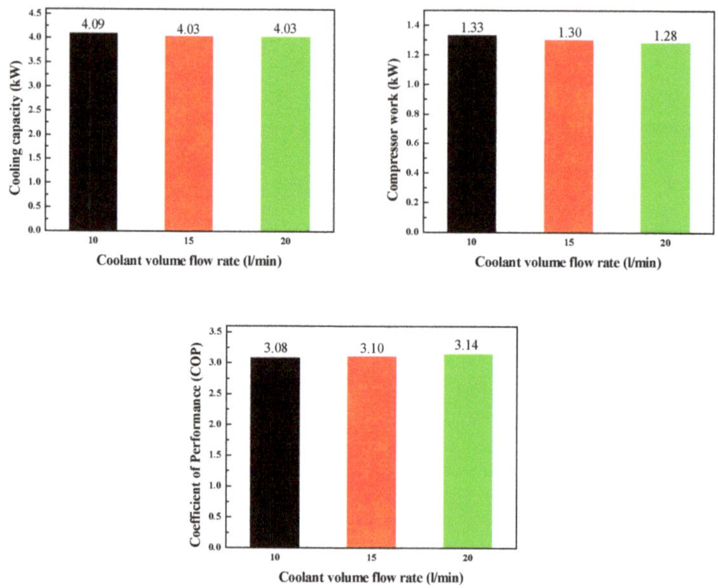

(a)

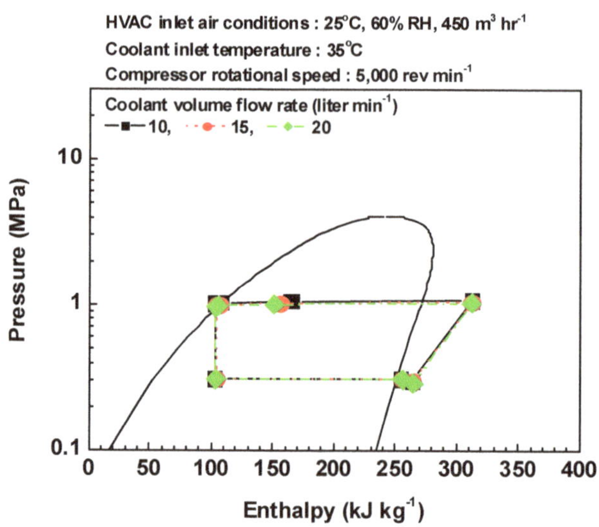

(b)

Figure 4. Effect of coolant volume flow rate on (**a**) cooling capacity, compressor work and coefficient of performance (COP) and (**b**) P-h diagram of heat pump system with chiller during the cooling mode.

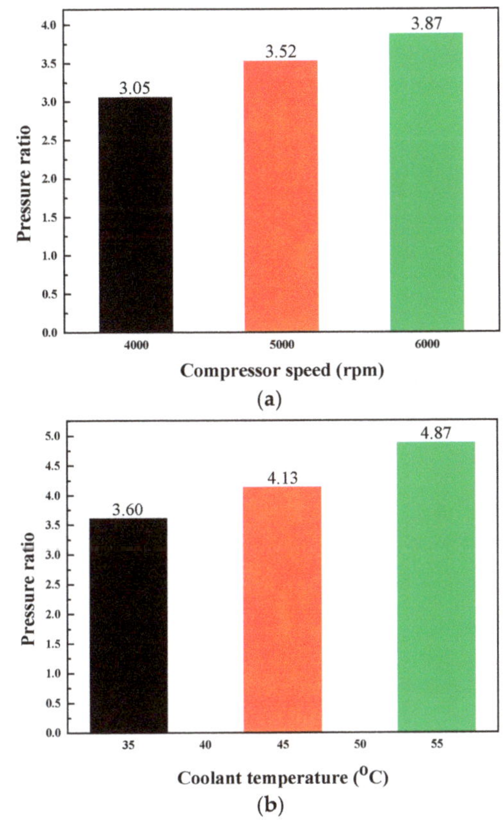

Figure 5. Pressure ratio characteristics with (**a**) compressor speed and (**b**) coolant temperature in cooling mode.

4.2. *Performance of Heat Pump System with Higher Pressure Side Chiller in Heating Mode*

This section elaborates the performance characteristics of the heat pump system with a high-pressure side chiller, namely, heat dissipation, chiller heat transfer rate, compressor power consumption, system efficiency, heater core performance and pressure characteristics for the steady state conditions of various compressor speeds, coolant temperatures and air velocities in heating mode. In addition, a brief discussion about the transient behavior of heat pump system with a high-pressure side chiller under heating mode is presented in this section.

4.2.1. Heat Dissipation at Various Air Velocities

The heat dissipation performance of the outdoor heat exchanger (condenser) of the heat pump system with a high-pressure side chiller for various air velocities is presented in Figure 6. The performance of the heat exchanger with air velocity was experimentally evaluated at the inlet temperature of 80 °C, inlet pressure of 1600 kPa, inlet SH of 25 and outlet SC of 8. With the increase in the air velocity, the mass flow rate of air increased, hence the air carried more heat from the outdoor heat exchanger, which resulted in the increase in the heat dissipation of the outdoor heat exchanger [11]. The heat dissipation of the outdoor heat exchanger increased linearly with the increase in the air velocity, as shown in Figure 6. The heat dissipation of the outdoor heat exchanger increased by 19% from 9.3 kW to 11.1 kW with the increase in the air velocity from 3 m/s to 4 m/s, whereas the heat dissipation increased by 13.7% from 11.1 W to 12.6 kW when the air velocity increased from 4 m/s to 5 m/s.

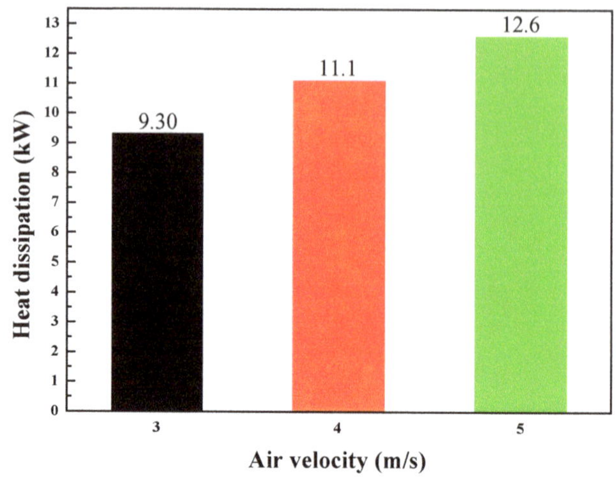

Figure 6. Heat dissipation performance of the outdoor heat exchanger (condenser) of heat pump system with chiller for various air velocities.

4.2.2. Chiller Heat Transfer Rate at Various Compressor Speeds and Coolant Temperatures

The effect of compressor speed and coolant temperature on the heat transfer rate of chiller is presented in Figure 7. Chiller heat transfer rate for various compressor speeds and coolant temperatures was experimentally recorded for ambient and air temperatures of −6.7 °C, air volume flow rate of 300 m^3/h, coolant flow rate of 10 L/min, and maximum temperature and maximum pressure of 120 °C and 2500 kPa. The chiller heat transfer rate increased with the increase in the compressor speed from 2000 rpm to 6000 rpm for the coolant temperatures of −6.7 °C, 10 °C, 20 °C, 30 °C, 40 °C and 50 °C. However, the chiller heat transfer rate increased for the increase in the compressor speed from 2000 rpm to 4000 rpm and reduced with the increase in the compressor speed from 4000 rpm to 6000 rpm for the coolant temperature of 0 °C. The compressor speed range became narrow with the increase in the coolant temperature to present the variation of chiller heat transfer rate. The variation of the chiller heat transfer rate is presented over the compressor speed range of 2000 rpm to 6000 rpm for coolant temperatures of−6.7 °C and 0 °C, that of 2000 rpm to 5000 rpm for the coolant temperatures of 10 °C, 20 °C and 30 °C and that of 2000 rpm to 4000 rpm for the coolant temperatures of 40 °C and 50 °C. With the increase in the compressor speed, the percentage increase in the chiller heat transfer rate decreased for coolant temperatures of −6.7 °C, 10 °C, 20 °C, 30 °C, 40 °C and 50 °C. The chiller heat transfer rate increased by 11.4%, 55.6%, 40.7%, 70.6%, 54.0% and 55.0% for the coolant temperatures of −6.7 °C, 10 °C, 20 °C, 30 °C, 40 °C and 50 °C, respectively with the increase in the compressor speed from 2000 rpm to 3000 rpm. The chiller heat transfer rate increased by 11.7%, 20.6%, 16.2%, 17.8%, 21.9% and 22.6% for the coolant temperatures of −6.7 °C, 10 °C, 20 °C, 30 °C, 40 °C and 50 °C, respectively, with the increase in the compressor speed from 3000 rpm to 4000 rpm. The chiller heat transfer rate increased by 11.6%, 1.56%, 7.29% and 11.0% for the coolant temperatures of −6.7 °C, 10 °C, 20 °C and 30 °C, respectively, with the increase in the compressor speed from 4000 rpm to 5000 rpm. Finally, the chiller heat transfer rate increased by 1.83% for the coolant temperatures of −6.7 °C with an increase in the compressor speed from 4000 rpm to 5000 rpm. In the case of the coolant temperature of 0 °C, when the compressor speed increased from 2000 rpm to 3000 rpm and 3000 rpm to 4000 rpm correspondingly, the chiller heat transfer rate increased by 21.6% and 11.4%, whereas the chiller heat transfer rate decreased by 2.90% and 3.20% with the increase in the compressor speed from 4000 rpm to 5000 rpm and 5000 rpm to 6000 rpm, respectively. The maximum chiller heat transfer rates of 4.91 kW, 4.73 kW, 3.83 kW, 3.84 kW, 3.37 kW and 3.05 kW were experimentally evaluated for the coolant temperatures of −6.7 °C, 10 °C, 20 °C, 30 °C, 40 °C and 50 °C, respectively at the highest compressor speeds of respective ranges, whereas the maximum chiller heat transfer rate of 3.50 W was experimentally evaluated at the middle of compressor speed range for a coolant temperature of 0 °C.

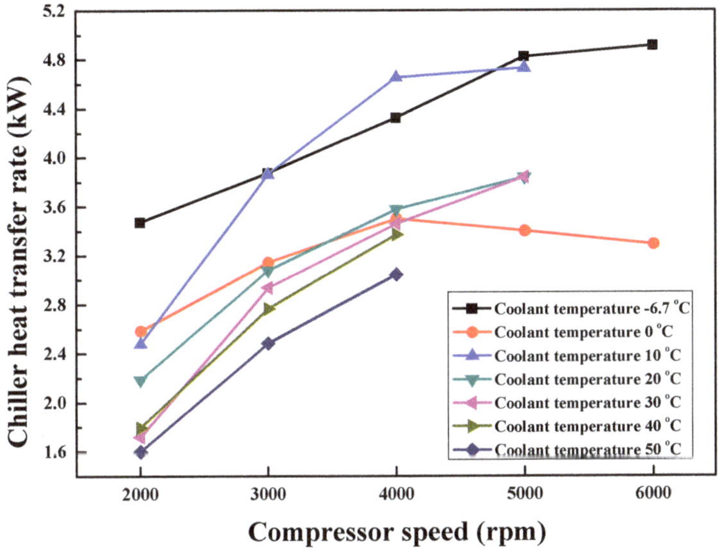

Figure 7. Effect of compressor speed and coolant temperature on heat transfer rate of chiller.

4.2.3. Compressor Power Consumption at Various Compressor Speeds and Coolant Temperatures

The behavior of compressor power consumption for coolant temperatures of −6.7 °C, 0 °C, 10 °C, 20 °C, 30 °C, 40 °C and 50 °C and compressor speeds of 2000 rpm, 3000 rpm, 4000 rpm, 5000 rpm and 6000 rpm is presented in Figure 8. The behavior of compressor power consumption for various compressor speeds and coolant temperatures were experimentally evaluated for ambient and air temperatures of −6.7 °C, air volume flow rate of 300 m^3/h, coolant flow rate of 10 L/min, maximum temperature of 120 °C and maximum pressure of 2500 kPa. For each coolant temperature, compressor power consumption increased with the increase in the compressor speed. In addition, the percentage increase in the compressor power consumption decreased with the increase in the compressor speed for each coolant temperature except 40 °C. The variation range of compressor power consumption with compressor speed decreased as the coolant temperature increased from −6.7 °C to 50 °C. The variation of compressor power consumption is presented over the compressor speed range of 2000 rpm to 6000 rpm for coolant temperatures of −6.7 °C and 0 °C. For coolant temperatures of 10 °C, 20 °C and 30 °C, the variation of compressor power consumption is presented over the compressor speed range of 2000 rpm to 5000 rpm. The variation of compressor power consumption is presented over the compressor speed range of 2000 rpm to 4000 rpm for coolant temperatures of 40 °C and 50 °C. The compressor power consumption increased by 76.3%, 64.2%, 76.5%, 51.8%, 44.0%, 35.7% and 35.7% for coolant temperatures of −6.7 °C, 0 °C, 10 °C, 20 °C, 30 °C, 40 °C and 50 °C, respectively, when the compressor speed increased from 2000 rpm to 3000 rpm. The compressor power consumption increased by 33.2%, 29.7%, 35.5%, 36.5%, 26.7%, 43.3% and 22.8% for coolant temperatures of −6.7 °C, 0 °C, 10 °C, 20 °C, 30 °C, 40 °C and 50 °C, respectively, when the compressor speed increased from 3000 rpm to 4000 rpm. The compressor power consumption increased by 21.6%, 17.4%, 22.8%, 19.7% and 18.1% for coolant temperatures of −6.7 °C, 0 °C, 10 °C, 20 °C and 30 °C, respectively, when the compressor speed increased from 4000 rpm to 5000 rpm. The compressor power consumption increased by 11.1% and 5.30% for coolant temperatures of −6.7 °C and 0 °C, respectively when the compressor speed increased from 5000 rpm to 6000 rpm. The minimum compressor power consumption of 0.48 kW,

0.44 kW, 0.60 kW, 0.70 kW, 0.86 kW, 0.95 kW and 1.05 kW were experimentally evaluated at the coolant temperatures of −6.7 °C, 0 °C, 10 °C, 20 °C, 30 °C, 40 °C and 50 °C, respectively, for a compressor speed of 2000 rpm. The maximum compressor power consumption of 1.51 kW and 1.16 kW were evaluated at the coolant temperatures of −6.7 °C and 0 °C, respectively, for compressor speed of 6000 rpm. For coolant temperatures of 10 °C, 20 °C and 30 °C, the maximum compressor power consumptions were evaluated as 1.77 kW, 1.75 kW and 1.85 kW, respectively, at the compressor speed of 5000 rpm. The maximum compressor power consumptions of 1.84 kW and 1.75 kW were evaluated at the coolant temperatures of 40 °C and 50 °C, respectively, for compressor speed of 4000 rpm.

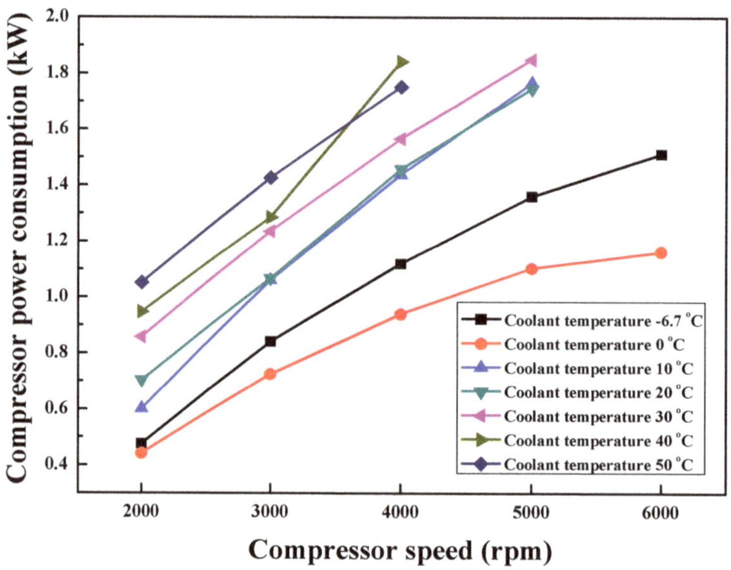

Figure 8. Behavior of compressor power consumption for coolant temperatures of −6.7 °C, 0 °C, 10 °C, 20 °C, 30 °C, 40 °C and 50 °C and compressor speeds of 2000 rpm, 3000 rpm, 4000 rpm, 5000 rpm and 6000 rpm.

4.2.4. System Efficiency at Various Compressor Speeds and Coolant Temperatures

The variation of system efficiency for compressor speed range of 2000 rpm to 6000 rpm and coolant temperature range of −6.7 °C to 50 °C is shown in Figure 9. The variation of system efficiency with compressor speed and coolant temperature was experimentally evaluated at ambient and air temperatures of −6.7 °C, maximum temperature of 120 °C, maximum pressure of 2500 kPa, air volume flow rate of 300 m^3/h and coolant flow rate of 10 L/min. The compressor speed range became narrow for the variation of system efficiency as the coolant temperature increased. The variation of system efficiency is presented over the compressor speed range of 2000 rpm to 6000 rpm for the coolant temperatures of −6.7 °C and 0 °C, that of 2000 rpm to 5000 rpm for coolant temperatures of 10 °C, 20 °C and 30 °C and that of 2000 rpm to 4000 rpm for coolant temperatures of 40 °C and 50 °C. For coolant temperatures of −6.7 °C, 0 °C, 10 °C and 20 °C, the system efficiency decreased with the increase in the compressor speed, whereas, for the coolant temperatures of 30 °C, 40 °C and 50 °C, the system efficiency increased, reached maximum value at compressor speed of 3000 rpm and decreased with further increase in the compressor speed. The system efficiency decreased by 35.5%, 25.5%, 10.8% and 7.51% for the coolant temperatures of −6.7 °C, 0 °C, 10 °C and 20 °C, respectively, and increased by 16.0%, 14.9% and 15.3% for the coolant temperatures of 30 °C, 40 °C and 50 °C, respectively, with the increase in the compressor speed from 2000 rpm to 3000 rpm. The system efficiency decreased by 16.2%, 15.6%, 11.2%, 15.4%, 6.34%, 15.5% and 1.49% with the increase in compressor speed from 3000 rpm to 4000 rpm for the coolant temperatures of −6.7 °C, 0 °C, 10 °C, 20 °C, 30 °C, 40 °C and 50 °C, respectively. The system efficiency decreased by 8.63%, 17.4%, 17.6%, 11.3% and 6.62% with the increase in compressor speed from 4000 rpm to 5000 rpm for the coolant temperatures of −6.7 °C, 0 °C, 10 °C, 20 °C and 30 °C, respectively. The system efficiency decreased by 7.72% and 8.82% with the increase in the compressor speed from 5000 rpm to 6000 rpm for the coolant temperatures of −6.7 °C and 0 °C, respectively. The system efficiency decreased as the coolant temperature increased from −6.7 °C to 50 °C for the same compressor speed. The coolant temperature of −6.7 °C showed the maximum system efficiency for the same compressor speed. The maximum system efficiencies of 7.21%, 5.81%, 4.08% and 3.08% were experimentally evaluated at the compressor speed of 2000 rpm for the coolant temperatures of −6.7 °C, 0 °C, 10 °C and 20 °C, respectively, whereas the maximum system efficiencies of 2.37%, 2.17% and 1.76% were experimentally evaluated at the compressor speed of 3000 rpm for the coolant temperatures of 30 °C, 40 °C and 50 °C, respectively.

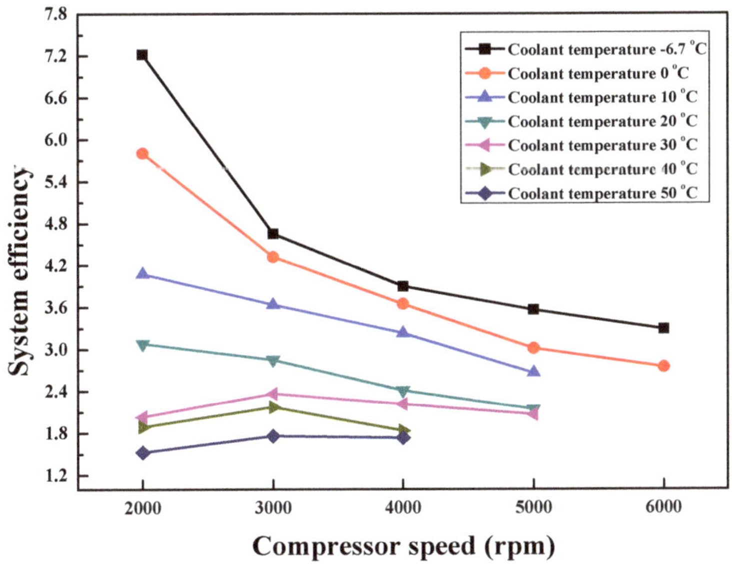

Figure 9. Variation of system efficiency for compressor speed range of 2000 rpm to 6000 rpm and coolant temperature range of −6.7 °C to 50 °C.

4.2.5. Heater Core Performance at Various Compressor Speeds and Coolant Temperatures

The variation of heater core performance for various compressor speeds and coolant temperatures is shown in Figure 10. The variation is presented at the ambient and air temperatures of −6.7 °C, air volume flow rate of 300 m^3/h, coolant flow rate of 10 L/min, maximum temperature of 120 °C and maximum pressure of 2500 kPa. As the coolant temperature increased from −6.7 °C to 50 °C, the compressor speed range of 2000 rpm to 6000 rpm decreased for the variation of heater core performance. The heater core performance was evaluated at the maximum compressor speed of 6000 rpm for coolant temperatures of −6.7 °C and 0 °C, that of 5000 rpm for coolant temperatures of 10 °C, 20 °C and 30 °C and that of 4000 rpm for coolant temperatures of 40 °C and 50 °C. With the increase in the compressor speed, the heater core performance increased for the coolant temperatures of −6.7 °C, 10 °C, 20 °C, 30 °C, 40 °C and 50 °C, whereas, for the coolant temperature of 0 °C the heater core performance showed parabolic variation with compressor speed. With the increase in the compressor speed from 2000 rpm to 3000 rpm, the heater core performance increased by 230%, 15.4%, 12.0%, 10.6%, 6.03%, 2.38% and 2.43% for the coolant temperatures of −6.7 °C, 0 °C, 10 °C, 20 °C, 30 °C, 40 °C and 50 °C, respectively. With the increase in the compressor speed from 3000 rpm to 4000 rpm, the heater core performance enhanced by 35.4%, 7.34%, 5.62%, 4.38%, 1.60%, 2.98% and 0.98% for the coolant temperatures of −6.7 °C, 0 °C, 10 °C, 20 °C, 30 °C, 40 °C and 50 °C, respectively. The heater performance enhanced by 28.0%, 1.95%, 2.36%, 1.52% and 2.43% for the coolant temperatures of −6.7 °C, 0 °C, 10 °C, 20 °C and 30 °C, respectively, when compressor speed increased from 4000 rpm to 5000 rpm. As the compressor speed increased from 5000 rpm to 6000 rpm, the heater core performance increased by 5.25% for coolant temperature of −6.7 °C and decreased by 5.80% for the coolant temperature of 0 °C. For the same compressor speed, the heater core performance enhanced with the increase in the coolant temperature. The heater core performance showed maximum behavior at the coolant temperature of 50 °C for same compressor speed. The maximum heater core performance of 0.36 kW was evaluated at compressor speed of 6000 rpm for the coolant temperature of −6.7 °C and that of 0.77 W at compressor speed of 5000 rpm for the coolant temperature of 0 °C. For the coolant temperatures of 10 °C, 20 °C and 30 °C, the corresponding maximum heater core performances of 1.85 kW, 2.75 kW and 3.58 kW were evaluated at the compressor speed of 5000 rpm. The maximum heater core performances of 4.35 kW and 4.98 kW were evaluated for coolant temperatures of 40 °C and 50 °C, respectively, at the compressor speed of 4000 rpm.

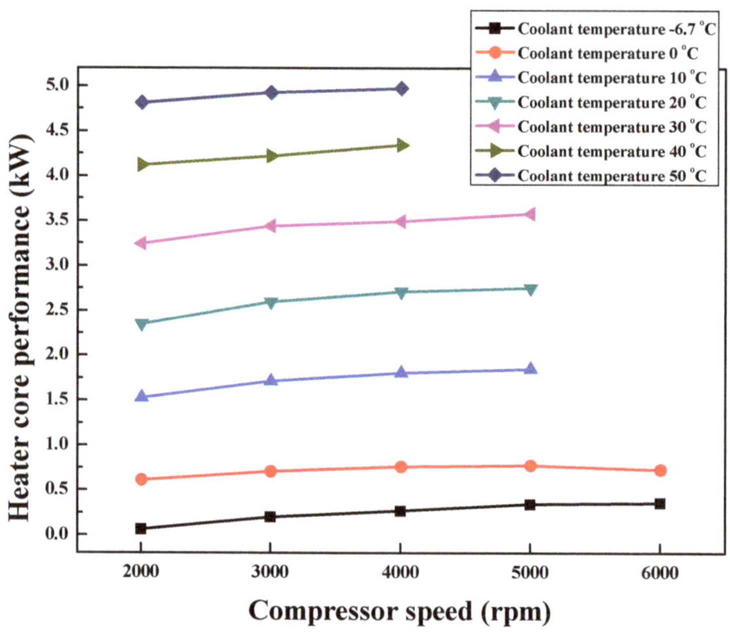

Figure 10. Variation of heater core performance for various compressor speeds and coolant temperatures.

4.2.6. Performance of Heat Pump System with a High-Pressure Side Chiller in Transient State

The transient performance of heat pump system with a high-pressure side chiller during the heating mode was experimentally evaluated for the ambient and air temperature of −6.7 °C, air volume flow rate of 300 m^3/h, coolant temperature and flow rate of −6.7 °C and 10 L/min, and compressor speed range of 3000 rpm to 6000 rpm. The variation in coolant temperature with time for compressor speeds of 3000 rpm, 4000 rpm, 5000 rpm and 6000 rpm is presented in Figure 11. Starting with the coolant temperature of −6.7 °C, for all compressor speeds, coolant temperature increased with time and converged, as shown in Figure 11. The increase in coolant temperature followed a smooth curve for compressor speeds of 3000 rpm, 4000 rpm and 5000 rpm, but, for the compressor speed of 6000 rpm, the increasing coolant temperature curve fluctuated due to experimental or environmental error. However, the curve converged accurately at the same time with other smooth curves. The convergence temperature was higher for a compressor speed of 6000 rpm followed by convergence temperatures corresponding to compressor speeds of 5000 rpm, 4000 rpm and 3000 rpm. However, the convergence temperature was obtained almost at the same time for all compressor speeds. The highest coolant convergence temperature of 22 °C was experimentally evaluated at the compressor speed of 6000 rpm followed by 20 °C, 16 °C and 13 °C at the compressor speeds of 5000 rpm, 4000 rpm and 3000 rpm, respectively. Even when starting with the lower coolant temperature of −6.7 °C, it took a longer time of 1 h 30 min to attain the convergence coolant temperature for all compressor speeds. This meant the time invested in cabin heating was longer for all compressor speeds. With reference to the ambient temperature of −6.7 °C, the temperature differences of 28.7 °C, 26.7 °C, 22.7 °C and 19.7 °C were evaluated at convergence time for compressor speeds of 6000 rpm, 5000 rpm, 4000 rpm and 3000 rpm, respectively.

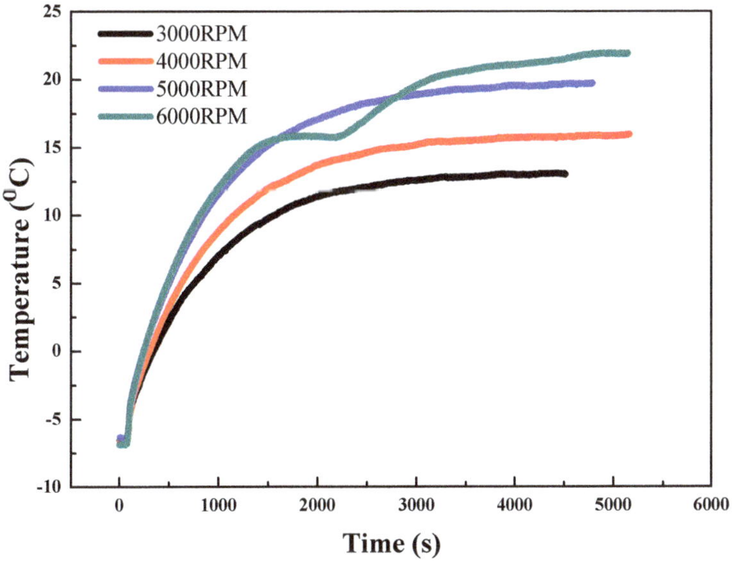

Figure 11. Variation in coolant temperature with time for compressor speeds of 3000 rpm, 4000 rpm, 5000 rpm and 6000 rpm.

4.2.7. Pressure Characteristics

Pressure ratio in heating mode was analyzed like cooling mode in the tested system, as shown in Figure 12. The variations of pressure ratio with coolant operating conditions and compressor speed were experimentally recorded for HVAC inlet air conditions of temperature of −6.7 °C, air flow rate of 300 m^3/h, and coolant flow rate of 10 L/min. In all cases of coolant temperature variation, as the compressor speed increased, pressure ratio had the same trend with cooling mode. However, pressure ratio in heating mode was higher than cooling mode by two to three times. In heating mode, because the tested system was exposed to temperatures under −6.7 °C, low pressure got down to below 100 kPa, similar to vacuum pressure, and high pressure led to applied coolant temperature increases up to 1500 kPa at the coolant temperature of 50 °C. As a result of that, pressure ratio had a wide range from 2.67 to 12.4, which led to the increase of compressor power consumption depicted in Figure 8.

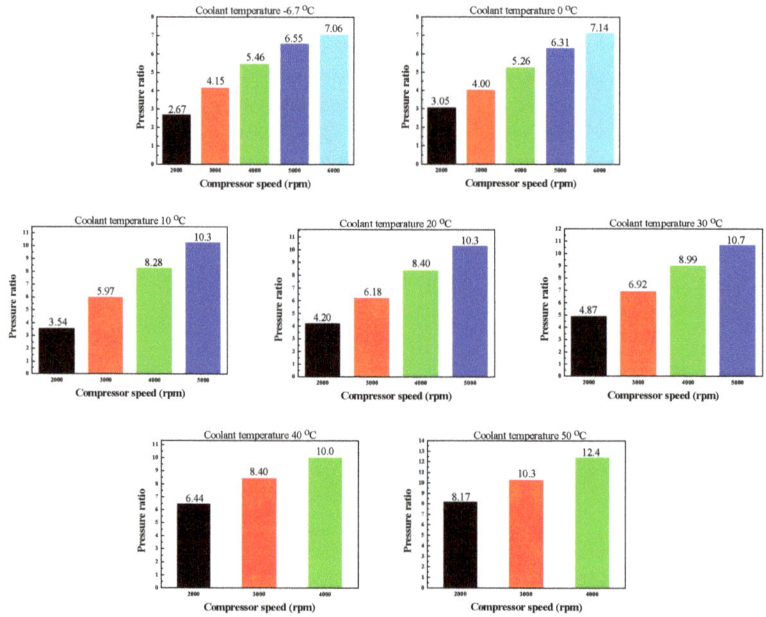

Figure 12. Pressure ratio characteristics with compressor speed and coolant temperature in heating mode.

5. Conclusions

The cooling and heating performance characteristics of the heat pump system with a high-pressure side chiller are experimentally investigated under real road driving conditions for light-duty commercial electric vehicles. The critical findings from the experimental investigations into the heat pump system with a high-pressure side chiller in cooling and heating modes are summarized as below.

(a) In the cooling mode of heat pump system with a high-pressure side chiller, the system efficiency decreases by 16.4% on an average and the cooling capacity and compressor work enhances by 8.0% and 27.0%, respectively, on average with variation in compressor speed.

(b) In the heating mode, the coolant gets heated by the discharged refrigerant of the electric compressor in the high-pressure side chiller and higher temperature coolant transfers heat to the inlet air of heater core so that the cabin heats up. In the high-pressure side chiller, the heat transfer rate is higher at the lower coolant temperature due to the higher temperature difference between the coolant and the refrigerant and because the coolant could absorb more heat from refrigerant at lower temperatures. However, with respect to the heater core to heat up the cabin because the coolant temperature is relatively low despite heat gain from the refrigerant, heat transfer rate in heater core is quantitatively low. On the contrary, when the coolant temperature is higher than ambient conditions, −6.7 °C, system efficiency of the tested heat pump is decreased due to the increase in compressor power consumption with higher pressure ratio.

(c) In the transient mode, the coolant temperature converges to 22 °C, 20 °C, 16 °C and 13 °C after 1 h 30 min for the compressor speeds of 6000 rpm, 5000 rpm, 4000 rpm and 3000 rpm, respectively. The attained temperature differences are 28.7 °C, 26.7 °C, 22.7 °C and 19.7 °C for 6000 rpm, 5000 rpm, 4000 rpm and 3000 rpm, respectively, at the convergent stage.

Author Contributions: Conceptualization, M.-Y.L.; K.S.G.; H.-B.J. and H.-S.L.; methodology, M.-Y.L.; K.S.G.; H.-B.J. and H.-S.L.; formal analysis, M.-Y.L.; K.S.G.; H.-B.J. and H.-S.L.; investigation, M.-Y.L.; H.-B.J. and H.-S.L.; resources, M.-Y.L.; K.S.G.; H.-B.J. and H.-S.L.; data curation, M.-Y.L. and H.-S.L.; writing—original draft preparation, M.-Y.L.; K.S.G. and H.-S.L.; writing—review and editing, M.-Y.L.; K.S.G. and H.-S.L.; visualization, M.-Y.L. and H.-S.L.; supervision, M.-Y.L. and H.-S.L.; project administration, H.-S.L.; funding acquisition, H.-S.L. All authors have read and agreed to the published version of the manuscript.

Funding: This work was funded by the Ministry of Trade, Industry & Energy(MOTIE), Korea Evaluation Institute of Industrial Technology (KEIT) through the Industrial Technology Innovation Program (20003988, Development of 4 kW heating and cooling modularization technology using waste heat recovery and alternative refrigerant on electric vehicle's exclusive platform).

Acknowledgments: This work was supported by the Ministry of Trade, Industry & Energy(MOTIE), Korea Institute for Advancement of Technology (KIAT) through the Automobile Parts Cluster Construction Program for the Environment-Friendly Vehicles (P0000760, Development of Integrated Smart Air-Conditioning System for Electric Commercial Vehicles), Korea Evaluation Institute of Industrial Technology(KEIT) through the Industrial Technology Innovation Program (20003988, Development of 4 kW heating and cooling modularization technology using waste heat recovery and alternative refrigerant on electric vehicle's exclusive platform) and the National Research Foundation of Korea (NRF) grant funded by the Korea government(MSIT) (No. 2020R1A2C1011555).

Conflicts of Interest: The authors declare no conflict of interest.

References

1. Lee, M.-Y.; Seo, J.-H.; Lee, H.-S.; Garud, K.S. Power generation, efficiency and thermal stress of thermoelectric module with leg geometry, material, segmentation and two-stage arrangement. *Symmetry* **2020**, *12*, 786. [CrossRef]
2. Kang, B.H.; Lee, H.J. A review of recent research on automotive HVAC systems for EVs. *Int. J. Air-Cond. Refrig.* **2017**, *25*, 1730003. [CrossRef]
3. Qi, Z. Advances on air conditioning and heat pump system in electric vehicles–A review. *Renew. Sustain. Energy Rev.* **2014**, *38*, 754–764. [CrossRef]
4. Torregrosa, B.; Payá, J.; Corberán, J.M. Modelling of mobile airconditioning systems for electric vehicles. In Proceedings of the 4th European Workshop—Mobile Air Conditioning and Vehicle Thermal Systems, Torino, Italy, 1–2 December 2011; pp. 1–2.
5. Zhang, Z.; Li, W.; Shi, J.; Chen, J. A study on electric vehicle heat pump systems in cold climates. *Energies* **2016**, *9*, 881. [CrossRef]
6. Qin, F.; Zhang, G.; Xue, Q.; Zou, H.; Tian, C. Experimental investigation and theoretical analysis of heat pump systems with two different injection portholes compressors for electric vehicles. *Appl. Energy* **2017**, *185*, 2085–2093. [CrossRef]
7. Qin, F.; Xue, Q.; Zhang, G.; Zou, H.; Tian, C. Experimental investigation on heat pump for electric vehicles with different refrigerant injection compressors. *Energy Procedia* **2015**, *75*, 1490–1495. [CrossRef]
8. Qin, F.; Xue, Q.; Velez GM, A.; Zhang, G.; Zou, H.; Tian, C. Experimental investigation on heating performance of heat pump for electric vehicles at −20 °C ambient temperature. *Energy Convers. Manag.* **2015**, *102*, 39–49. [CrossRef]
9. Qin, F.; Shao, S.; Tian, C.; Yang, H. Experimental investigation on heating performance of heat pump for electric vehicles in low ambient temperature. *Energy Procedia* **2014**, *61*, 726–729. [CrossRef]
10. Zhou, G.; Li, H.; Liu, E.; Li, B.; Yan, Y.; Chen, T.; Chen, X. Experimental study on combined defrosting performance of heat pump air conditioning system for pure electric vehicle in low temperature. *Appl. Therm. Eng.* **2017**, *116*, 677–684. [CrossRef]
11. Ahn, J.H.; Kang, H.; Lee, H.S.; Jung, H.W.; Baek, C.; Kim, Y. Heating performance characteristics of a dual source heat pump using air and waste heat in electric vehicles. *Appl. Energy* **2014**, *119*, 1–9. [CrossRef]
12. Jung, J.; Jeon, Y.; Lee, H.; Kim, Y. Numerical study of the effects of injection-port design on the heating performance of an R134a heat pump with vapor injection used in electric vehicles. *Appl. Therm. Eng.* **2017**, *127*, 800–811. [CrossRef]
13. Zhang, L.; Hashimoto, K.; Hasegawa, H.; Saikawa, M. Performance analysis of a heat pump system with integrated desiccant for electric vehicles. *Int. J. Refrig.* **2018**, *86*, 154–162. [CrossRef]
14. Choi, Y.U.; Kim, M.S.; Kim, G.T.; Kim, M.; Kim, M.S. Performance analysis of vapor injection heat pump system for electric vehicle in cold startup condition. *Int. J. Refrig.* **2017**, *80*, 24–36. [CrossRef]

15. Lee, D.Y.; Cho, C.W.; Won, J.P.; Park, Y.C.; Lee, M.Y. Performance characteristics of mobile heat pump for a large passenger electric vehicle. *Appl. Therm. Eng.* **2013**, *50*, 660–669. [CrossRef]
16. Kwon, C.; Kim, M.S.; Choi, Y.; Kim, M.S. Performance evaluation of a vapor injection heat pump system for electric vehicles. *Int. J. Refrig.* **2017**, *74*, 138–150. [CrossRef]
17. Liu, C.; Zhang, Y.; Gao, T.; Shi, J.; Chen, J.; Wang, T.; Pan, L. Performance evaluation of propane heat pump system for electric vehicle in cold climate. *Int. J. Refrig.* **2018**, *95*, 51–60. [CrossRef]
18. Li, W.; Liu, R.; Liu, Y.; Wang, D.; Shi, J.; Chen, J. Performance evaluation of R1234yf heat pump system for an electric vehicle in cold climate. *Int. J. Refrig.* **2020**, *115*, 117–125. [CrossRef]
19. Ahn, J.H.; Lee, J.S.; Baek, C.; Kim, Y. Performance improvement of a dehumidifying heat pump using an additional waste heat source in electric vehicles with low occupancy. *Energy* **2016**, *115*, 67–75. [CrossRef]
20. Bellocchi, S.; Guizzi, G.L.; Manno, M.; Salvatori, M.; Zaccagnini, A. Reversible heat pump HVAC system with regenerative heat exchanger for electric vehicles: Analysis of its impact on driving range. *Appl. Therm. Eng.* **2018**, *129*, 290–305. [CrossRef]
21. Lee, H.S.; Lee, M.Y. Steady state and start-up performance characteristics of air source heat pump for cabin heating in an electric passenger vehicle. *Int. J. Refrig.* **2016**, *69*, 232–242. [CrossRef]
22. Li, S.; Wang, S.; Ma, Z.; Jiang, S.; Zhang, T. Using an air cycle heat pump system with a turbocharger to supply heating for full electric vehicles. *Int. J. Refrig.* **2017**, *77*, 11–19. [CrossRef]
23. Cho, C.W.; Lee, H.S.; Won, J.P.; Lee, M.Y. Measurement and evaluation of heating performance of heat pump systems using wasted heat from electric devices for an electric bus. *Energies* **2012**, *5*, 658–669. [CrossRef]
24. Lee, M.Y.; Lee, H.S.; Won, H.P. Characteristic evaluation on the cooling performance of an electrical air conditioning system using R744 for a fuel cell electric vehicle. *Energies* **2012**, *5*, 1371–1383. [CrossRef]
25. Lee, H.S.; Won, J.P.; Cho, C.W.; Kim, Y.C.; Lee, M.Y. Heating performance characteristics of stack coolant source heat pump using R744 for fuel cell electric vehicles. *J. Mech. Sci. Technol.* **2012**, *26*, 2065–2071. [CrossRef]
26. Shi, Y.; Guo, X.; Zhang, X. Study on economized vapor injection heat pump system using refrigerant R32. *Int. J. Air-Cond. Refrig.* **2016**, *24*, 1650006. [CrossRef]
27. Raj, A.K.; Kunal, G.; Srinivas, M.; Jayaraj, S. Performance analysis of a double-pass solar air heater system with asymmetric channel flow passages. *J. Therm. Anal. Calorim.* **2019**, *136*, 21–38. [CrossRef]
28. Arun, K.R.; Kunal, G.; Srinivas, M.; Kumar, C.S.; Mohanraj, M.; Jayaraj, S. Drying of untreated Musa nendra and Momordica charantia in a forced convection solar cabinet dryer with thermal storage. *Energy* **2020**, *192*, 116697. [CrossRef]
29. Garud, K.S.; Seo, J.H.; Cho, C.P.; Lee, M.Y. Artificial Neural Network and Adaptive Neuro-Fuzzy Interface System Modelling to Predict Thermal Performances of Thermoelectric Generator for Waste Heat Recovery. *Symmetry* **2020**, *12*, 259. [CrossRef]
30. Garud, K.S.; Seo, J.H.; Patil, M.S.; Bang, Y.M.; Pyo, Y.D.; Cho, C.P.; Lee, M.Y. Thermal–electrical–structural performances of hot heat exchanger with different internal fins of thermoelectric generator for low power generation application. *J. Therm. Anal. Calorim.* **2020**, 1–33. [CrossRef]
31. Hamad, A.J.; Khalifa, A.H.N.; Khalaf, D.Z. The Effect of Compressor Speed Variation and Vapor Injection on the Performance of Modified Refrigeration System. *Int. Rev. Mech. Eng. IREME* **2018**, *12*, 285–292. [CrossRef]

symmetry

MDPI

Article

Power Generation, Efficiency and Thermal Stress of Thermoelectric Module with Leg Geometry, Material, Segmentation and Two-Stage Arrangement

Moo-Yeon Lee [1], Jae-Hyeong Seo [1], Ho-Seong Lee [2] and Kunal Sandip Garud [1,*]

[1] School of Mechanical Engineering, Dong-A University, 37 Nakdong-Daero 550, Saha-gu, Busan 49315, Korea; mylee@dau.ac.kr (M.-Y.L.); cheonchw@donga.ac.kr (J.-H.S.)
[2] Thermal Management Research Center, KATECH, 74 Younjung-Ri, Pungse-Myun, Chonan 330-912, Korea; leehs@katech.re.kr
* Correspondence: 1876936@donga.ac.kr; Tel.: +82-51-200-7659

Received: 29 March 2020; Accepted: 12 April 2020; Published: 8 May 2020

Abstract: The objective of this study was to investigate the power generation, efficiency, and thermal stress of a thermoelectric module with leg geometry, material, segmentation, and two-stage arrangement. The effects of leg geometry, material, segmentation, and two-stage arrangement on maximum power, maximum efficiency, and maximum stress under various temperature differences and voltage load conditions were investigated. The performance parameters of the thermoelectric module were evaluated based on a numerical approach using ANSYS 19.1 commercial software. An analytical approach based on theoretical equations of the thermoelectric module was used to verify the accuracy and reliability of the numerical approach. The numerically predicted values for maximum power and maximum efficiency of the thermoelectric module were validated as ±5% and those for the maximum thermal stress of the thermoelectric module as ±7% with the corresponding calculated theoretical values. In addition, the predicted values of maximum power and maximum stress of the thermoelectric module were validated as ±2% and ±5%, respectively, with studies reported by Ma et al. and Al-Merbati et al. Of all the combinations, the single stage segmented arrangement with cylindrical leg geometry and SiGe+Bi_2Te_3 material was suggested as the best combination with maximum power of 0.73 W, maximum efficiency of 13.2%, and maximum thermal stress of 0.694 GPa.

Keywords: efficiency; power generation; thermal stress; thermoelectric module

1. Introduction

The world is currently facing global environmental challenges related to energy crisis, depletion of fossil fuel, and global warming [1,2]. To overcome these challenges, numerous research works have been carried out on high efficiency energy conversion technologies, as well as on the effective utilization of energy resources. One of the major sources of energy waste is automobile vehicles, where more than 50% of fuel energy is exhausted as heat [1]. Hence, the recovery and conversion of automobile exhaust heat using suitable conversion technology is a prerequisite in reducing fuel consumption as well as excessive demand of fossil fuel and natural resources. A thermoelectric generator is the most promising and convenient technology to convert waste heat into electrical energy [3].

Thermoelectric generators convert energy without degrading the environment. In addition, no mechanical moving parts are involved in the operation, and it is hence free of noise and vibration [4]. However, low conversion efficiency restricts their globalization in various application fields [5]. This low conversion efficiency is due to two major challenges: the lower Figure of Merit ($Z\overline{T}$) of the thermoelectric materials and various equipment-level challenges. Several new materials with higher $Z\overline{T}$ have been developed or the $Z\overline{T}$ of existing materials improved to overcome the low conversion efficiency. However, research works on thermoelectric equipment level challenges are comparatively less, especially those that seek to improve the performance of thermoelectric generators.

The thermo-electrical and thermo-mechanical behavior of four thermoelectric leg geometries wer compared by Erturun et al. for the non-segmented and segmented configurations [1]. For effective utilization of heat sources or sinks that are circular in shape, Shen et al. presented the concept and superiority of an annular segmented thermoelectric generator over an annular non-segmented thermoelectric generator [2]. Jin predicted the thermal stresses and deformations induced in a multilayer thin film thermoelectric module by developing the laminate and strength of material models [3]. Ma et al. showed the superiority of the performance of a multifunctional gradient thermoelectric module over a single functional gradient thermoelectric module [4]. Patil et al. investigated the performance characteristics of symmetrical segmented as well as asymmetrical segmented thermoelectric generators for various geometrical parameters and operating conditions [5]. Erturun et al. reported the influence of rectangular and cylindrical leg geometries, leg height, leg width/diameter, and leg spacing on power output, conversion efficiency, and thermal stresses. The predicted power, efficiency, and stress values using statistical models show accuracies of 8.85%, 1.22%, and 6.56%, respectively, with the corresponding simulated values using finite element analysis [6]. Wu et al. numerically investigated the effects of thermoelectric leg spacing, thickness of conducting plate, thickness of soldering layers, and thickness of ceramic plate on the power, efficiency, and thermal stresses of a thermoelectric generator under various heat flux conditions [7]. Erturun et al. proved that, compared to cylindrical legs, cylindrical coaxial legs show 7% higher efficiency and 10% reduced stresses, whereas rotated cylindrical legs showed 1.2% lower stresses and 0.3% higher efficiency [8]. Jia et al. concluded that a linear-shaped thermoelectric generator had superior performance compared to conventional π shaped thermoelectric generators [9]. Yilbas et al. conducted thermal stress analysis on a thermoelectric module with horizontally arranged rectangular as well as tapered leg geometries. The tapered leg geometry showed a lower value of thermal stress compared to the rectangular leg geometry [10]. Merbati et al. proved that thermoelectric legs with the cold side area twice the area of the hot side show lower values of thermal stress with small improvement in thermoelectric performance [11]. Ali et al. investigated the power generation and conversion efficiency of pin-shaped thermoelectric legs for various temperature ratios as well as external load resistance ratios [12]. Erturun et al. compared six different thermoelectric leg geometries and concluded that the cylindrical legs show less thermal stress and deformation with enhancement in thermoelectric power generation [13]. Yilbas et al. showed improvements in the first and second law efficiencies of the thermoelectric generator with double tapered leg geometry under various conditions of external load resistance and temperature ratio [14]. Gao et al. conducted thermal stress analysis on a single stage thermoelectric generator constructed with cubical leg geometry and Bi_2Te_3 material in order to optimize its mechanical performance [15]. Two-stage and three-stage thermoelectric generators were designed by Zhang et al. [16] and Kanimba et al. [17] in order to achieve high conversion efficiency. Wilbrecht et al. presented a two-stage thermoelectric module with 44.2% higher efficiency and 22.8% lower power output compared to the single stage thermoelectric module for diesel electric locomotives [18]. For a cryogenic nitrogen-based waste cold heat recovery system, Weng et al. presented four two-layer cascaded thermoelectric generators with power generation of 0.93 W [19]. Chen et al. concluded that multistage thermoelectric generators show better performance than single stage thermoelectric generators [20]. Kaibe et al. developed a cascaded thermoelectric generator with conversion efficiency of 12.1% [21]. Thermoelectric generators arranged thermally as well as electrically in three different series/parallel connections were compared by Almeida et al. based on the equivalent $Z\overline{T}$ [22]. Ibrahim et al. optimized the shape and length size of thermoelectric legs to enhance the power output and conversion efficiency of a thermoelectric generator [23]. Under steady state temperature conditions, Turenne et al. showed that thermal stresses in larger thermoelectric generators consisting of a number of thermoelectric modules were reduced due to the provision of soldering layers [24]. Kunal et al. showed the applicability of thermoelectric modules in waste heat recovery for lower power generation [25].

The comparison of single stage, segmented, and two-stage arrangements of thermoelectric modules with various combinations of leg geometries and materials based on their electrical and structural performances has not been explored yet. Therefore, the objective of this study was to numerically compare single stage, two-stage, and single stage segmented arrangements with three different leg geometries, namely, square prism legs, cylindrical legs, and trapezoidal legs, constructed with SiGe, Bi_2Te_3, and a combination of SiGe+Bi_2Te_3 materials in ANSYS 19.1 commercial software under various boundary conditions of temperature difference and voltage load. The comparison was made for maximum power, maximum efficiency, and maximum thermal stress. Additionally, in order to achieve higher power, higher efficiency, and lower thermal stress, an optimum configuration was suggested with leg geometry, material, segmentation, and two-stage arrangement.

2. Methodology

The single stage, two-stage, and single stage segmented arrangements with square prism legs, cylindrical legs, and trapezoidal legs, as well as SiGe, Bi_2Te_3, and combination of SiGe+Bi_2Te_3 materials were modelled in order to investigate the behavior of power generation, efficiency, and stress of the thermoelectric module. Modelled configurations of the thermoelectric module were analyzed numerically using the thermal electric and static structure solvers of ANSYS 19.1 under various temperature difference and voltage load conditions.

2.1. Geometry and Material Specifications

Among the three arrangements of the thermoelectric module, the single stage was baseline and the two-stage and single stage segmented arrangements were considered as modified cases of the single stage arrangement. Figure 1 shows the schematic diagrams of the single stage, two-stage, and single stage segmented arrangements of a thermoelectric module. A thermoelectric module with single stage arrangement comprised of one pair of p- and n-type thermoelectric legs and three conducting plates. The two-stage arrangement comprised of two single stage arrangements, positioned in the form of steps with an intermediate plate between them. The two pairs of thermoelectric legs in the two-stage arrangement were either made up of the same or different materials [17,18], whereas the single stage segmented arrangement was obtained by dividing the thermoelectric legs of the single stage arrangement into two equal halves. The upper half was named Segment 1 and the lower half Segment 2. Both segments were made of different materials [1,4].

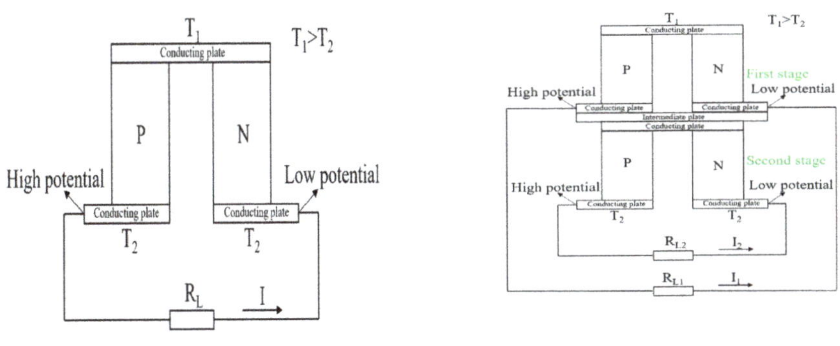

(**a**) Schematic of single stage arrangement (**b**) Schematic of two-stage arrangement

Figure 1. *Cont.*

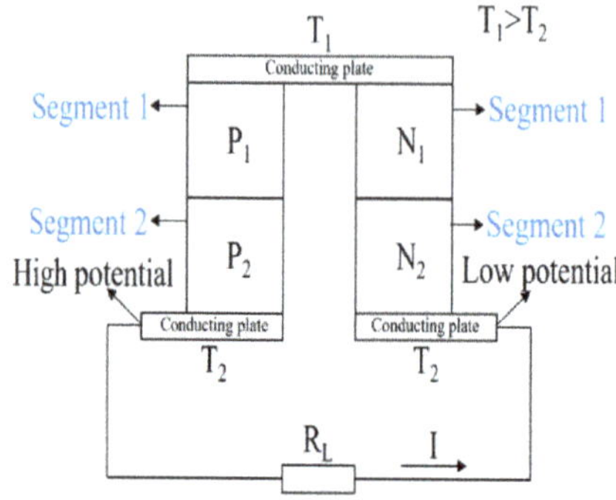

(**c**) Schematic of single stage segmented arrangement

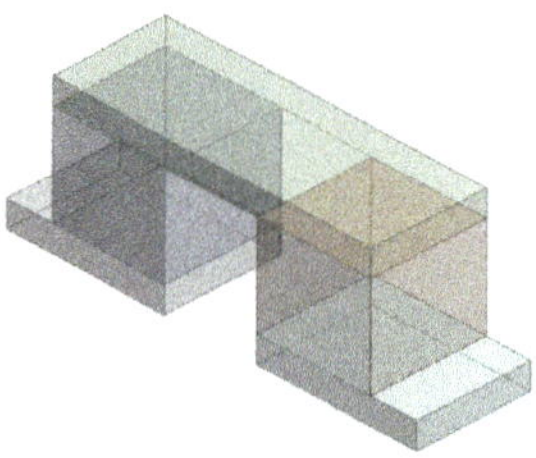

(**d**) single stage square

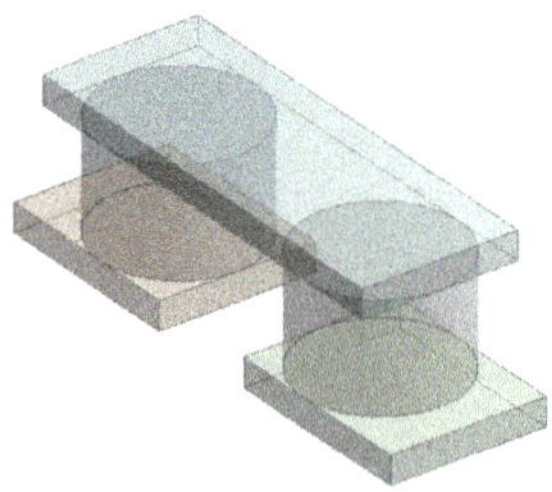

(**e**) single stage cylindrical

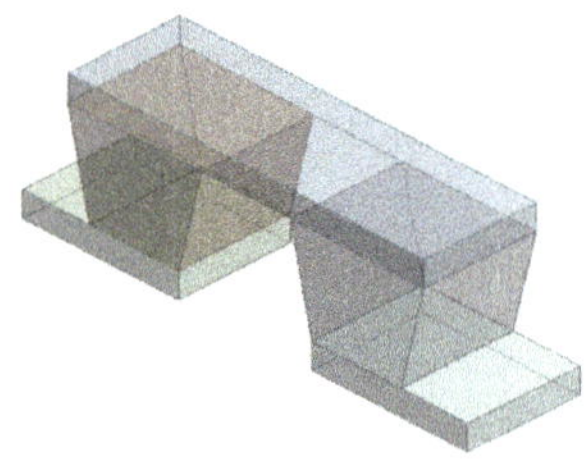

(**f**) single stage trapezoidal Alegs, hotside> Alegs, coldside

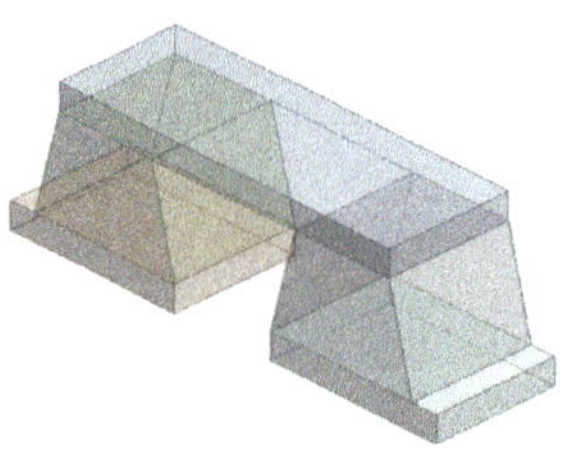

(**g**) single stage trapezoidal legs Alegs, coldside> Alegs, hotside

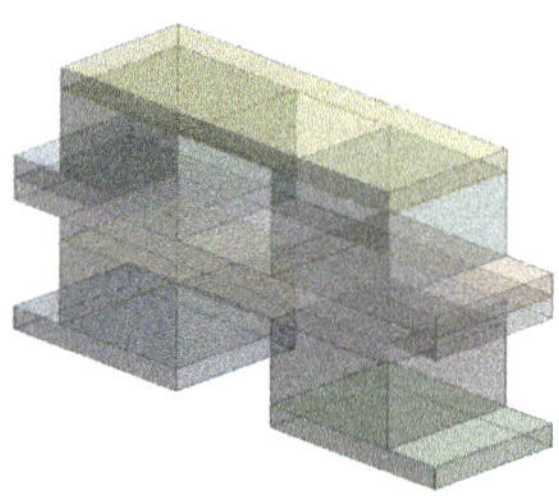

(**h**) two-stage square

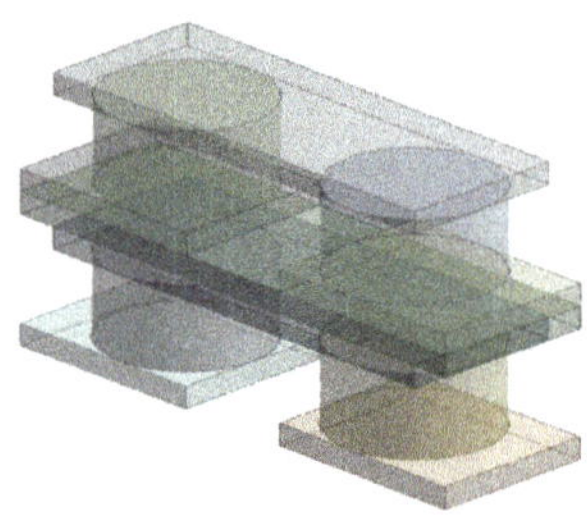

(**i**) two-stage cylindrical

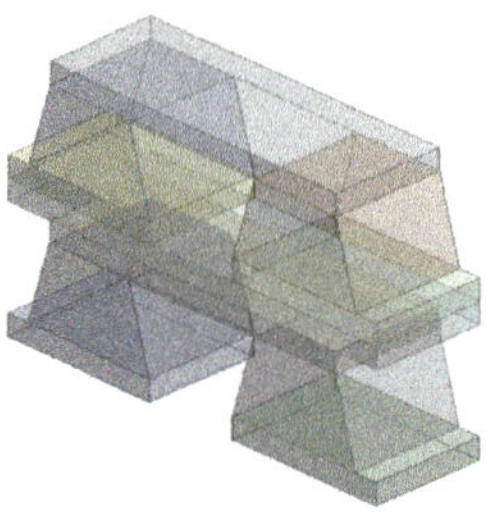

(**j**) two-stage trapezoidal $A_{legs, hotside}$> $A_{legs, coldside}$

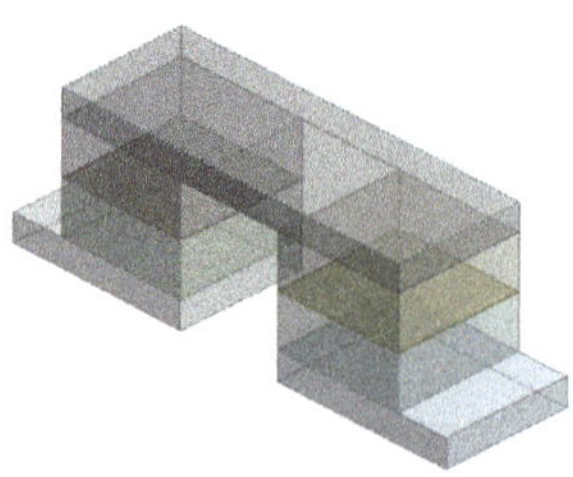

(**k**) single stage segmented square

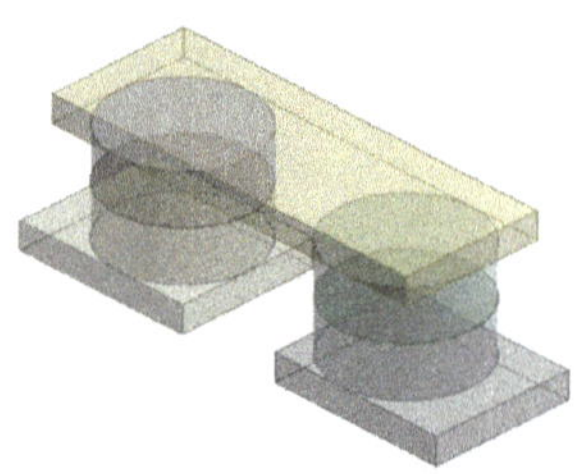

(**l**) single stage segmented cylindrical

Figure 1. *Cont.*

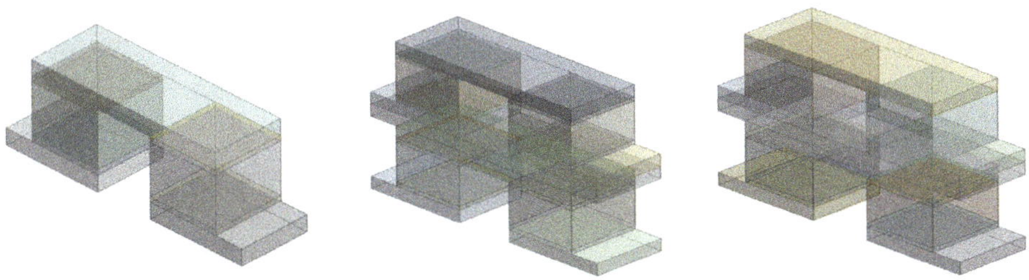

(m) single stage square with soldering layer
(n) two-stage square with soldering layer
(o) two-stage square—second stage with soldering layer

Figure 1. Various configurations of the thermoelectric module with different combinations of arrangements and leg geometries.

The single stage and two-stage arrangements of the thermoelectric module were modelled with square prism legs, cylindrical legs, and trapezoidal legs; whereas the single stage segmented arrangement of the thermoelectric module was modelled with square prism and cylindrical leg geometries. The square prism legs were modelled with a 1.0 mm × 1.0 mm cross-section area (base area) and 0.96 mm height [4]. The spacing between the legs was kept constant at 0.8 mm [4]. By having the same cross-section area (base area), height, spacing, and volume as the square prism legs, cylindrical legs with 1.13 mm diameter were modelled. Trapezoidal legs with two different leg configurations, $A_{legs, coldside} > A_{legs, hotside}$ and $A_{legs, hotside} > A_{legs, coldside}$ were also modelled. The larger side area had dimensions of 1.2 mm × 1.0 mm and the smaller side area had dimensions of 0.8 mm × 1.0 mm. The trapezoidal legs were also modelled with same cross-section area, height, spacing, and volume as that of the square prism legs and cylindrical legs. However, the cross-section areas at the base of the square prism and cylindrical legs were identical to the cross-section area at the center of the trapezoidal legs. The spacing in square prism and cylindrical legs was constant for leg height, which varied in the trapezoidal legs with 0.8 mm spacing at the larger side area, continuously increasing toward the smaller side area. Each pair of p- and n- type thermoelectric legs was sandwiched between three copper conducting plates—two at the bottom and one at the top. The bottom two plates had the same dimensions: 1.4 mm × 1.4 mm area and 0.25 mm thickness. In order to construct various leg geometries with the same cross-section area, spacing, and volume, the area of the top conducting plate was kept different in each case. The top conducting plates with square prism legs, cylindrical legs, and trapezoidal legs with $A_{legs, hotside} > A_{legs, coldside}$ and $A_{legs, coldside} > A_{legs, hotside}$ had areas of 2.8 mm × 1.0 mm [4], 3.06 mm × 1.0 mm, 3.2 mm × 1.0 mm and 2.8 mm × 1.0 mm, respectively. However, the thickness of the top conducting plate was 0.25 mm in each case. The thermoelectric legs in the single stage configuration were either made of SiGe or Bi_2Te_3 material in the two-stage configurations; the thermoelectric legs of both the stages were made of SiGe or Bi_2Te_3 or a combination of SiGe+Bi_2Te_3 materials. For the segmented configuration, the thermoelectric legs were made of a combination of SiGe+Bi_2Te_3 materials. In the two-stage arrangement with combination of SiGe+Bi_2Te_3, the upper stage was made of SiGe material and the bottom stage of Bi_2Te_3 material. Similarly, in the segmented legs, the upper half was made of SiGe material and the lower half of Bi_2Te_3 material. For a combination of SiGe+Bi_2Te_3 materials, the SiGe material was used near the hot side and the Bi_2Te_3 material near the cold side due to their melting point temperatures. The configurations of the thermoelectric module were modelled and compared without considering soldering layers. However, in order to show the effect of soldering layers on the performance characteristics of the thermoelectric module, the single stage and two-stage arrangements with square prism legs and Bi_2Te_3 material as well as the two-stage arrangements with square prism legs and SiGe+Bi_2Te_3 material were modelled separately with the soldering layers. Soldering layers were provided between the thermoelectric legs and the hot plate as well as the thermoelectric legs and the cold plates with an area of 1.0 mm × 1.0 mm and thickness of 0.08 mm [4]. Hence, when soldering layers were provided in those cases, the height of the legs reduced to 0.8 mm. The soldering layers were made up of 63Sn-37Pb material. The two-stage configuration carried an intermediate plate between both the stages of the thermoelectric modules. An intermediate plate in each case of the two-stage arrangement was made up of ceramic material having an area of 3.6 mm × 1.0 mm and thickness of 0.25 mm. Figure 1 shows the various configurations of the thermoelectric module with different combinations of arrangements and leg geometries. Table 1 shows the material specifications used in the numerical analysis.

Table 1. Material specifications used in the numerical analysis [1,4,26,27].

Material	SiGe	Bi_2Te_3	Copper	63Sn-37Pb	Al_2O_3
T_M (°C)	1177	585	1083	183	2072
α (μV/K)	115 (p-type) −115 (n-type)	227 (p-type) −227 (n-type)	-	-	-
Z (K^{-1})	2.38×10^{-4}	1.53×10^{-3}	-	-	-
ρ (Ω m)	1×10^{-5}	1.95×10^{-5}	1.84×10^{-8}	1.45×10^{-7}	1×10^{12}
k (W/m K)	5.56	1.73	394.5	50	17.2–37.2
E (GPa)	155	49.7	100–119	12–36	366–380
v	0.3	0.28	0.31	0.40	0.26
α_T (10^{-6}/K)	4.2	16.8	16.7–17.6	24	4.89–6.68

2.2. Mesh Convergence

Mesh dependency of maximum power, maximum efficiency, and maximum stress for the single stage thermoelectric module constructed with square prism legs and SiGe material is shown in Figure 2a–c, respectively. Maximum power and maximum efficiency showed negligible variations within ±0.00008% and ±0.00002%, respectively, for the selected mesh numbers; however, maximum stress showed a variation within ±0.7% above mesh number of 115,401. Hence, the mesh with 115,401 nodes was selected for the single stage thermoelectric module with square prism legs and SiGe material to evaluate maximum power, maximum efficiency, as well as maximum stress. Similarly, mesh dependency was conducted for the thermoelectric module with combinations of leg geometries, materials, and arrangements. The number of nodes for the same arrangement of thermoelectric module with materials and leg geometries are the same, but the number of nodes for varying arrangements are different. Therefore, the number of nodes for various combinations of leg geometries and materials after mesh dependency was selected as 115,401 for the single stage arrangement, 540,706 for the two-stage arrangement, and 242,138 for the single stage segmented arrangement. Maximum power, maximum efficiency, and maximum stress were converged as ±1% above 115,401 (number of nodes) for the single stage arrangement, 540,706 for the two-stage arrangement, and 242,138 for the single stage segmented arrangement of the thermoelectric module with leg geometries and materials [28]. Meshing for the single stage, two-stage, and segmented arrangements of the thermoelectric module with square prism leg geometry is shown in Figure 3a–c, respectively.

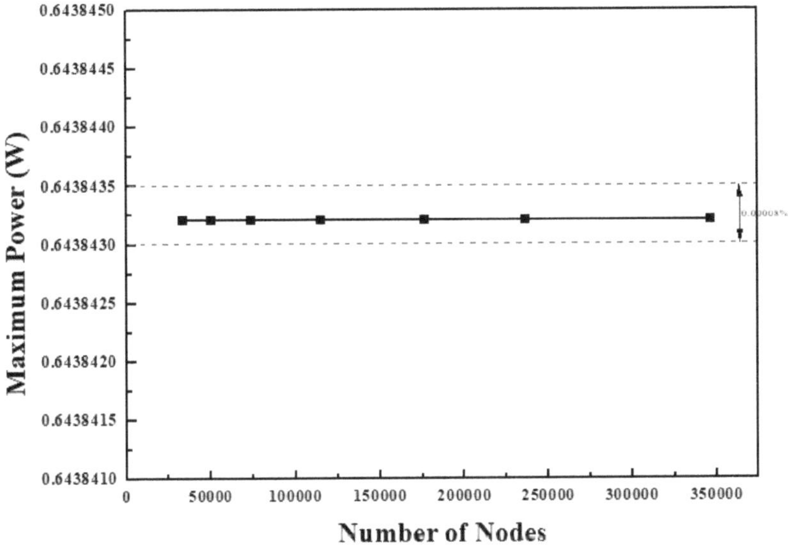

(**a**) Maximum power

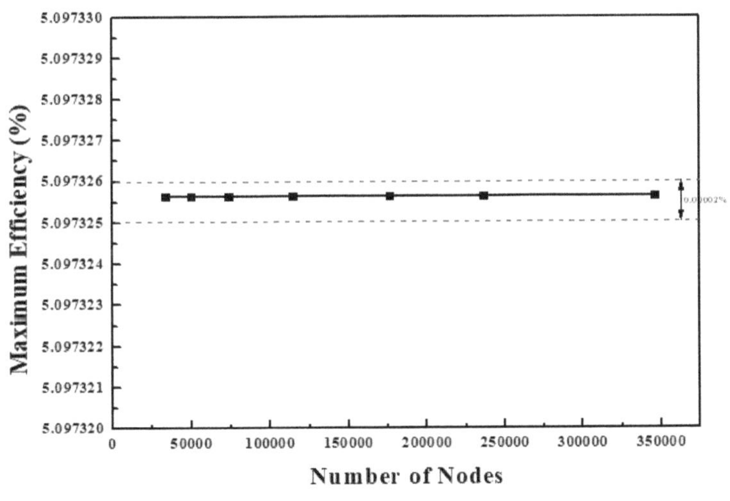

(**b**) Maximum efficiency

Figure 2. *Cont.*

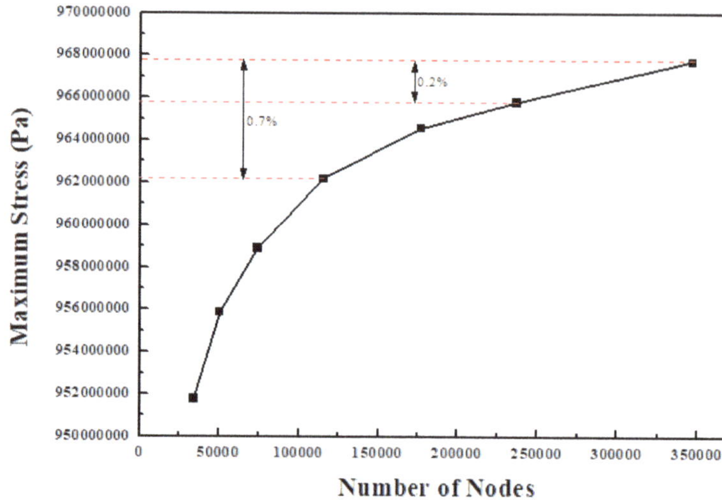

(**c**) Maximum stress

Figure 2. Mesh independency test for (**a**) maximum power, (**b**) maximum efficiency, and (**c**) maximum stress.

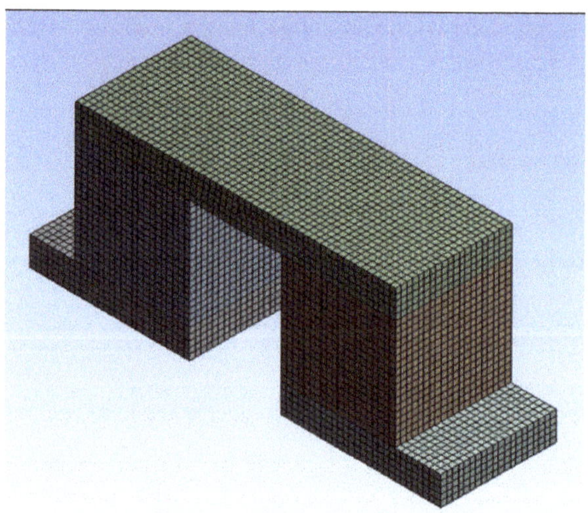

(**a**) Single stage arrangement

Figure 3. *Cont.*

(b) Two-stage arrangement

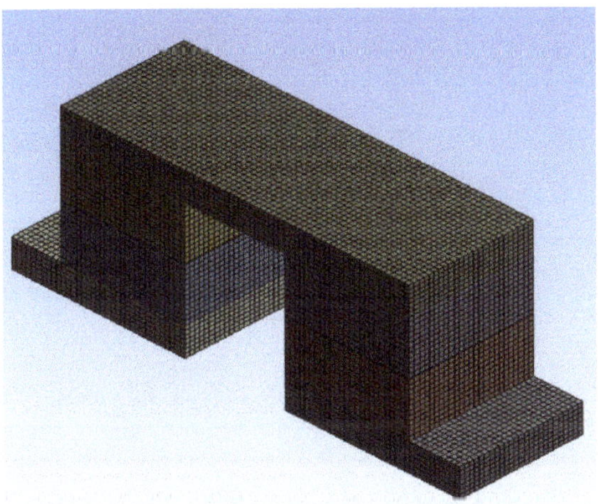

(c) Single stage segmented

Figure 3. Meshing for (**a**) single stage, (**b**) two-stage, and (**c**) single stage segmented arrangements with square prism leg geometries.

2.3. Boundary Conditions

In order to predict temperature distribution, power, conversion efficiency, and thermal stress, a thermoelectric module with various configurations was investigated under four boundary conditions, temperatures of hot and cold junctions, and voltage conditions at high and low potential sides. These boundary conditions were applied to the four different faces, as shown in Figure 1a–c. Temperature of the cold junction was kept constant at 20 °C and voltage at the low potential side was fixed at 0 V. Temperature of the hot junction and voltage at the high potential were not constant for the various configurations of the thermoelectric module. The boundary conditions for the hot junction temperature and the high potential voltage of leg geometries, materials, and arrangements were shown in Table 2. The hot side temperature was varied with a step size of 50 °C and the high potential voltage with a step size of 0.001 V. The maximum temperatures for the hot junction of various configurations were decided based on the melting point temperature of the corresponding material of construction. The high potential voltages varied from 0 V to voltage value at which current and power became zero. During variations in high potential voltages, there existed an optimum voltage, which showed maximum power and maximum efficiency. Optimum voltage values were different with different configurations of the thermoelectric module because the variation ranges of the voltage were different for different configurations. In addition, the optimum voltage varied with the temperature difference for each configuration of the thermoelectric module. Maximum power and maximum efficiency were simulated at the optimum voltage for each configuration of the thermoelectric module at different temperature difference conditions. Apart from these four boundary conditions, the convection boundary condition with heat transfer coefficient of 1×10^{-6} W/mm^2 °C was applied to all remaining faces of each configuration of the thermoelectric module. The heat transfer from the surfaces with the convection boundary condition to its surroundings was negligible.

Table 2. Boundary conditions used in the numerical analysis.

Material	Arrangement	Thermoelectric Legs	Boundary Conditions
			Temperature Conditions
SiGe	Single stage, two-stage	Square prism, cylindrical, trapezoidal	50 °C–1000 °C
Bi_2Te_3 (without soldering layer)	Single stage, two-stage	Square prism, cylindrical, trapezoidal	50 °C–500 °C
Bi_2Te_3 (with soldering layer)	Single stage, two-stage	Square prism	50 °C–150 °C
SiGe+Bi_2Te_3 (without soldering layer)	Two-stage	Square prism, cylindrical, trapezoidal	50 °C–1000 °C
SiGe+Bi_2Te_3 (without soldering layer)	Segmented	Square prism, cylindrical	50 °C–1000 °C
SiGe+Bi_2Te_3 (with soldering layer)	Two-stage	Square prism	50 °C–150 °C
			Voltage Conditions
All configurations	Single stage, two-stage, segmented	Square prism, cylindrical, trapezoidal	0 V to voltage value at which current and power become zero

2.4. Governing Equations

The thermoelectric module works on the conjugate physics of thermal and electrical effects; hence, the coupled equations related to heat transfer (temperature), and the electric field has to be solved in order to analyze the thermoelectrical characteristics of the thermoelectric modules [4,29].

The thermoelectric coupling equations are presented as:

$$\nabla \cdot \left(p \cdot \vec{J} \right) - \nabla \cdot (k \cdot \nabla T) = \vec{J} . \vec{E} \tag{1}$$

$$\nabla \cdot (\sigma \cdot \alpha \cdot \nabla T) + \nabla \cdot (\sigma \cdot \nabla \varnothing) = 0 \tag{2}$$

The term $\vec{J} . \vec{E}$ in Equation (1) represents volumetric joule heating.

For all configurations of the thermoelectric module, the thermoelectric coupling equations were solved in the thermal electric solver of ANSYS 19.1 software using the Galerkin finite element method and the corresponding boundary conditions in order to predict their thermoelectric behavior. The dimensionless displacement-strain relation needed to be solved in order to deal with the coupled thermal stress effect [7,30]:

$$\bar{\varepsilon}_{xx} = \frac{\partial \bar{u}}{\partial \bar{x}}, \qquad \bar{\varepsilon}_{yy} = \frac{\partial \bar{v}}{\partial \bar{y}}, \qquad \bar{\varepsilon}_{zz} = \frac{\partial \bar{w}}{\partial \bar{z}} \tag{3}$$

$$\bar{\varepsilon}_{xy} = 0.5 \left(\frac{\partial \bar{u}}{\partial \bar{y}} + \frac{\partial \bar{v}}{\partial \bar{x}} \right), \qquad \bar{\varepsilon}_{yz} = 0.5 \left(\frac{\partial \bar{w}}{\partial \bar{y}} + \frac{\partial \bar{v}}{\partial \bar{z}} \right), \qquad \bar{\varepsilon}_{zx} = 0.5 \left(\frac{\partial \bar{w}}{\partial \bar{x}} + \frac{\partial \bar{u}}{\partial \bar{z}} \right) \tag{4}$$

The exact dimensionless relationship between the stress and the strain is presented in terms of the non-dimensional Jacobian matrix, which is deducted from Equations (3) and (4) using Newton's method [7,30].

$$\begin{bmatrix} \overline{\sigma}_{xx} \\ \overline{\sigma}_{yy} \\ \overline{\sigma}_{zz} \\ \overline{\sigma}_{yz} \\ \overline{\sigma}_{zx} \\ \overline{\sigma}_{xy} \end{bmatrix} = \frac{E}{(1+v)(1-2v)} \times \begin{bmatrix} 1-v & v & v & 0 & 0 & 0 \\ v & 1-v & v & 0 & 0 & 0 \\ v & v & 1-v & 0 & 0 & 0 \\ 0 & 0 & 0 & 1-2v & 0 & 0 \\ 0 & 0 & 0 & 0 & 1-2v & 0 \\ 0 & 0 & 0 & 0 & 0 & 1-2v \end{bmatrix} \begin{bmatrix} \overline{\varepsilon}_{xx} \\ \overline{\varepsilon}_{yy} \\ \overline{\varepsilon}_{zz} \\ \overline{\varepsilon}_{yz} \\ \overline{\varepsilon}_{zx} \\ \overline{\varepsilon}_{xy} \end{bmatrix} - \begin{bmatrix} 1 \\ 1 \\ 1 \\ 0 \\ 0 \\ 0 \end{bmatrix} \frac{\alpha_T E \overline{T}}{1-2v} \quad (5)$$

The non-symmetrical Jacobian matrix was solved in the static structure solver of ANSYS 19.1 software for all configurations of the thermoelectric module using the corresponding boundary conditions in order to predict the stress and strain generated within them.

2.5. Data Reduction

2.5.1. Physical Assumptions

The heat transfer from the hot side to the cold side of the thermoelectric module was caused by the conduction mode based on Fourier's law using the inherent thermal properties of thermoelectric materials [4]. In addition, the conduction heat transfer along the thermoelectric legs was assumed to be one-dimensional. Heat transfer by the convection and the radiation modes was neglected [4]. The Seebeck and Peltier effects were taken into consideration in the present study. However, because the properties of materials are independent of temperature, the Thomson effect was neglected [12]. The thermal resistance of the conducting plates was also neglected [17].

2.5.2. Theoretical Analysis

In order to confirm the validity of the numerically predicted values of all performance characteristics of the thermoelectric module, the same performance characteristics were evaluated with a theoretical approach. The voltage at maximum power indicates optimum voltage and is calculated from Equation (6) [1,31,32]. Here, α is the Seebeck coefficient, which is measured as the amount of thermoelectric voltage induced in any material due to temperature difference across that material. Each thermoelectric material has its own unique value of Seebeck coefficient. The Seebeck coefficient is also recognized as thermopower, thermoelectric power, and thermoelectric sensitivity:

$$V_{theoretical,opt} = \alpha\ (T_H - T_C) \quad (6)$$

Current depends on voltage, external load resistance, and internal resistance—Equation (7) [10]. $R = R_L$ is the condition to obtain optimal current wherein power is maximum [4,31]:

$$I_{theoretical,opt} = \frac{V_{theoretical,opt}}{R + R_L} = \frac{\alpha\ (T_H - T_C)}{2R} \quad (7)$$

By manipulating the equations of $I = \frac{\alpha\ (T_H - T_C)}{R+R_L}$ and $P = I^2(R + R_L)$, the power equation was obtained. To obtain maximum power, constraint of $\frac{\partial P}{\partial R_L} = 0$ was applied on the power equation, which gave the maximum power condition of $R = R_L$. When $R = R_L$ is used in the power equation, Equation (8) is derived, which calculates the theoretical maximum power of the thermoelectric module [4,33]:

$$P_{theoretical,max} = \frac{\propto^2}{4R}(T_H - T_C)^2 \quad (8)$$

The theoretical maximum efficiency was calculated using Equation (9) under the same condition of $R = R_L$ in the theoretical efficiency equation of $\eta_{theoretical} = \frac{\Delta T \cdot R_L}{T_H(R+R_L) - \frac{\Delta T \cdot R}{2} + \frac{(R+R_L)^2}{ZR}}$ [4].

$$\eta_{theoretical,max} = \Delta T \frac{\sqrt{1+Z\overline{\overline{T}}} - 1}{T_H\sqrt{1+Z\overline{\overline{T}}} + T_C} \quad (9)$$

Resistance for square prism and cylindrical legs was calculated with Equation (10) [1].

$$R = \frac{(\rho_p + \rho_n)L}{A} \quad (10)$$

Resistance of the trapezoidal legs cannot be calculated using Equation (10) because of the non-uniform cross-sectional area, which varies along the height of the thermoelectric legs. In the trapezoidal legs, an unevenly distributed area along the centerline of leg in height was converted to an evenly distributed area along the same centerline of the leg such that volume of the leg was same in both cases. Using the uniform area approach, the resistance for the trapezoidal legs was calculated with Equation (11) [11].

$$R = \frac{\sigma_p + \sigma_n}{2\sigma_p\sigma_n \frac{A_0}{L}\left(\frac{R_A - 1}{R_A + 1}\right)} \ln(R_A) \quad (11)$$

where R_A is the ratio of the top side area of the trapezoidal leg (A_T) to the bottom side area of the trapezoidal leg (A_B) from Equation (12). A_0 is the equivalent uniform cross-section area of the trapezoidal leg:

$$R_A = \frac{A_T}{A_B} \quad (12)$$

In order to evaluate the maximum power and maximum efficiency of the thermoelectric module with the two-stage arrangement, it is necessary to derive the intermediate temperature between the two stages. This intermediate temperature can be obtained using voltage and temperature difference correlation equation of $V = \alpha\ (T_H - T_i)$ or $\alpha\ (T_i - T_C)$. In the present study, the voltages applied to both the stages were different; hence, the equations were modified to $V_{Top} = \alpha\ (T_H - T_{i1})$ and $V_{Bottom} = \alpha\ (T_{i2} - T_C)$. T_{i1} is the cold side temperature for the first stage, whereas T_{i2} is the hot side temperature of the second stage. The difference of $T_{i1} - T_{i2}$ shows the temperature loss due to the thermal resistance of the intermediate plate. The voltage boundary conditions applied to the high potential sides of both the stages were inserted in the corresponding voltage equations, obtaining T_{i1} and T_{i2} temperatures. These temperatures are used in Equations (8) and (9) in order to calculate the maximum power and maximum efficiency of the two-stage arrangement of the thermoelectric module. The segmented thermoelectric module consists of a combination of two materials in a single thermoelectric leg. Hence, the concept of equivalent Seebeck coefficient, equivalent resistivity, equivalent thermal conductivity, and equivalent Figure of Merit was used. Using this equivalent parameter concept, maximum power and maximum efficiency of the segmented thermoelectric module were calculated theoretically from the following equations [22,34]:

$$P_{theoretical,max} = \frac{{\alpha_{equivalent}}^2}{R_{equivalent}}(T_H - T_C)^2 \quad (13)$$

$$\alpha_{equivalent} = \frac{\alpha_2 k_1 + \alpha_1 k_2}{k_1 + k_2} \quad (14)$$

$$\eta_{theoretical,max} = \Delta T \frac{\sqrt{1+Z_{equivalent}\overline{\overline{T}}} - 1}{T_H\sqrt{1+Z_{equivalent}\overline{\overline{T}}} + T_C} \quad (15)$$

$$Z_{equivalent} = \frac{\alpha_{equivalent}{}^2}{(\rho_1 + \rho_2) * \left(\frac{k_1 k_2}{k_1 + k_2}\right)} \tag{16}$$

Thermal stress induced in the thermoelectric modules due to various temperature conditions, as well as due to mismatch in thermal properties of different materials, were calculated in the form of von Mises stress. This is the equivalent of the second deviatory stresses and computed using the following equation [1,3]:

$$\sigma_v = \sqrt{\frac{(\sigma_{11} - \sigma_{22})^2 + (\sigma_{22} - \sigma_{33})^2 + (\sigma_{33} - \sigma_{11})^2 + 6(\sigma_{12}{}^2 + \sigma_{23}{}^2 + \sigma_{31}{}^2)}{2}} \tag{17}$$

Normal stress in the longitudinal direction can be calculated using the following equation. In this equation, normal strain $\varepsilon_n = 0$ and curvature K = 0 for the present study [3] are:

$$\sigma_{xx} = \frac{E}{1 - v^2}(\varepsilon_n + Z_c K) - \frac{E\alpha_T}{1 - v}(T_H - T_C) = -\frac{E\alpha_T}{1 - v}(T_H - T_C) \tag{18}$$

Thermal stress in the direction of temperature gradient can be calculated using the following equations, which are based on temperature difference and thermal properties.

$$\sigma_T = \alpha_T E (T_H - T_C) \tag{19}$$

$$\sigma_T = \alpha_T E \Delta T \tag{20}$$

Thermal strain is the ratio of thermal stress to the Young modulus, which is represented by Equation (21).

$$\varepsilon = \frac{\sigma_T}{E} \tag{21}$$

2.5.3. Numerical Analysis

Optimum voltage at maximum power and efficiency exists for the given temperature difference condition. The current corresponding to the optimum voltage is called optimum current. Using the simulated optimum current value corresponding to the employed optimum voltage value, the maximum power based on the numerical approach was calculated using Equation (22) [13].

$$P_{numerical,max} = V_{numerical,opt} I_{numerical,opt}. \tag{22}$$

The maximum efficiency was calculated using the numerical maximum power and heat absorbed [13]. The heat absorbed by the thermoelectric module from the hot side was simulated:

$$n_{numerical,max} = \frac{P_{numerical,opt}}{H_a} \tag{23}$$

Similarly, thermal stress in terms of the equivalent stress or von-Mises stress was simulated from the numerical analysis with the static structure solver of ANSYS under various temperature conditions.

The maximum power and maximum efficiency were taken into consideration with the thermal-electric solver under various boundary conditions for the hot side and cold side temperatures, as well as voltage loads of the low and high potential sides. The thermal stress analysis of various configurations of the thermoelectric module was carried out in the static structure solver using the same temperature boundary conditions as the thermal-electric solver.

3. Results and Discussion

The results and discussion of this study includes the validation of numerically predicted values of maximum power, maximum efficiency, and maximum stress with data published by Ma et al. [4] and Al-Merbati et al. [11], as well as the values calculated using the theoretical approach. The comparison of different configurations of the thermoelectric module based on maximum power, maximum efficiency, and maximum stress is considered and the effect of soldering layers on the performance of the thermoelectric module is discussed.

3.1. Validation

The maximum power and maximum stress predicted in this study using ANSYS codes were validated with the corresponding data published by Ma et al. [4] and Al-Merbati et al. [11]. The validation of codes for maximum power and stress are presented and the comparison between the numerical and theoretical values of the various performance parameters are discussed.

3.1.1. Code Validation

The maximum power values evaluated in the present study using the thermal electric solver were found to be ±2% of the corresponding maximum power values reported by Ma et al. [4]. Similarly, the maximum stress and centerline stress values evaluated numerically in the present study using the static structure solver were validated at ±5% with the corresponding maximum stress and centerline stress values reported by Al-Merbati et al. [11]. The maximum power and maximum stress evaluated using ANSYS codes in the present study were found to be in the error range of ±5% with the corresponding data of Ma et al. [4] and Al-Merbati et al. [11]. Hence, the application of numerical approach using the validated ANSYS codes is justified for the further analysis. Figure 4a,b show the maximum power and stress validation with the corresponding previous studies of Ma et al. [4] and Al-Merbati et al. [11].

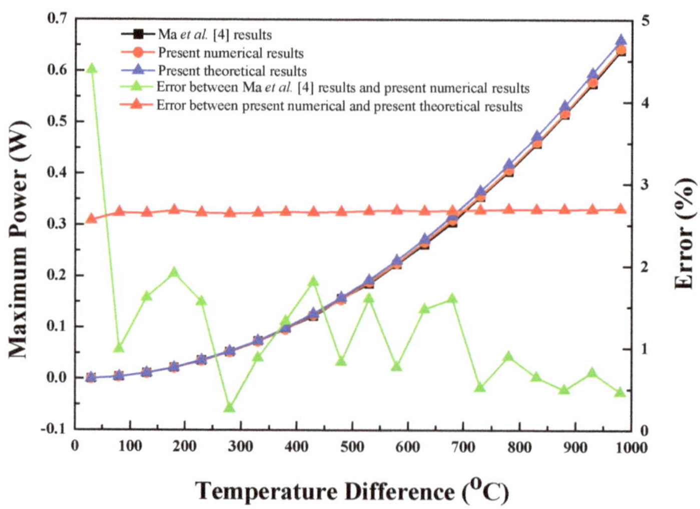

(**a**) Maximum power

Figure 4. *Cont.*

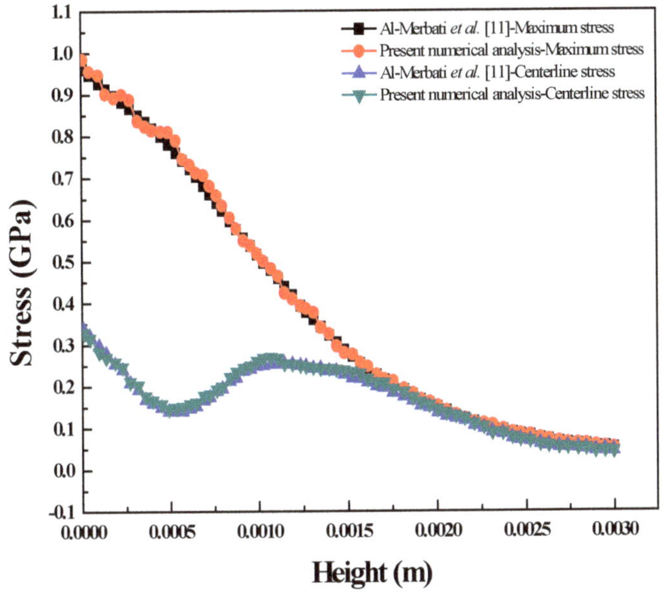

(**b**) Stress

Figure 4. Validation with previous study for (**a**) maximum power and (**b**) stress.

3.1.2. Validation of Numerical Results with Theoretical Results

The theoretical maximum power for the thermoelectric module with leg geometries, materials, and arrangements were calculated using correlations with Table 3. The maximum power for various configurations of the thermoelectric module were calculated using Equations (8) and (13). Similarly, the maximum efficiency for various configurations of the thermoelectric module was calculated theoretically using Equations (9) and (15). The simulated values of maximum power as well as maximum efficiency were calculated using Equations (22) and (23), respectively, which were compared with the corresponding theoretical values of maximum power and maximum efficiency. Figure 5a,b show the comparison between the numerical and theoretical results of maximum power and maximum efficiency, respectively, for the various configurations of the thermoelectric module. The numerical values of maximum power as well as maximum efficiency were validated with the corresponding theoretical values at ±5% with the entire temperature difference range for all configurations of the thermoelectric module.

The maximum stress values were calculated theoretically using Equations (18)–(20). The calculated theoretical values of the maximum stress were compared with the corresponding predicted values. The comparison of theoretical and numerical results of maximum stress for the single stage arrangement of the thermoelectric module with three leg geometries and two materials is shown in Figure 5c. The equations show that the theoretical values of stress are not dependent on leg geometry; hence, the stress values are same for all leg geometries. However, different materials have different theoretical values of maximum stress. For the same material, the numerical stress values for cylindrical legs show closer agreement, whereas square prism legs as well as trapezoidal legs with two-area configurations show almost equal error with the corresponding theoretical stress values. For the same material, the numerically predicted maximum stress for all the leg geometries was found at ±7% with the corresponding theoretical stress with the entire temperature difference range.

Table 3. Maximum power correlations derived for various leg geometries, materials, and arrangements.

Arrangements	Material	Leg Geometry	Maximum Power (W)
Single stage	SiGe	Square Cylindrical	$6.89 \times 10^{-7} \times (\Delta T)^2$
Single stage	SiGe	Trapezoidal ($A_{leg,hotside} > A_{leg,coldside}$)	$6.80 \times 10^{-7} \times (\Delta T)^2$
Single stage	SiGe	Trapezoidal ($A_{leg,coldside} > A_{leg,hotside}$)	$6.80 \times 10^{-7} \times (\Delta T)^2$
Single stage	Bi_2Te_3	Square Cylindrical	$1.38 \times 10^{-6} \times (\Delta T)^2$
Single stage	Bi_2Te_3	Trapezoidal	$1.36 \times 10^{-6} \times (\Delta T)^2$
Single stage (SL)	Bi_2Te_3	Square	$1.65 \times 10^{-6} \times (\Delta T)^2$
Two-stage—1st stage	SiGe	Square Cylindrical	$6.89 \times 10^{-7} \times (\Delta T)^2$
Two-stage—2nd stage	SiGe	Square Cylindrical	$6.89 \times 10^{-7} \times (\Delta T)^2$
Two-stage—1st stage	SiGe	Trapezoidal	$6.80 \times 10^{-7} \times (\Delta T)^2$
Two-stage—2nd stage	SiGe	Trapezoidal	$6.80 \times 10^{-7} \times (\Delta T)^2$
Two-stage—1st stage	Bi_2Te_3	Square Cylindrical	$1.38 \times 10^{-6} \times (\Delta T)^2$
Two-stage—2nd stage	Bi_2Te_3	Square Cylindrical	$1.38 \times 10^{-6} \times (\Delta T)^2$
Two-stage—1st stage	Bi_2Te_3	Trapezoidal	$1.36 \times 10^{-6} \times (\Delta T)^2$
Two-stage—2nd stage	Bi_2Te_3	Trapezoidal	$1.36 \times 10^{-6} \times (\Delta T)^2$
Two-stage—1st stage	SiGe	Square Cylindrical	$6.89 \times 10^{-7} \times (\Delta T)^2$
Two-stage—2nd stage	Bi_2Te_3	Square Cylindrical	$1.38 \times 10^{-6} \times (\Delta T)^2$
Two-stage—1st stage	SiGe	Trapezoidal	$6.80 \times 10^{-7} \times (\Delta T)^2$
Two-stage—2nd stage	Bi_2Te_3	Trapezoidal	$1.36 \times 10^{-6} \times (\Delta T)^2$
Two-stage—1st stage (SL)	Bi_2Te_3	Square	$1.65 \times 10^{-6} \times (\Delta T)^2$
Two-stage—2nd stage (SL)	Bi_2Te_3	Square	$1.65 \times 10^{-6} \times (\Delta T)^2$
Two-stage—1st stage	SiGe	Square	$6.89 \times 10^{-7} \times (\Delta T)^2$
Two-stage—2nd stage (SL)	Bi_2Te_3	Square	$1.65 \times 10^{-6} \times (\Delta T)^2$
Segmented	SiGe+Bi_2Te_3	Square Cylindrical	$1.41 \times 10^{-6} \times (\Delta T)^2$

SL means soldering layer.

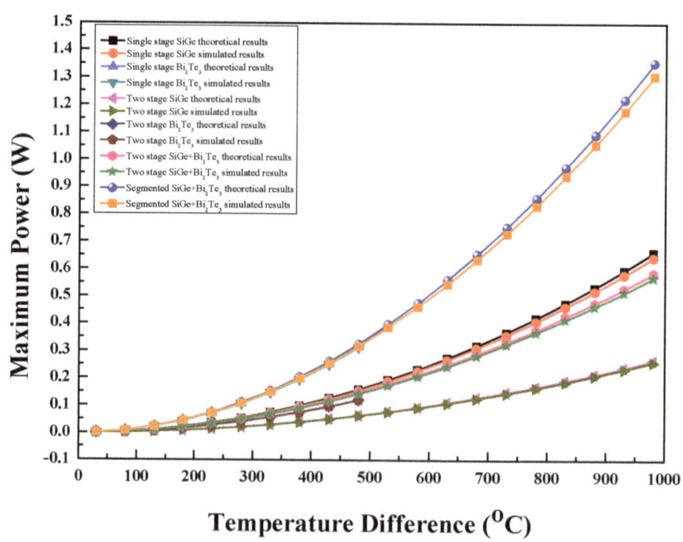

(**a**) Maximum power of all the leg geometries

Figure 5. *Cont.*

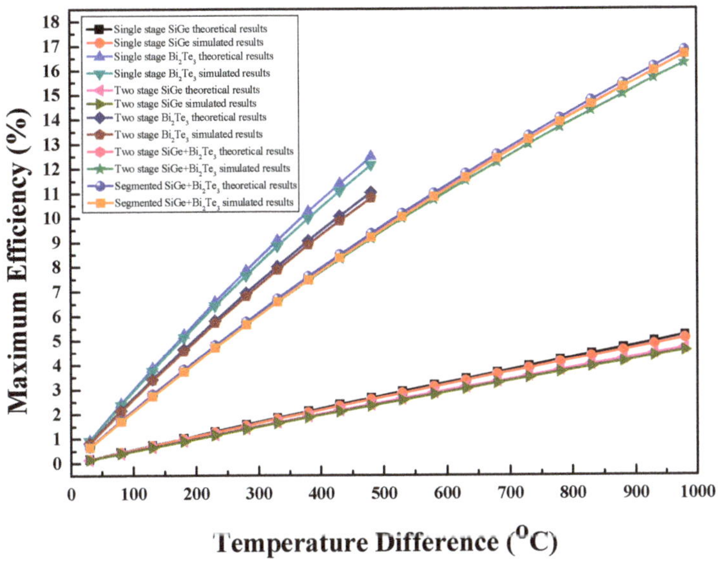

(**b**) Maximum efficiency of all the leg geometries

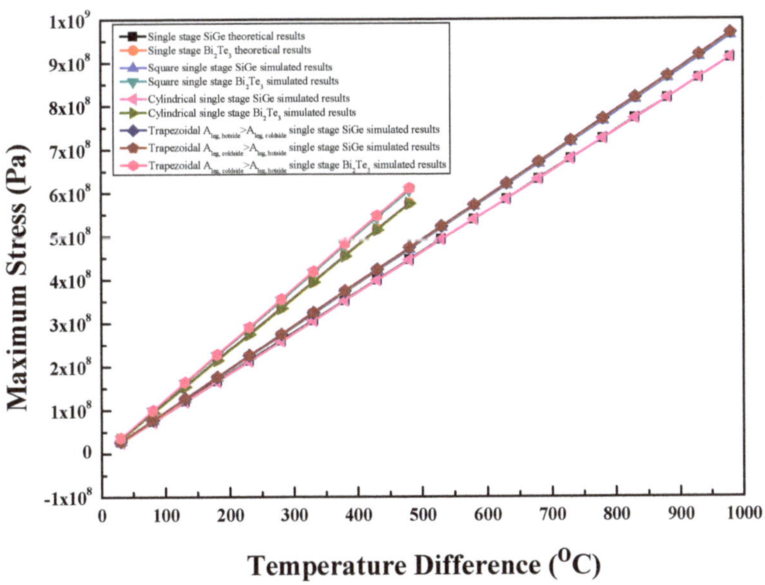

(**c**) Maximum stress for single stage arrangement

Figure 5. Comparison of numerical and theoretical results for (**a**) maximum power, (**b**) maximum efficiency, and (**c**) maximum stress.

3.2. Optimum Temperature

The single stage as well as two-stage arrangements with either SiGe or Bi_2Te_3 material were operated at maximum temperature differences of 980 °C and 480 °C, respectively, lower than the melting point temperature of SiGe or Bi_2Te_3 material. If the two-stage and segmented arrangements of the thermoelectric module with SiGe+Bi_2Te_3 material are operated at the maximum temperature difference of 980 °C, the operating temperature of SiGe material is below its melting point temperature, but the operating temperature of Bi_2Te_3 material is higher than its melting point temperature limit. Therefore, optimum temperature was introduced for the two-stage and segmented arrangements of the thermoelectric module with SiGe+Bi_2Te_3 material in order to operate them below the melting point temperature limit without failure. The optimum temperature is the hot junction temperature of the second stage for the two-stage arrangement of the thermoelectric module with SiGe+Bi_2Te_3 material and the interface temperature between two segments for the segmented arrangement of the thermoelectric module with SiGe+Bi_2Te_3 material [4,17]. The maximum value of the optimum temperature is the melting point temperature of Bi_2Te_3 material, which is around 585 °C. Figure 6 shows the optimum temperature values of the two-stage and segmented arrangements of the thermoelectric module with SiGe+Bi_2Te_3 material at various temperature difference conditions. The optimum temperature increases linearly with the temperature difference, as shown in the figure. As presented in the figure, in order to maintain optimum temperature of 585 °C or below, the two-stage arrangement with square prism legs are operated at the maximum temperature difference of 880 °C or lower, the two-stage arrangement with cylindrical and trapezoidal legs are operated at the maximum temperature difference of 830 °C or lower, and the segmented arrangement with square prism and cylindrical legs are operated at the maximum temperature difference of 730 °C or lower.

For the single stage arrangement of the thermoelectric module with SiGe material, trapezoidal legs with $A_{legs, hotside} > A_{legs, coldside}$ showed higher average stress compared to trapezoidal legs with $A_{legs, coldside} > A_{legs, hotside}$ [11], which is discussed in detail in Section 3.5.1. Therefore, out of the two configurations of trapezoidal legs, only trapezoidal legs with $A_{legs, coldside} > A_{legs, hotside}$ were considered for maximum power, maximum efficiency, and maximum stress analyses for single stage arrangement with Bi_2Te_3 material and two-stage arrangement with SiGe, Bi_2Te_3, and SiGe+Bi_2Te_3 materials.

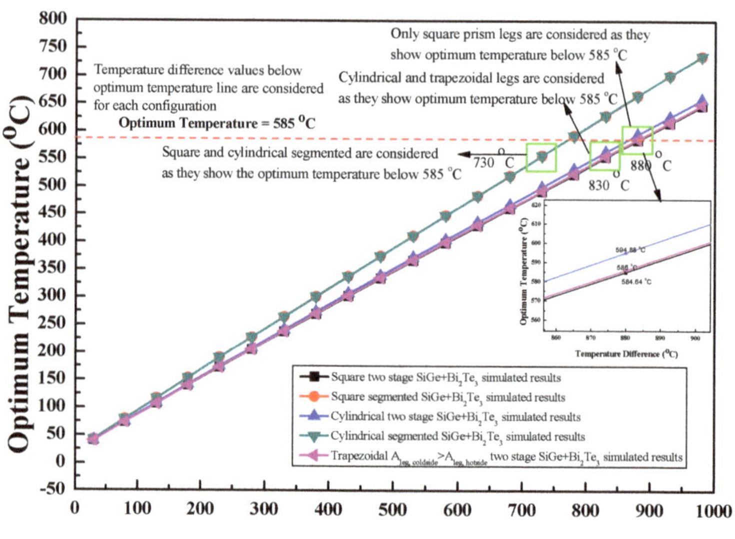

Figure 6. Optimum temperature variation with respect to temperature difference.

3.3. *Maximum Power*

The comparison of maximum power for the single stage, two-stage, and single stage segmented arrangements of the thermoelectric module with various leg geometries and materials are presented in this section.

3.3.1. Single Stage Arrangement

The comparison of maximum power with temperature difference for the single stage arrangement of the thermoelectric module with three leg geometries and two semiconductor materials is shown in Figure 7a. For the same material, leg geometry had no significant effect on maximum power because of the same internal resistance as well as the same optimum voltage values [1,6,10]. At the same temperature difference, the optimum voltage of the thermoelectric module was the same and leg geometry had the same volume and base area with equal internal resistance for all the legs. However, materials with high $Z\overline{T}$ showed higher maximum power than those with lower $Z\overline{T}$ [4,17]. Hence, for the same leg geometry, the Bi_2Te_3 material showed higher maximum power than the SiGe material at the same temperature difference. This behavior was observed up to a temperature difference of 480 °C as the Bi_2Te_3 material could not be operated at a higher temperature difference because operation above that temperature difference increased the hot side temperature above its melting point. Hence, beyond the temperature difference of 480 °C, the SiGe material showed an increase in maximum power with the highest value at the temperature difference of 980 °C. For all the materials and leg geometries with the single stage arrangement of the thermoelectric module, the maximum power increased with the temperature difference. All the leg geometries with the SiGe material showed maximum power of 0.65 W at a temperature difference of 980 °C, while all the leg geometries with the Bi_2Te_3 material showed maximum power of 0.31 W at a temperature difference of 480 °C.

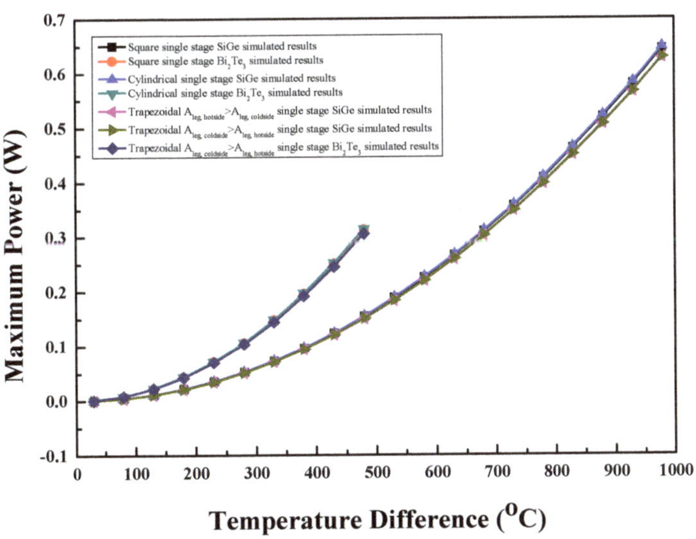

(**a**) Single stage arrangement

Figure 7. *Cont.*

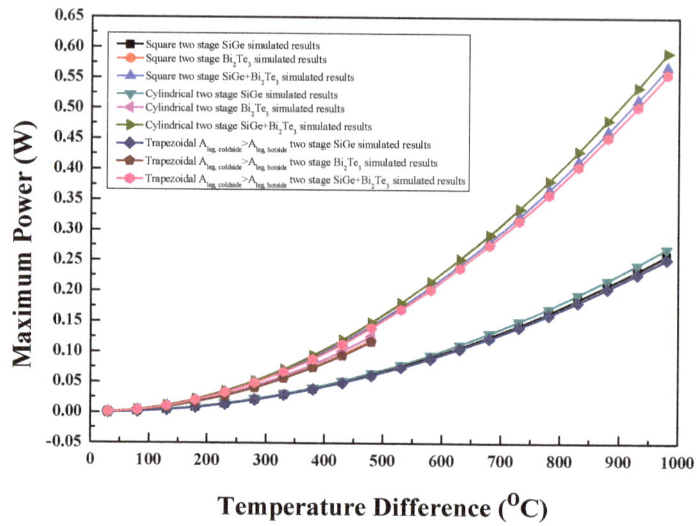

(**b**) Two-stage arrangement

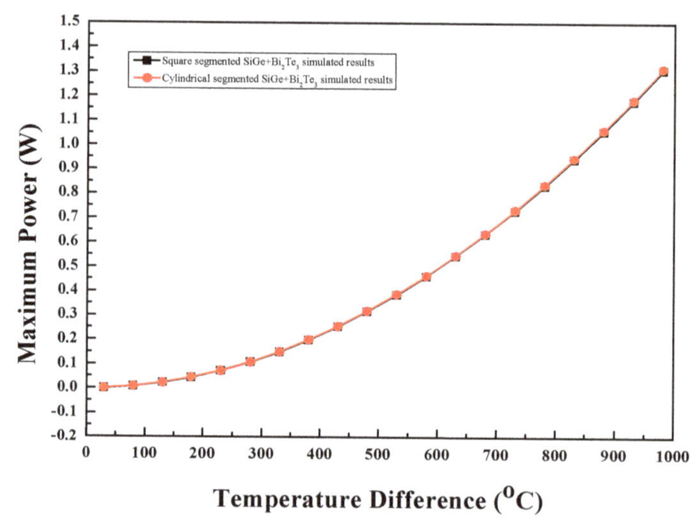

(**c**) Single stage segmented arrangement

Figure 7. Maximum power for (**a**) single stage arrangement, (**b**) two-stage arrangement, and (**c**) single stage segmented arrangement.

3.3.2. Two-Stage Arrangement

The variation of maximum power with the temperature difference for the two-stage thermoelectric module with various leg geometries and materials is shown in Figure 7b. In the two-stage arrangement of the thermoelectric module with the same leg geometry, a combination of materials in the first stage made up of SiGe material and the second stage made up of Bi_2Te_3 material showed significant enhancement in maximum power, compared to both stages either made of SiGe or Bi_2Te_3 alone [17,18].

In the case of SiGe, Bi_2Te_3, and SiGe+Bi_2Te_3 material, the square prism and trapezoidal legs with $A_{legs,\ coldside} > A_{legs,\ hotside}$ showed similar maximum power with the temperature difference but the cylindrical legs showed enhancement in maximum power at the corresponding same temperature difference, although the degree of increase was less [1,6,10]. In the case of the SiGe+Bi_2Te_3 material, maximum power of 0.46 W was obtained for square prism legs at a temperature difference of 880 °C, whereas the cylindrical legs showed 0.43 W and trapezoidal legs showed 0.41 W maximum power at a temperature difference of 830 °C. For the SiGe material and temperature difference of 980 °C, the square prism and trapezoidal legs showed maximum power of 0.25 W and the cylindrical legs showed maximum power of 0.27 W. For the Bi_2Te_3 material and temperature difference of 480 °C, the square prism legs and trapezoidal legs showed maximum power of 0.12 W and the cylindrical legs showed maximum power of 0.13 W. The maximum power increased with the temperature difference in all the cases, as shown in Figure 7b.

3.3.3. Single Stage Segmented Arrangement

The maximum power variation with the temperature difference is shown in Figure 7c for the segmented thermoelectric module constructed with two leg geometries and the SiGe+Bi_2Te_3 material. Maximum power increased with the temperature difference for both the combinations. The leg geometry had no effect on the maximum power [1,6,10], as shown in Figure 7c, due to same internal resistance and same optimum voltage condition. The square prism as well as the cylindrical legs showed maximum power of 0.73 W at a temperature difference of 730 °C.

3.4. *Maximum Efficiency*

The comparison of maximum efficiency for the single stage, two-stage, and single stage segmented arrangements of the thermoelectric module with various leg geometries and materials are presented in this section.

3.4.1. Single Stage Arrangement

The variation of maximum efficiency with the temperature difference for the single stage thermoelectric module with three leg geometries and two materials is shown in Figure 8a. Maximum efficiency is the ratio of maximum power to the heat absorbed [13]. Similar to that of maximum power, the leg geometries had no effect on maximum efficiency for the same material [1,6,10]. For all the leg geometries and materials, as the temperature difference increased, the maximum efficiency also increased. For the same leg geometry, the Bi_2Te_3 material showed superior increase in maximum efficiency compared to the SiGe material over the entire temperature difference range due to high $Z\overline{T}$ [4]. Although the Bi_2Te_3 material was operated up to a temperature difference of 480 °C due to its low melting point temperature, it still showed the highest value of maximum efficiency at a temperature difference of 480 °C, compared to the maximum efficiency value for the SiGe material at a temperature difference of 980 °C. At temperature difference of 480 °C, the Bi_2Te_3 material had a higher maximum power, thus showing higher maximum efficiency than SiGe. From a temperature difference of 480 to 980 °C, the SiGe material showed increase in maximum power and the heat absorbed also increased due to increase in temperature difference, which resulted in lower maximum efficiency for the SiGe material than that of the Bi_2Te_3 material. Maximum efficiency of 12.2% was observed at a temperature difference of 480 °C for all the leg geometries with the Bi_2Te_3 material. However, maximum efficiency of 5.1% was obtained for all the leg geometries with the SiGe material at a temperature difference of 980 °C.

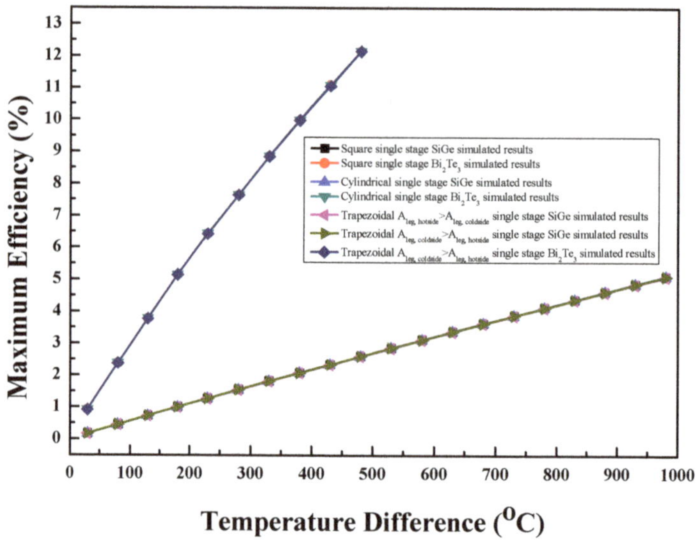

(**a**) Single stage arrangement

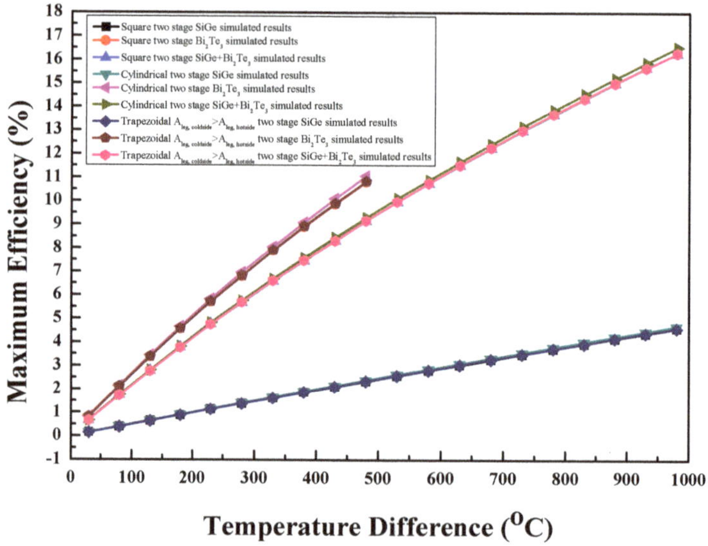

(**b**) Two-stage arrangement

Figure 8. *Cont.*

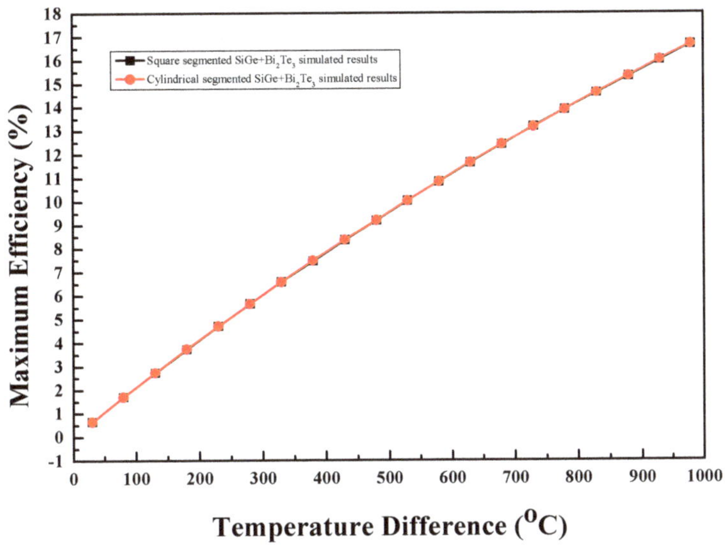

(**c**) Single stage segmented arrangement

Figure 8. Maximum efficiency for (**a**) single stage arrangement (**b**) two-stage arrangement, and (**c**) single stage segmented arrangement.

3.4.2. Two-Stage Arrangement

Figure 8b shows maximum efficiency variations with temperature difference for the two-stage arrangement of the thermoelectric module with various leg geometries and materials. For the same leg geometry, the Bi_2Te_3 material showed the highest values of maximum efficiency with temperature difference, compared to the SiGe+Bi_2Te_3 and SiGe materials but up to a temperature difference of 480 °C. However, from temperature difference of 480 °C, the SiGe+Bi_2Te_3 material showed increase in maximum efficiency up to the temperature difference corresponding to the optimum temperature condition. The SiGe material showed the lowest values of maximum efficiency until the temperature difference of 980 °C. Although Bi_2Te_3 as well as SiGe+Bi_2Te_3 materials are operated till a particular temperature difference because of the limitation of their melting point temperatures, they still showed comparatively higher values of maximum efficiencies than the SiGe material. For the same material, leg geometry had very less effect on maximum efficiency because of the same optimum voltage load and same internal resistance [1,6,10]. In the case of SiGe, Bi_2Te_3 and SiGe+Bi_2Te_3 materials, square prism, cylindrical, and trapezoidal legs showed almost the same maximum efficiency variation with temperature difference. At a temperature difference of 980 °C and for the SiGe material, the square prism legs and cylindrical legs showed maximum efficiency of 4.6% and the cylindrical legs showed maximum efficiency of 4.7%. For the Bi_2Te_3 material and temperature difference of 480 °C, the square prism legs and cylindrical legs showed maximum efficiency of 10.8% and the cylindrical legs showed maximum efficiency of 11.1%. In the case of the SiGe+Bi_2Te_3 material, the square prism legs showed maximum efficiency of 15% at a temperature difference of 880 °C and the cylindrical as well as trapezoidal legs showed maximum efficiencies of 14.6% and 14.4%, respectively, at a temperature difference of 830 °C. The maximum efficiency of the two-stage arrangement of the thermoelectric module with leg geometries and materials increased over the entire temperature difference range.

3.4.3. Single Stage Segmented Arrangement

The maximum efficiency of the single stage segmented arrangement of the thermoelectric module with two leg geometries and combination of SiGe+Bi_2Te_3 material increased with the temperature difference. The maximum efficiency variation with the temperature difference is shown in Figure 8c. Like maximum power, the leg geometries had no effect on the maximum efficiency of the single stage segmented arrangement of the thermoelectric module with two leg geometries and combination of the SiGe+Bi_2Te_3 material [1,6,10]. At a temperature difference of 730 °C, the square prism and cylindrical legs showed maximum efficiency of 13.2%.

3.5. *Maximum Stress*

The comparison of maximum stress for the single stage, two-stage, and single stage segmented arrangements of the thermoelectric module with various leg geometries and materials is presented in this section. Stress variation along the height of various thermoelectric legs of different materials and for various arrangements is also presented.

3.5.1. Single Stage Arrangement

The maximum stress variation with temperature difference for the single stage thermoelectric module with various leg geometries and various materials is shown in Figure 9a. For the same leg geometry, the Bi_2Te_3 material showed higher values of maximum stress compared to the SiGe material till a temperature difference of 480 °C; above that, the SiGe material showed increase in maximum stress with the highest value at a temperature difference of 980 °C due to the increase in temperatures. The coefficient of the thermal expansion of the Bi_2Te_3 material was higher than that of the SiGe material. Therefore, the thermal stress induced in the Bi_2Te_3 material was higher than that induced in the SiGe material at the same temperature difference condition. Further, for the same material, the cylindrical legs showed fewer maximum stress than the other two leg geometries [1,10] and square prism legs, trapezoidal legs with $A_{legs,\ coldside} > A_{legs,\ hotside}$, and $A_{legs,\ hotside} > A_{legs,\ coldside}$ showed almost equal values of maximum stress due to the same geometrical structure with sharp corner edges. Square prism and trapezoidal legs have sharp edges, which are absent in cylindrical legs; therefore, the latter show lower thermal stress. Square prism legs, trapezoidal legs with $A_{legs,\ coldside} > A_{legs,\ hotside}$ and trapezoidal legs with $A_{legs,\ hotside} > A_{legs,\ coldside}$ with the SiGe material showed average stress of approximately 39 MPa, 38 MPa, and 41 MPa, respectively. The intensity of stress was high near the hot junction plate [1,10,11] and in the case of trapezoidal legs with $A_{legs,\ coldside} > A_{legs,\ hotside}$, the area of legs exposed to the hot side plate was less compared to the area of legs exposed to the hot side plate by the square prism legs and trapezoidal legs with $A_{legs,\ hotside} > A_{legs,\ coldside}$. Therefore, the average stress induces in the trapezoidal legs with $A_{legs,\ coldside} > A_{legs,\ hotside}$ was lower than that of the square prism and trapezoidal legs with $A_{legs,\ hotside} > A_{legs,\ coldside}$ [11]. Based on the area of legs exposed to the hot junction plate and the average stress values, the square prism legs and trapezoidal legs with $A_{legs,\ coldside} > A_{legs,\ hotside}$ was preferred over the trapezoidal legs with $A_{legs,\ hotside} > A_{legs,\ coldside}$. For the SiGe material and at a temperature difference of 980 °C, the square prism legs and trapezoidal legs with $A_{legs,\ coldside} > A_{legs,\ hotside}$ and $A_{legs,\ hotside} > A_{legs,\ coldside}$ showed maximum stress of 0.96 GPa and the cylindrical legs showed maximum stress of 0.91 GPa. Similarly, for the Bi_2Te_3 material and at a temperature difference of 480 °C, the square prism legs and trapezoidal legs with $A_{legs,\ coldside} > A_{legs,\ hotside}$ showed a maximum stress value of 0.61 GPa, whereas the cylindrical legs showed maximum stress of 0.58 GPa. The maximum stress of the single stage arrangement of the thermoelectric module with all leg geometries and materials increased linearly with the temperature difference.

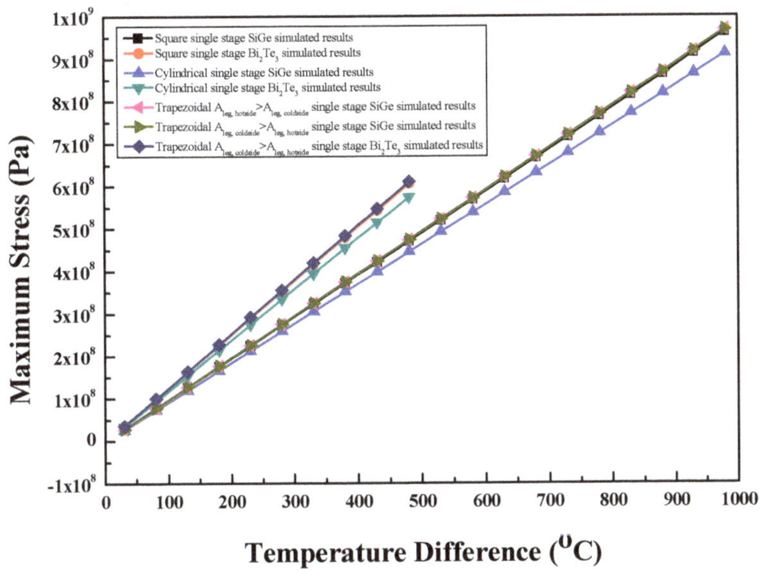

(**a**) Single stage arrangement

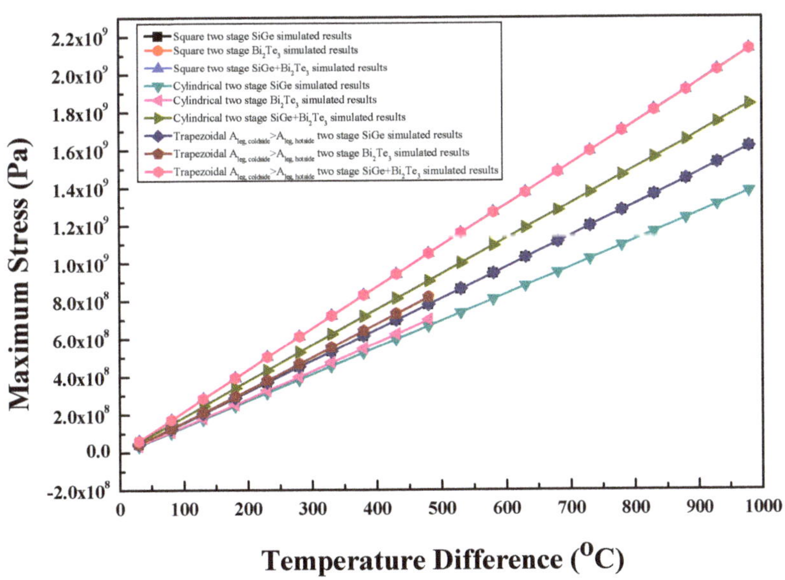

(**b**) Two-stage arrangement

Figure 9. *Cont.*

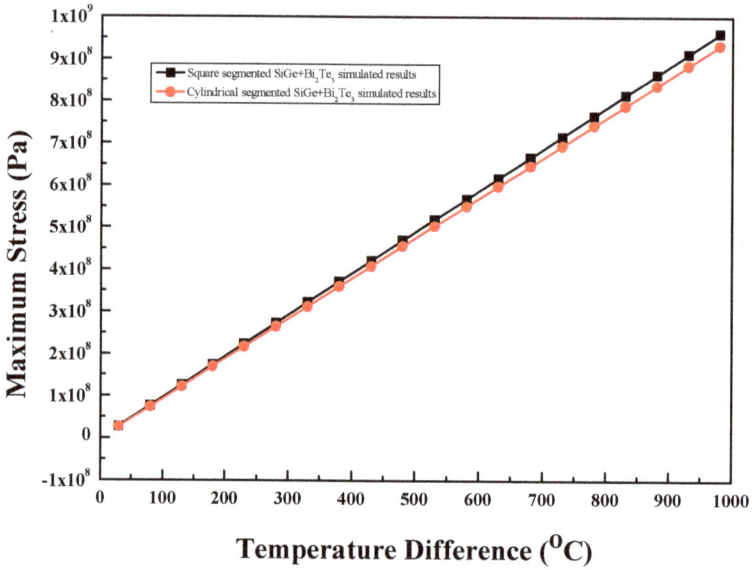

(c) Single stage segmented arrangement

Figure 9. Maximum stress for (**a**) single stage arrangement (**b**) two-stage arrangement, and (**c**) single stage segmented arrangement.

3.5.2. Two-Stage Arrangement

The maximum stress shows linear variation with the temperature difference for the two-stage thermoelectric module with various leg geometries and materials, as shown in Figure 9b. In the case of SiGe, Bi_2Te_3 and SiGe+Bi_2Te_3 materials, the square prism and trapezoidal legs showed almost same maximum stress variation with temperature difference, which were higher than the corresponding maximum stress values for the cylindrical legs [1,10]. The cylindrical legs have a smooth geometrical structure; hence, they presented lower maximum stress compared to the other two leg geometries. For the same leg geometry, the SiGe+Bi_2Te_3 material showed higher maximum stress, followed by the Bi_2Te_3 and SiGe materials. The Bi_2Te_3 material showed higher thermal stress than the SiGe material due to its higher coefficient of thermal expansion. The SiGe+Bi_2Te_3 material showed higher thermal stress than the Bi_2Te_3 material because when two different materials with different thermal properties are connected at higher temperature conditions, it results in higher stress generation [1]. For the SiGe material and at temperature difference of 980 °C, the square prism and trapezoidal legs showed maximum stress of 1.62 GPa and the cylindrical legs showed maximum stress of 1.38 GPa. Similarly, for the Bi_2Te_3 material and at a temperature difference of 480 °C, the square prism and trapezoidal legs showed maximum stress of 0.82 GPa and the cylindrical legs showed maximum stress of 0.7 GPa. For the SiGe+Bi_2Te_3 material, the square prism legs showed maximum stress of 1.91 GPa at a temperature difference of 880 °C and the cylindrical as well as trapezoidal legs showed maximum stress of 1.56 GPa and 1.81 GPa, respectively, at a temperature difference of 830 °C.

3.5.3. Single Stage Segmented Arrangement

Maximum stress varies linearly with temperature difference for the single stage segmented arrangement of the thermoelectric module with both the leg geometries and the SiGe+Bi_2Te_3 material, as shown in Figure 9c. The cylindrical leg geometry showed the lowest values of maximum stress, compared to the square prism legs [1,10] over the entire temperature difference range due to no sharp edges and a smooth geometrical structure. At a temperature difference of 730 °C and for the SiGe+Bi_2Te_3 material, the square prism legs showed maximum stress of 0.72 GPa and the cylindrical legs showed maximum stress of 0.69 GPa.

3.5.4. Stress Variation along Thermoelectric Leg Height

Stress variation along the selected locations of the thermoelectric module with various combinations of leg geometries, materials, and arrangements are discussed here. The selected centerline locations in the vertical direction for various arrangements of the thermoelectric module with square prism legs are shown in Figure 10a–c. In Figure 10a–c, the dotted line of locations on the thermoelectric legs shows the direction from the bottom of the thermoelectric legs to the top. Similar locations are selected in the cylindrical and trapezoidal leg geometries. The variation of stress for the selected locations were considered at the maximum operating temperature difference for each combination.

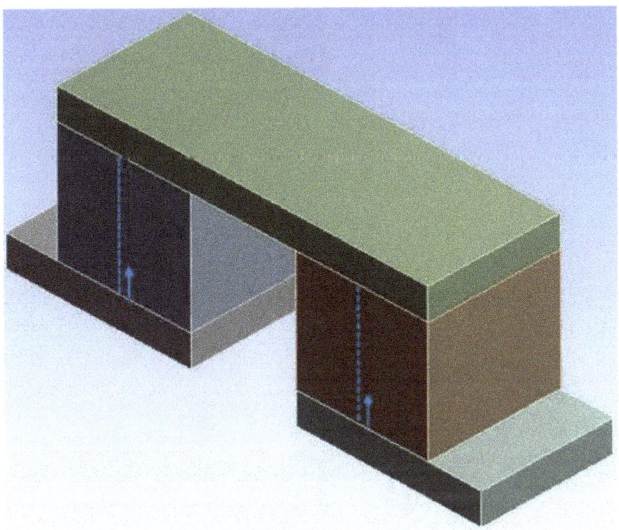

(**a**) Single stage square legs

Figure 10. *Cont.*

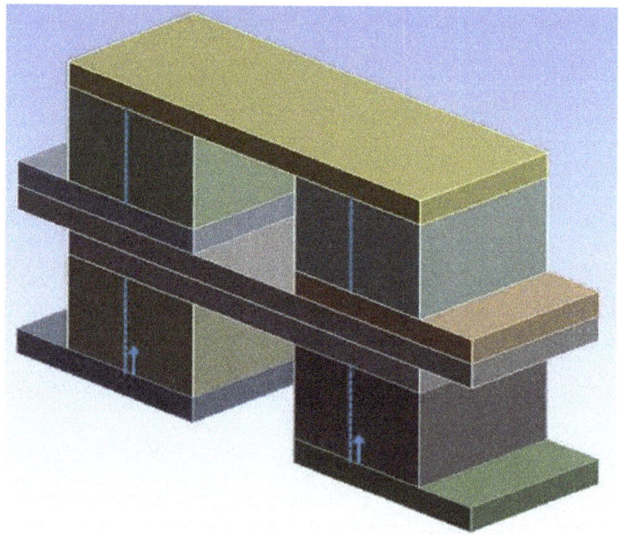

(**b**) Two-stage square legs

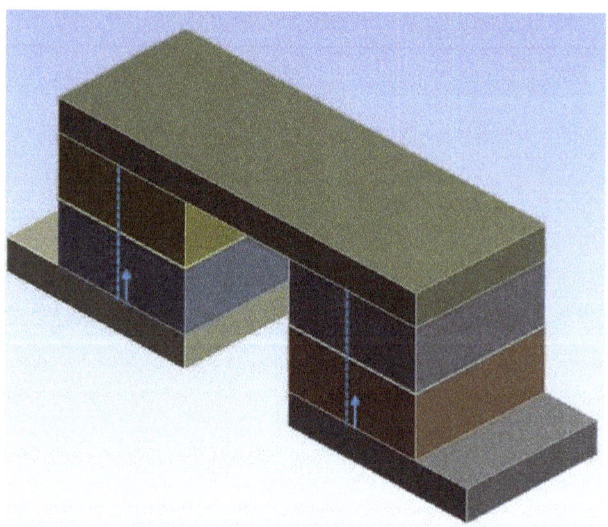

(**c**) Single stage segmented stage square legs

Figure 10. *Cont.*

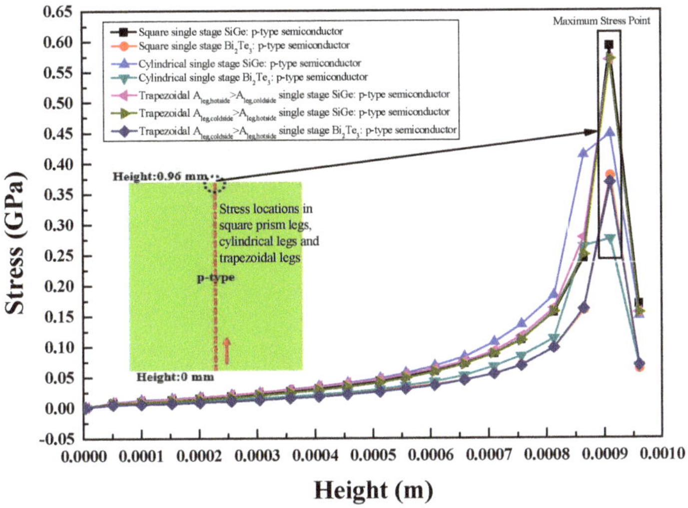

(**d**) p-type semiconductors in single stage arrangement

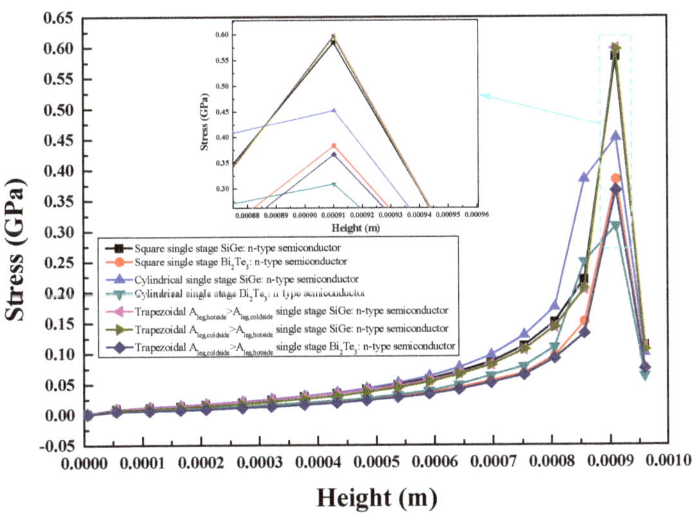

(**e**) n-type semiconductors in single stage arrangement

Figure 10. *Cont.*

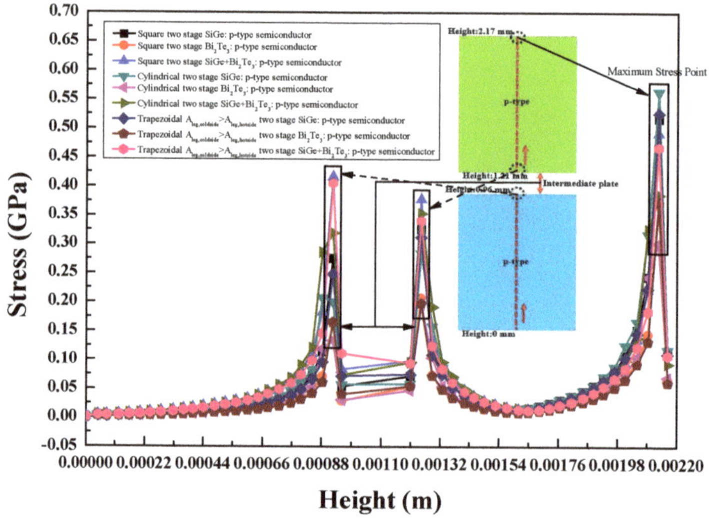

(**f**) p-type semiconductors in two-stage arrangement

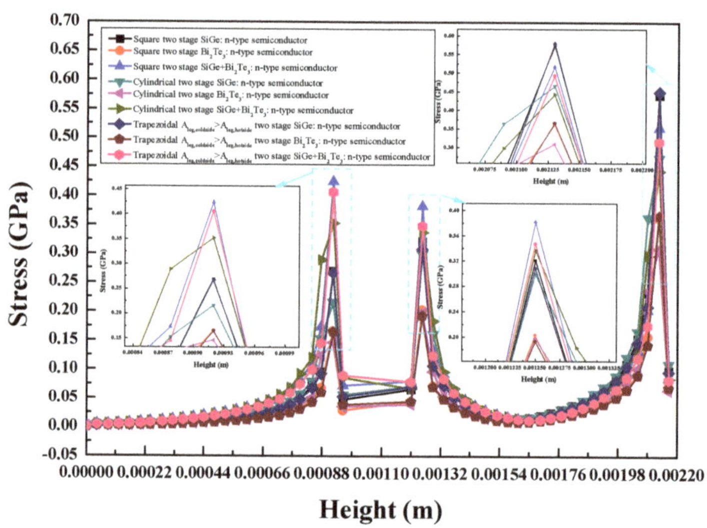

(**g**) n-type semiconductors in two-stage arrangement

Figure 10. *Cont.*

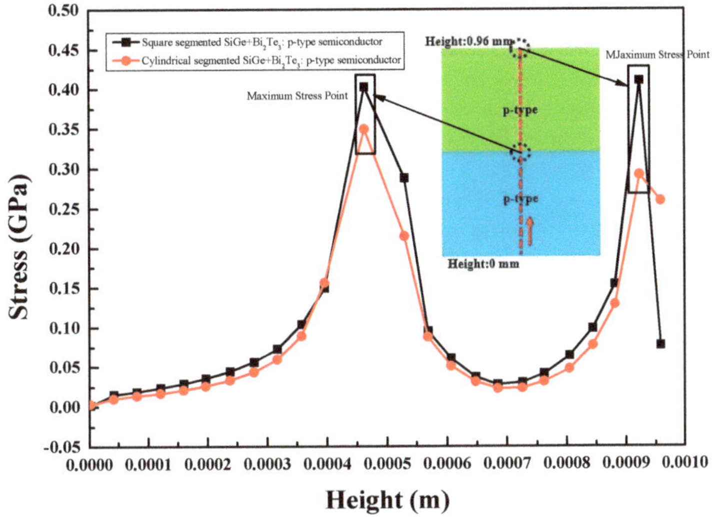

(**h**) p-type semiconductors in single stage segmented arrangement

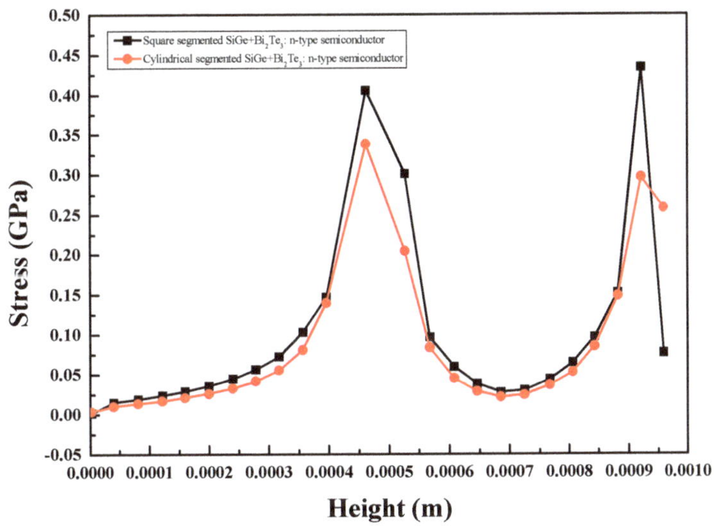

(**i**) n-type semiconductors in single stage segmented arrangement

Figure 10. Selected centerline locations in the vertical direction for (**a**) single stage square legs (**b**), two-stage square legs (**c**), single stage segmented stage square legs and variation of the stress along the selected centerline locations of (**d**) p type semiconductors in single stage arrangement, (**e**) n-type semiconductors in single stage arrangement, (**f**) p type semiconductors in two-stage arrangement, (**g**) n-type semiconductors in two-stage arrangement, (**h**) p type semiconductors in single stage segmented arrangement, and (**i**) n-type semiconductors in single stage segmented arrangement.

Single Stage Arrangement

For the single stage arrangement of the thermoelectric module, the variation of stress along the selected centerline locations of various thermoelectric legs with different materials is shown in Figure 10d,e. Figure 10d shows the variations of stress in different p-type semiconductor materials of various legs geometries and Figure 10e shows the variations of stress in different n-type semiconductor materials of various legs geometries. For each geometry, same p-type and n-type semiconductor material showed similar behavior with small variations in values due to same thermal properties [1]. However, the same p-type and n-type semiconductor material had different electrical properties. The intensity of stress was high at the intersection of the thermoelectric legs and the hot plate (top of the thermoelectric legs) for the single stage arrangement of the thermoelectric module with materials and leg geometries, as shown in Figure 10d,e because the materials of the thermoelectric legs and hot plate were different, and the top of the legs was exposed to a higher temperature. The stress reduced from the hot side of the thermoelectric legs to the cold side of the thermoelectric legs with the minimum value at the bottom of the leg near the cold junction plate because of reduction in temperature [1,10,11]. The materials of the thermoelectric legs and the cold side plate were also different, but the temperature at this location was low to generate lower stress. For the SiGe material, the square and trapezoidal legs showed higher stress compared to the cylindrical legs. Similarly, for the Bi_2Te_3 material, the cylindrical legs presented lower stress than the other leg geometries. The higher temperature difference showed higher thermal stress. Therefore, the Bi_2Te_3 material showed lower stress than the SiGe material for all three leg geometries because stress variation along the selected locations was presented at a temperature difference of 480 °C for the Bi_2Te_3 material and 980 °C for the SiGe material.

Two-Stage Arrangement

Figure 10f,g show the stress variation along the selected center line locations in the two-stage arrangement with various thermoelectric leg geometries and various materials. The variation of stress in various leg geometries with different p-type semiconductor materials is shown in Figure 10f. Figure 10g shows the stress variation in different leg geometries with various n-type semiconductor materials. In the two-stage arrangement of the thermoelectric module, the variation of stress was the same for the same p-type and n-type semiconductor material with a small variation in their values due to same thermal properties [1]. In the second stage of the two-stage arrangement, the variation trend of stress was similar as of the single stage arrangement case, the minimum value of stress was at the bottom of the leg near the cold junction, and increased towards the hot side of the legs. The intensity of stress for the second stage of the two-stage arrangement with various materials and leg geometries was high near the intersection of the thermoelectric legs of the second stage and the intermediate plate. A height of 0.00096 m to 0.001210 m points to the intermediate plate, and stress variation in the intermediate plate is not considered here, as this section deals only with stress variation in the thermoelectric legs. The variation trend of stress in the first stage of the two-stage arrangement is different compared to the second stage of the two-stage arrangement. In the first stage of the two-stage arrangement, higher values of stress occurred at two locations—the first was at the interconnection of the intermediate plate with the thermoelectric legs of the first stage, and the second one was at the interconnection of the thermoelectric legs of the first stage and the hot side plate. When materials with different properties are connected with each other at a higher temperature, it results in higher stress [1]. However, the values of stress in the thermoelectric legs near the hot side plate were higher than that near the intermediate plate because of a higher temperature of the hot side plate than that of the intermediate plate. As shown in Figure 10f,g, the minimum values of stress for the first stage of the two-stage arrangement occurred at the middle of the thermoelectric legs—at a location between 0.00154 m to 0.00176 m. For the same leg geometry, the SiGe+Bi_2Te_3 and SiGe materials showed higher values of stress than the Bi_2Te_3 material because the stress variation of the latter for the selected locations was at a temperature difference of 480 °C, which is lower than the temperature difference at which stress variation were presented for the SiGe+Bi_2Te_3 and SiGe materials. In the case of the second stage of the two-stage arrangement, the square prism and trapezoidal legs showed higher stress than the cylindrical legs for the same material, while in the case of the first stage of the two-stage arrangement, the cylindrical legs showed higher stress than the square prism and trapezoidal legs for the same material. Thus, for the selected centerline locations on the thermoelectric legs, the intensity of stress was high in the cylindrical legs for the first stage of the two-stage arrangement, but it was low in the cylindrical legs for the second stage of the two-stage arrangement. As the stress effect is considered for the whole leg geometry of the two-stage arrangement, the cylindrical legs were found to have lower stress than the other two leg geometries.

Single Stage Segmented Arrangement

For the single stage segmented arrangement of the thermoelectric module with two leg geometries and the SiGe+Bi_2Te_3 material, the variation of stress along the selected center line locations in the p-type and n-type semiconductors is presented in Figure 10h,i. For the same leg geometry, the same behavior of the stress variation was observed for the p-type and n-type semiconductors [1] with the SiGe+Bi_2Te_3 material due to same thermal properties. The maximum values of stress occur at the interconnection of both the materials (at the middle height) as well as at the interconnection of the thermoelectric legs and the hot side plate [1] because of the interconnections of different materials at both the locations. The hot side plate and thermoelectric legs with the different materials were interconnected at a higher temperature with a higher stress at the interconnection. In addition, two different materials with different thermal properties are interconnected at the middle height of the thermoelectric legs at a higher temperature with higher stress at the middle height of the legs. The maximum stress at the middle height and the hot side of the thermoelectric legs are almost equal. Minimum stress was observed at the cold side of the thermoelectric legs, as shown in Figure 10h,i, because the temperature in this location was low, resulting in lower stress. The square legs showed higher values of stress compared to the cylindrical legs due to its sharp edges.

Figure 11 shows the stress distribution contours of the thermoelectric legs of different configurations in the thermoelectric module. Figure 11 supports the graphical presentation and discussion of stress described for various configurations of the thermoelectric module in Section 3.5. Figure 11 also describes that the intensity of stress is high at the intersection of the hot side plate and the thermoelectric legs for all configurations of the thermoelectric module. In addition, for various leg geometries and materials, the intensity of stress was high at the interconnection and the intermediate plate of thermoelectric legs for the two-stage arrangement and at the middle height of thermoelectric legs for the single stage segmented arrangement. The strain generated in various configurations of the thermoelectric module, as presented in Equation (21), is a replica of the stress. Namely, the behavior of the strain is same as the stress for various configurations of the thermoelectric module. Therefore, detailed discussions on this strain are not considered further in this study.

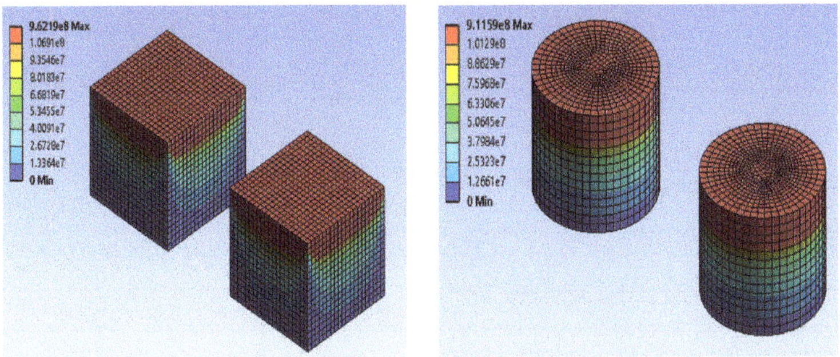

(**a**) Single stage square prism legs (**b**) Single stage cylindrical legs

Figure 11. *Cont.*

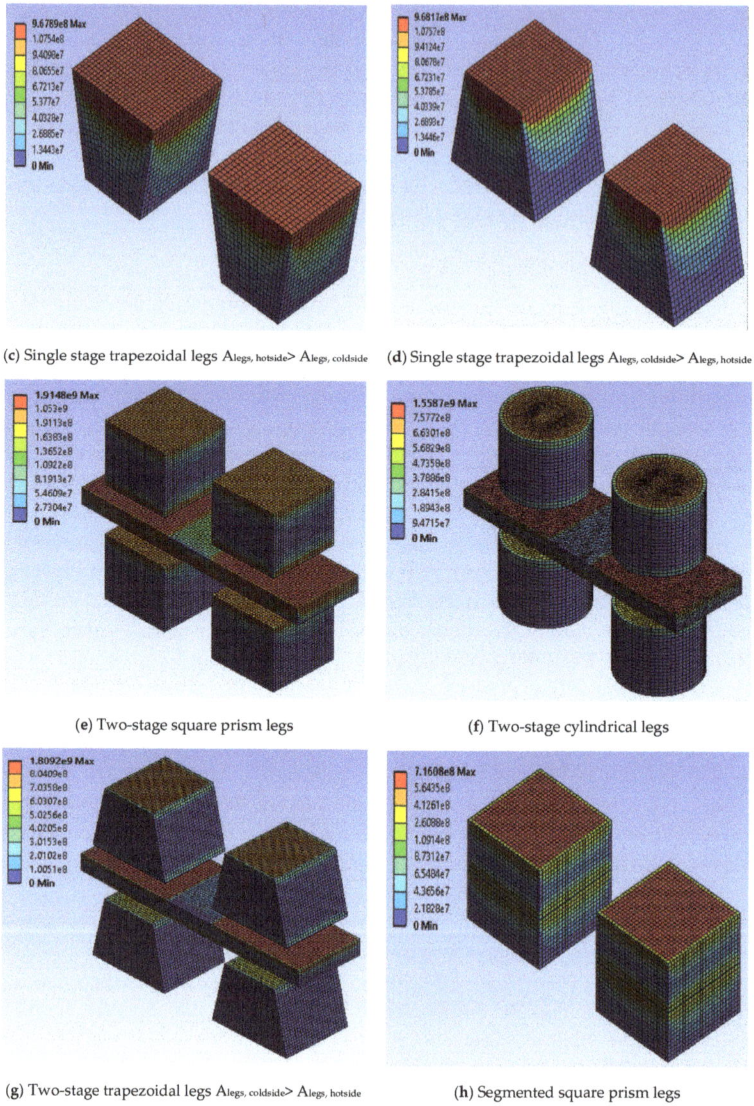

(**c**) Single stage trapezoidal legs $A_{legs, hotside} > A_{legs, coldside}$ (**d**) Single stage trapezoidal legs $A_{legs, coldside} > A_{legs, hotside}$

(**e**) Two-stage square prism legs (**f**) Two-stage cylindrical legs

(**g**) Two-stage trapezoidal legs $A_{legs, coldside} > A_{legs, hotside}$ (**h**) Segmented square prism legs

Figure 11. *Cont.*

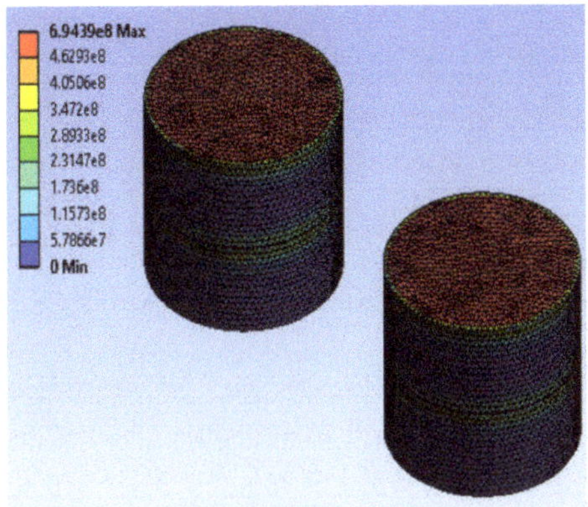

(i) Segmented cylindrical legs

Figure 11. Maximum thermal stress contours in (**a**) single stage square prism legs, (**b**) single stage cylindrical legs, (**c**) single stage trapezoidal legs $A_{legs,\ hotside} > A_{legs,\ coldside}$, (**d**) single stage trapezoidal legs $A_{legs,\ coldside} > A_{legs,\ hotside}$, (**e**) two-stage square prism legs, (**f**) two-stage cylindrical legs, (**g**) two-stage trapezoidal legs $A_{legs,\ coldside} > A_{legs,\ hotside}$, (**h**) single stage segmented square prism legs, and (**i**) single stage segmented cylindrical legs.

3.6. Selection of Optimum Configuration for the Thermoelectric Module

A comparison of the thermoelectric module with combinations of leg geometry, material, segmentation, and two-stage arrangement based on maximum power, maximum efficiency, and maximum stress at the corresponding maximum operating temperature difference is discussed in this section. Optimum configurations of the thermoelectric module with the combinations of leg geometry, material, segmentation, and two-stage arrangement were selected based on three performance parameters of maximum power, maximum efficiency, and maximum stress. Therefore, higher values of maximum power and maximum efficiency, and lower values of maximum stress of all combinations for the thermoelectric module were suggested as optimum configuration.

A graphical presentation of the comparison of various configurations of the thermoelectric module based on maximum temperature difference, maximum efficiency, maximum power, and maximum stress is shown in Figure 12. In Figure 12 and Table 4, numbers 1 to 18 on the abscissa presents various combinations of the thermoelectric module and the thermoelectric module constructed with the segmented arrangement, cylindrical legs, and combination of the SiGe+Bi_2Te_3 material show the optimum values of maximum power, maximum efficiency, and maximum stress. The single stage segmented arrangement of the thermoelectric module with the cylindrical leg geometry and the SiGe+Bi_2Te_3 material showed a combination of higher maximum power, higher maximum efficiency, and lower maximum thermal stress [35]. Therefore, based on the overall effect of power, efficiency, and stress, a thermoelectric module constructed with the segmented arrangement, cylindrical legs, and combination of the SiGe+Bi_2Te_3 material is suggested as the optimum configuration of the thermoelectric module.

The comparison of all 18 configurations of the thermoelectric module was carried out based on computational time/single case of simulation, which is presented in Table 5. From the table, it can be concluded that the two-stage arrangement has a higher computational time, followed by the segmented arrangement and single stage arrangement, respectively. The computational time for the selected optimum configuration of the thermoelectric module constructed with the segmented arrangement, cylindrical legs, and combination of SiGe+Bi_2Te_3 materials is 2100 s.

Table 4. Comparison of various combinations of thermoelectric module based on maximum temperature difference, maximum power, maximum efficiency, and maximum stress.

Combination	Arrangements	Material	Leg Geometry	Maximum Temperature Difference (°C)	Maximum Power (W)	Maximum Efficiency (%)	Maximum Stress (GPa)
1	Single stage	SiGe	Square	980	0.6438	5.0973	0.962
2	Single stage	SiGe	Cylindrical	980	0.6472	5.1058	0.912
3	Single stage	SiGe	Trapezoidal ($A_{leg,hotside} > A_{leg,coldside}$)	980	0.6275	5.0910	0.968
4	Single stage	SiGe	Trapezoidal ($A_{leg,coldside} > A_{leg,hotside}$)	980	0.6280	5.0964	0.968
5	Single stage	Bi_2Te_3	Square	480	0.3132	12.156	0.607
6	Single stage	Bi_2Te_3	Cylindrical	480	0.3145	12.163	0.575
7	Single stage	Bi_2Te_3	Trapezoidal	480	0.3056	12.156	0.611
8	Two-stage	SiGe	Square	980	0.2590	4.5902	1.62
9	Two-stage	SiGe	Cylindrical	980	0.2710	4.6881	1.38
10	Two-stage	SiGe	Trapezoidal	980	0.2537	4.5982	1.62
11	Two-stage	Bi_2Te_3	Square	480	0.1192	10.811	0.822
12	Two-stage	Bi_2Te_3	Cylindrical	480	0.1258	11.060	0.701
13	Two-stage	Bi_2Te_3	Trapezoidal	480	0.1169	10.833	0.821
14	Two-stage	SiGe+Bi_2Te_3	Square	880	0.4627	15.011	1.91
15	Two-stage	SiGe+Bi_2Te_3	Cylindrical	830	0.4302	14.551	1.56
16	Two-stage	SiGe+Bi_2Te_3	Trapezoidal	830	0.4053	14.365	1.81
17	Segmented	SiGe+Bi_2Te_3	Square	730	0.7293	13.172	0.716
18	Segmented	SiGe+Bi_2Te_3	Cylindrical	730	0.7325	13.165	0.694

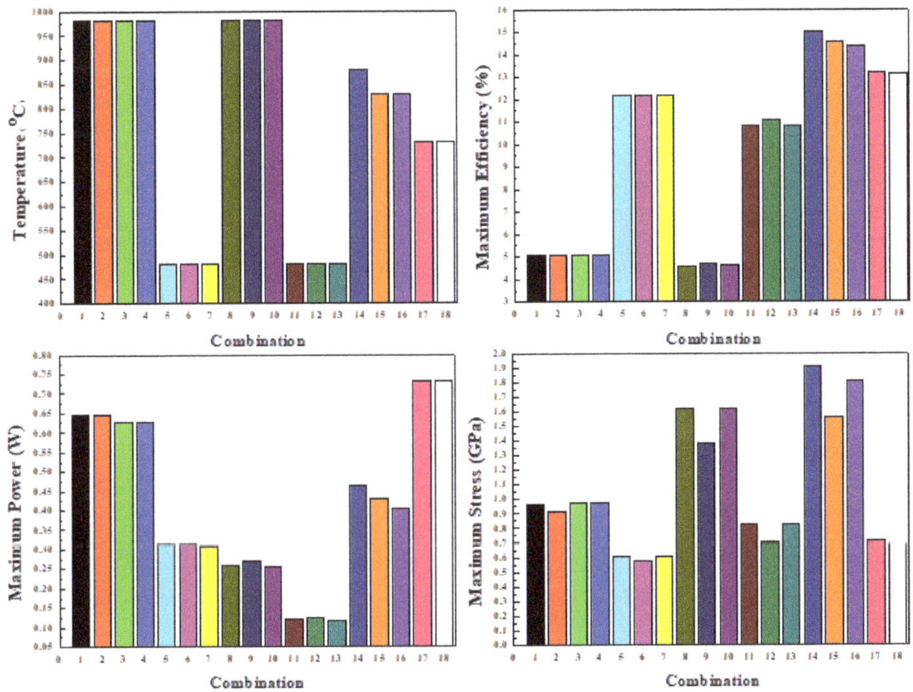

Figure 12. Comparison of various configurations of thermoelectric module based on maximum temperature difference, maximum efficiency, maximum power, and maximum stress.

Table 5. Comparison of various combinations of thermoelectric module based on computational time/single case of simulation.

Combination	Arrangements	Material	Leg Geometry	Computational Time (s)/Single Case
1	Single stage	SiGe	Square	1200
2	Single stage	SiGe	Cylindrical	1560
3	Single stage	SiGe	Trapezoidal ($A_{leg,hotside}$> $A_{leg,coldside}$)	1500
4	Single stage	SiGe	Trapezoidal ($A_{leg,coldside}$> $A_{leg,hotside}$)	1440
5	Single stage	Bi_2Te_3	Square	1260
6	Single stage	Bi_2Te_3	Cylindrical	1380
7	Single stage	Bi_2Te_3	Trapezoidal	1320
8	Two-stage	SiGe	Square	3300
9	Two-stage	SiGe	Cylindrical	6000
10	Two-stage	SiGe	Trapezoidal	4500
11	Two-stage	Bi_2Te_3	Square	3540
12	Two-stage	Bi_2Te_3	Cylindrical	5940
13	Two-stage	Bi_2Te_3	Trapezoidal	8940
14	Two-stage	SiGe+Bi_2Te_3	Square	3780
15	Two-stage	SiGe+Bi_2Te_3	Cylindrical	6240
16	Two-stage	SiGe+Bi_2Te_3	Trapezoidal	7140
17	Segmented	SiGe+Bi_2Te_3	Square	1920
18	Segmented	SiGe+Bi_2Te_3	Cylindrical	2100

3.7. *Effect of Soldering Layers*

Table 6 shows the effect of soldering layers on maximum power, maximum efficiency, and maximum thermal stress of the thermoelectric module. For the same range of operating temperature difference, the Bi_2Te_3 material showed higher thermal stress compared to the SiGe material. Hence, soldering layers were only provided on the thermoelectric legs constructed with the Bi_2Te_3 material [4]. In order to investigate the effect of the soldering layers, the single stage arrangement of the thermoelectric module with the square prism legs and the Bi_2Te_3 material as well as the two-stage arrangement of the thermoelectric module with the square prism legs and the Bi_2Te_3 material are compared with and without soldering layers. In addition, the two-stage arrangement with square prism legs with the SiGe+Bi_2Te_3 material is also compared with and without soldering layers. In the case of the SiGe+Bi_2Te_3 material with the soldering layers, the second stage of the Bi_2Te_3 material is provided with soldering layers. Addition of the soldering layers reduces the thermal stress compared to the corresponding same configuration without soldering layers. The soldering layers experience most of the deformation compared to the other parts of the thermoelectric module and sometimes undergo plastic deformation. Hence, excessive thermal stress on the thermoelectric legs are reduced because the larger effect of the thermal stress is absorbed by the soldering layers. The thickness of the soldering layers is very small; hence, the resistance offered is a small contribution to the total effective resistance of the thermoelectric module. Therefore, the soldering layers have no significant effect on maximum power or maximum efficiency. The melting point temperature of the soldering layers is 185 °C; hence, various configurations of the thermoelectric module with soldering layers are operated at a maximum hot junction temperature below 185 °C. Therefore, a comparison of configurations of the thermoelectric module provided with and without soldering layers was done for temperature difference of 150 °C. Maximum stress for the various configurations with the soldering layers were considered only for the thermoelectric legs, not the entire thermoelectric module, as is shown in Table 6.

Table 6. Effect of soldering layers on various performance parameters.

Arrangements	Material	Maximum Power-Theoretical (W)	Maximum Power-Numerical (W)	Maximum Efficiency-Numerical (%)	Maximum Stress-Numerical (GPa)
Single stage	Bi_2Te_3	0.0233	0.0230	3.7805	0.163
Single stage (SL)	Bi_2Te_3	0.0279	0.0270	3.7486	0.022
Two-stage	Bi_2Te_3	0.0091	0.0089	3.3790	0.212
Two-stage (SL)	Bi_2Te_3	0.0104	0.0101	3.2696	0.024
Two-stage	SiGe+Bi_2Te_3	0.0111	0.0109	2.7764	0.283
Two-stage (SL)	SiGe+Bi_2Te_3	0.0120	0.0115	2.6337	0.151

SL means soldering layer.

4. Conclusions

The behavior of maximum power, maximum efficiency, and maximum thermal stress was investigated numerically and theoretically for a thermoelectric module with various combinations of leg geometry, material, segmentation, and two-stage arrangement. From the conducted numerical and theoretical analysis, the following critical observations are summarized:

(a) For all configurations of the thermoelectric module, the numerically predicted values of maximum power and maximum efficiency were validated at ±5% error and the numerically predicted values of maximum stress were validated at ±7% error with their theoretical values, respectively.

(b) For the same arrangement and the same material, leg geometries with the same volume and same base area have a negligible effect on maximum power and maximum efficiency. However, the cylindrical legs showed lower values of maximum stress than that of the square prism and trapezoidal legs.

(c) For the same arrangement and leg geometry, the SiGe+Bi_2Te_3 material for the thermoelectric module showed higher maximum power and maximum efficiency than the other individual materials. The thermal stress of the thermoelectric module with the SiGe+Bi_2Te_3 material was higher than those of the thermoelectric module with the SiGe and Bi_2Te_3 materials. In addition, the intensity of stress in the thermoelectric module with the SiGe+Bi_2Te_3 material could be reduced by using soldering layers without affecting maximum power and maximum efficiency.

(d) The segmented arrangement of the thermoelectric module showed a higher maximum power and maximum efficiency and lower maximum stress in all combinations of the thermoelectric module. Therefore, the segmented arrangement of the thermoelectric module with cylindrical leg geometry and a combination of SiGe+Bi_2Te_3 materials was selected as the optimum configuration for the thermoelectric module with maximum power of 0.73 W, maximum efficiency of 13.2%, and maximum thermal stress of 0.69 GPa.

Author Contributions: Conceptualization, M.-Y.L.; J.-H.S., and K.S.G.; methodology, M.-Y.L., J.-H.S., H.-S.L., and K.S.G.; software, M.-Y.L., J.-H.S., and K.S.G.; validation, M.-Y.L., J.-H.S., and K.S.G.; formal analysis, M.-Y.L., J.-H.S., H.-S.L., and K.S.G.; investigation, M.-Y.L., J.-H.S., H.-S.L., and K.S.G.; resources, M.-Y.L., J.-H.S., and K.S.G.; data curation, M.-Y.L.; writing—original draft preparation, M.-Y.L., J.-H.S., and K.S.G.; writing—review and editing, M.-Y.L., J.-H.S., and K.S.G.; visualization, M.-Y.L.; supervision, M.-Y.L.; project administration, M.-Y.L.; funding acquisition, M.-Y.L. All authors have read and agreed to the published version of the manuscript.

Funding: This research received no external funding.

Acknowledgments: This work was supported by the Dong-A University research fund.

Conflicts of Interest: The authors declare no conflict of interest.

Nomenclature

A	Leg cross-sectional area (m^2)
E	Yong modulus (GPa)
$\vec{E}$	Electric field intensity (N/C or V/m)
H_a	Heat absorbed (W)
$I_{theoretical,opt}$	Theoretical optimum current (A)
$I_{numerical,opt}$	Numerical optimum current (A)
$\vec{J}$	Electric current intensity (A/ m^2)
K	Curvature (m^{-1})
k_1	Thermal conductivity of segment 1 (W/m °C)
k_2	Thermal conductivity of segment 2 (W/m °C)
L	Leg length (m)
$P_{theoretical,max}$	Theoretical maximum power (W)
$P_{numerical,max}$	Numerical maximum power (W)
p	Peltier coefficient (V)
R	Internal resistance (Ω)
$R_{equivalent}$	Equivalent resistance (Ω)
R_L	External load resistance (Ω)
T_H	Hot junction temperature (°C)
T_C	Cold junction temperature (°C)
$\overline{T}$	Average temperature (°C)
ΔT	Temperature difference (°C)
∇T	Temperature gradient
$V_{theoretical,opt}$	Theoretical optimum voltage (V)
$V_{numerical,opt}$	Numerical optimum voltage (V)
$Z\overline{T}$	Figure of Merit
Z_c	Coordinate (m)
$Z_{equivalent}\overline{T}$	Equivalent Figure of Merit
α	Seebeck coefficient (V/K)
$\eta_{theoretical,max}$	Theoretical maximum efficiency

$\eta_{numerical,max}$	Numerical maximum efficiency
ρ_p	Electrical resistivity of p-type (Ω m)
ρ_n	Electrical resistivity of n-type (Ω m)
σ_p	Electrical conductivity of p-type (Ω^{-1} m^{-1})
σ_n	Electrical conductivity of n-type (Ω^{-1} m^{-1})
$\alpha_{equivalent}$	Equivalent Seebeck coefficient (V/K)
α_1	Seebeck coefficient of segment 1 (V/K)
α_2	Seebeck coefficient of segment 2 (V/K)
ρ_1	Electrical resistivity of segment 1 (Ω m)
ρ_2	Electrical resistivity of segment 2 (Ω m)
σ_v	von-Mises stress (GPa)
σ_T *and* σ_{ij}	Thermal stress (i, j = 1,2,3) (GPa)
σ_{xx}	Normal stress in longitudinal direction (GPa)
α_T	Coefficient of thermal expansion (K^{-1})
ε	Strain
ε_n	Normal strain
v	Poisson's ratio
$\nabla\varnothing$	Electrical potential gradient

References

1. Erturun, U.; Erermis, K.; Mossi, K. Effect of various leg geometries on thermo-mechanical and power generation performance of thermoelectric devices. *Appl. Therm. Eng.* **2014**, *73*, 128–141. [CrossRef]
2. Shen, Z.G.; Liu, X.; Chen, S.; Wu, S.Y.; Xiao, L.; Chen, Z.X. Theoretical analysis on a segmented annular thermoelectric generator. *Energy* **2018**, *157*, 297–313. [CrossRef]
3. Jin, Z.H. Thermal stresses in a multilayered thin film thermoelectric structure. *Microelectron. Reliab.* **2014**, *54*, 1363–1368. [CrossRef]
4. Ma, Q.; Fang, H.; Zhang, M. Theoretical analysis and design optimization of thermoelectric generator. *Appl. Therm. Eng.* **2017**, *127*, 758–764. [CrossRef]
5. Patil, M.S.; Seo, J.H.; Lee, M.Y. Numerical Study on Geometric Parameter effects of Power Generation Performances for Segmented Thermoelectric Generator. *Int. J. Air-Cond. Refrig.* **2018**, *26*, 1850004. [CrossRef]
6. Erturun, U.; Erermis, K.; Mossi, K. Influence of leg sizing and spacing on power generation and thermal stresses of thermoelectric devices. *Appl. Energy* **2015**, *159*, 19–27. [CrossRef]
7. Wu, Y.; Ming, T.; Li, X.; Pan, T.; Peng, K.; Luo, X. Numerical simulations on the temperature gradient and thermal stress of a thermoelectric power generator. *Energy Conv. Manag.* **2014**, *88*, 915–927. [CrossRef]
8. Erturun, U.; Mossi, K. Thermoelectric devices with rotated and coaxial leg configurations: Numerical analysis of performance. *Appl. Therm. Eng.* **2015**, *85*, 304–312. [CrossRef]
9. Jia, X.; Gao, Y. Optimal design of a novel thermoelectric generator with linear-shaped structure under different operating temperature conditions. *Appl. Therm. Eng.* **2015**, *78*, 533–542. [CrossRef]
10. Yilbas, B.S.; Akhtar, S.S.; Sahin, A.Z. Thermal and stress analyses in thermoelectric generator with tapered and rectangular pin configurations. *Energy* **2016**, *114*, 52–63. [CrossRef]
11. Al-Merbati, A.S.; Yilbas, B.S.; Sahin, A.Z. Thermodynamics and thermal stress analysis of thermoelectric power generator: Influence of pin geometry on device performance. *Appl. Therm. Eng.* **2013**, *50*, 683–692. [CrossRef]
12. Ali, H.; Sahin, A.Z.; Yilbas, B.S. Thermodynamic analysis of a thermoelectric power generator in relation to geometric configuration device pins. *Energy Conv. Manag.* **2014**, *78*, 634–640. [CrossRef]
13. Erturun, U.; Mossi, K. A feasibility investigation on improving structural integrity of thermoelectric modules with varying geometry. In Proceedings of the ASME 2012 Conference on Smart Materials, Adaptive Structures and Intelligent Systems, Stone Mountain, GA, USA, 19–21 September 2012; American Society of Mechanical Engineers: Stone Mountain, GA, USA; pp. 939–945.
14. Yilbas, B.S.; Ali, H. Thermoelectric generator performance analysis: Influence of pin tapering on the first and second law efficiencies. *Energy Conv. Manag.* **2015**, *100*, 138–146. [CrossRef]
15. Gao, J.L.; Du, Q.G.; Zhang, X.D.; Jiang, X.Q. Thermal stress analysis and structure parameter selection for a Bi2Te3-based thermoelectric module. *J. Electron. Mater.* **2011**, *40*, 884–888. [CrossRef]

16. Zhang, L.; Tosho, T.; Okinaka, N.; Akiyama, T. Design of cascaded oxide thermoelectric generator. *Mater. Trans.* **2008**, *49*, 1675–1680. [CrossRef]
17. Kanimba, E.; Pearson, M.; Sharp, J.; Stokes, D.; Priya, S.; Tian, Z. A modeling comparison between a two-stage and three-stage cascaded thermoelectric generator. *J. Power Sources* **2017**, *365*, 266–272. [CrossRef]
18. Wilbrecht, S.; Beitelschmidt, M. The Potential of a Cascaded TEG System for Waste Heat Usage in Railway Vehicles. *J. Electron. Mater.* **2018**, *47*, 3358–3369. [CrossRef]
19. Weng, C.C.; Lin, M.C.; Huang, M.J. A waste cold recovery from the exhausted cryogenic nitrogen by using thermoelectric power generator. *Energy* **2016**, *103*, 385–396. [CrossRef]
20. Chen, G.; Mu, Y.; Zhai, P.; Li, G.; Zhang, Q. An investigation on the coupled thermal–mechanical–electrical response of automobile thermoelectric materials and devices. *J. Electron. Mater.* **2013**, *42*, 1762–1770. [CrossRef]
21. Kaibe, H.; Aoyama, I.; Mukoujima, M.; Kanda, T.; Fujimoto, S.; Kurosawa, T.; Ishimabushi, H.; Ishida, K.; Rauscher, L.; Hata, Y.; et al. Development of thermoelectric generating stacked modules aiming for 15% of conversion efficiency. In Proceedings of the ICT 2005. 24th International Conference on Thermoelectrics, Clemson, SC, USA, 19–23 June 2005; pp. 242–247.
22. Vargas-Almeida, A.; Olivares-Robles, M.A.; Camacho-Medina, P. Thermoelectric system in different thermal and electrical configurations: Its impact in the figure of merit. *Entropy* **2013**, *15*, 2162–2180. [CrossRef]
23. Ibrahim, A.; Rahnamayan, S.; Martin, M.V.; Yilbas, B. Multi-objective thermal analysis of a thermoelectric device: Influence of geometric features on device characteristics. *Energy* **2014**, *77*, 305–317. [CrossRef]
24. Turenne, S.; Clin, T.; Vasilevskiy, D.; Masut, R.A. Finite element thermomechanical modeling of large area thermoelectric generators based on bismuth telluride alloys. *J. Electron. Mater.* **2010**, *39*, 1926–1933. [CrossRef]
25. Garud, K.S.; Seo, J.H.; Patil, M.S.; Bang, Y.M.; Pyo, Y.D.; Cho, C.P.; Lee, M.Y. Thermal-Electrical-Structural Performances of Hot Heat Exchanger with Different Internal Fins of Thermoelectric Generator for Low Power Generation Application. *J. Therm. Anal. Calorim.* **2020**. [CrossRef]
26. Jones, M.H.; Jones, S.H. *The General Properties of Si, Ge, SiGe, SiO2 and Si3N4*; Virginia Semiconductor: Fredericksburg, VA, USA, 2002.
27. Bennis, A.; Leinenbach, C.; Raudzis, C.; Müller-Fiedler, R.; Kronmüller, S. Measurement technique for elastic and mechanical properties of polycrystalline silicon-germanium films using surface acoustic waves and projection masks. *arXiv* **2007**, arXiv:0711.3305.
28. Seo, J.H.; Patil, M.S.; Cho, C.P.; Lee, M.Y. Heat transfer characteristics of the integrated heating system for cabin and battery of an electric vehicle under cold weather conditions. *Int. J. Heat Mass Transf.* **2018**, *117*, 80–94. [CrossRef]
29. Garud, K.S.; Seo, J.H.; Cho, C.P.; Lee, M.Y. Artificial Neural Network and Adaptive Neuro-Fuzzy Interface System Modelling to Predict Thermal Performances of Thermoelectric Generator for Waste Heat Recovery. *Symmetry* **2020**, *12*, 259. [CrossRef]
30. Al-Merbati, A.S.; Yilbas, B.S.; Sahin, A.Z. A model study for cyclic thermal loading and thermal performance of a thermoelectric generator. *Int. J. Energy Res.* **2014**, *38*, 1351–1360. [CrossRef]
31. Elzalik, M.; Rezk, H.; Mostafa, R.; Thomas, J.; Shehata, E.G. An experimental investigation on electrical performance and characterization of thermoelectric generator. *Int. J. Energy Res.* **2020**, *44*, 128–143. [CrossRef]
32. Baheta, A.T.; Looi, K.K.; Oumer, A.N.; Habib, K. Thermoelectric Air-Conditioning System: Building Applications and Enhancement Techniques. *Int. J. Air-Cond. Refrig.* **2019**, *27*, 1930002. [CrossRef]
33. Luo, Y.; Kim, C.N. Effects of the cross-sectional area ratios and contact resistance on the performance of a cascaded thermoelectric generator. *Int. J. Energy Res.* **2019**, *43*, 2172–2187. [CrossRef]
34. Tritt, T.M. Thermoelectric phenomena, materials, and applications. *Ann. Rev. Mater. Res.* **2011**, *41*, 433–448. [CrossRef]
35. Ali, H.; Yilbas, B.S. Configuration of segmented leg for the enhanced performance of segmented thermoelectric generator. *Int. J. Energy Res.* **2017**, *41*, 274–288. [CrossRef]

symmetry

MDPI

Article

Artificial Neural Network and Adaptive Neuro-Fuzzy Interface System Modelling to Predict Thermal Performances of Thermoelectric Generator for Waste Heat Recovery

Kunal Sandip Garud [1], Jae-Hyeong Seo [1], Chong-Pyo Cho [2] and Moo-Yeon Lee [1,*]

1 School of Mechanical Engineering, Dong-A University, 37 Nakdong-Daero 550, Saha-gu, Busan 49315, Korea; kunalgarud1@gmail.com (K.S.G.); cheonchw@naver.com (J.-H.S.)

2 Energy Saving Technologies Laboratory, Korea Institute of Energy Research,152 Gajeong-ro, Yuseong-gu, Daejeon 34129, Korea; cnsever@kier.re.kr

* Correspondence: mylee@dau.ac.kr; Tel.: +82-51-200-7659

Received: 16 January 2020; Accepted: 5 February 2020; Published: 8 February 2020

check for updates

Abstract: The present study elaborates the suitability of the artificial neural network (ANN) and adaptive neuro-fuzzy interface system (ANFIS) to predict the thermal performances of the thermoelectric generator system for waste heat recovery. Six ANN models and seven ANFIS models are formulated by considering hot gas temperatures and voltage load conditions as the inputs to predict current, power, and thermal efficiency of the thermoelectric generator system for waste heat recovery. The ANN model with the back-propagation algorithm, the Levenberg–Marquardt variant, Tan-Sigmoidal transfer function and 25 number of hidden neurons is found to be an optimum model to accurately predict current, power and thermal efficiency. For current, power and thermal efficiency, the ANFIS model with pi-5 or gauss-5-membership function is recommended as the optimum model when the prediction accuracy is important while the ANFIS model with gbell-3-membership function is suggested as the optimum model when the prediction cost plays a crucial role along with the prediction accuracy. The proposed optimal ANN and ANFIS models present higher prediction accuracy than the coupled numerical approach.

Keywords: adaptive neuro-fuzzy interface system; artificial neural network; current; power; thermal efficiency; thermoelectric generator system; waste heat recovery

---

1. Introduction

In the last two decades, the researches on waste heat recovery technologies have been increased to diminish the global energy crisis [1]. The thermoelectric generators are the evolving technology for the waste heat recovery due to its non-polluting and silent operational characteristics [2]. The thermoelectric generators convert the thermal energy into the electricity using the Seebeck effect of semiconductor materials [2]. The automobile vehicles are the heavy consumers of the fossil fuel and approximately 30%–40% of heat is lost as the vehicle exhaust [3]. The significant research works have been done on the applications of the thermoelectric generators in the automobile vehicle for waste heat recovery. However, the low conversion efficiency of the thermoelectric generators restricts their commercialization in the automobile field [3]. To improve the performance of the thermoelectric generators in the waste heat recovery of the automobile vehicles, the hot heat exchanger is provided with various internal structures in the form of fins, inserts and protrusions.

Wang et al. have proved that the dimpled surface hot heat exchanger enhances the power output of the thermoelectric generator by 173.60% and reduces the pressure drop by 20.57% compared to inserted fins hot heat exchanger [3]. Niu et al. have recommended that the baffles should be installed in front of the thermoelectric modules and the baffle angle should be increased in the flow direction to enhance the performance of the thermoelectric generator and to reduce the pressure drop [4]. Liu et al. and Quan et al. have proved that the chaos shaped internal structure in the heat exchanger improves the power output and thermal performance of the automotive thermoelectric generators [5,6]. Luo et al. have improved the performance of the thermoelectric generators by proposing the heat exchanger with hotter side converges towards the inward [7]. Nithyanandam et al. have proposed that the metal foam-based heat exchanger shows 5.7 to 7.8 times higher power output than that without the metal foam for automobile waste heat recovery [8]. Cao et al. have proved that the heat pipe with insertion depth of 60 mm and gas flow direction of 15° enhances the open circuit voltage, maximum power and maximum power density of the automotive thermoelectric generator system by 7.5, 10.17 and 15.49%, respectively [9]. He et al. have shown that the plate type heat exchanger shows the maximum conversion efficiency of 5% for the louvered fins and 4.5% for the smooth and offset strip fins, respectively [10]. Lu et al. have proved that the hot heat exchanger configurations with uniform winglet vortex and non-uniform winglet vortex show higher power output of the thermoelectric generator than the hot heat exchanger without fins [11]. Rana et al. have the generated maximum power of 79.02 W by designing the heat exchanger with 0.08 m length, 1 m height, 4 mm gap size and 50 thermoelectric modules [12]. Suter et al. have proposed 1 kW thermoelectric stack with the counterflow parallel plate heat exchanger and 127 pairs of thermoelectric modules to convert the geothermal reservoir heat to electricity using the optimized stack volume of 0.0021 m^3 and optimized the conversion efficiency of 4.2% [13]. Zhao et al. have showed that the application of intermediate fluid improves the maximum power output and generation efficiency of the automotive thermoelectric generator system [14,15]. Lu et al. have shown that 1-inlet 2-outlet heat exchanger has improved the performance characteristics compared to 2-inlet 2-outlet and empty cavity heat exchangers [16].

The conducted literature review concludes that the numerous experimental and numerical studies have been demonstrated on the thermoelectric generator system for waste heat recovery. The experimental study on the thermoelectric generator system for waste heat recovery shows that the energy imbalance results into the excessive loss within the system and improper insulation results in heat loss from the system to the environment. In addition, the thermocouples embedded into the thermoelectric generator system for waste heat recovery show measuring errors in the temperatures of various parts of the system, and manufacturing complexity arises due to non-uniform material properties. The numerical study on the thermoelectric generator system for waste heat recovery requires the powerful computational devices which involve higher computational time and higher computational cost. In the last few years, the artificial intelligence techniques have a secured position as the effective prediction tools to predict and optimize the performances of the various physical system. The artificial intelligence techniques are the most efficient tools to accurately predict the performance of the thermoelectric generator system for waste heat recovery and diminish the limitations of the corresponding experimental and numerical approaches. Dheenamma et al. have shown the artificial neural network models to predict the overall heat transfer coefficient, friction factors of hot and cold fluids, and effectiveness of the plate type heat exchanger by considering the Reynolds number of hot and cold fluids, Prandtl numbers of hot and cold fluids and the concentration of the cold fluid as the input conditions [17]. Angeline et al. have formulated the artificial neural network to predict the performance parameters of open circuit voltage, maximum power and matched load resistance of the thermoelectric generator for various hot side temperatures. The predicted results from the artificial neural network are found within 3% error with the corresponding experimental results [18,19].

The application of the artificial neural network and adaptive neuro-fuzzy interface system for the performance prediction of the thermoelectric generator system for waste heat recovery has not been investigated. Therefore, six ANN models with various combinations of training variants, transfer functions and the number of hidden neurons as well as seven ANFIS models with various combinations of types of membership functions and the number of membership functions are formulated. The developed ANN and ANFIS models predict the current, power and thermal efficiency of the thermoelectric generator system for waste heat recovery under the hot gas inlet temperatures and voltage load conditions.

2. Experimental Set-Up

The experimental set-up of the thermoelectric generator system for waste heat recovery is shown in Figure 1. The thermoelectric generator system for waste heat recovery is designed with heat exchanger, four cold fluid channels and 12 thermoelectric modules. The hot gas passes through the heat exchanger and the cold-water flows through the cold fluid channels. The thermoelectric modules are arranged between the heat exchanger body and the cold fluid channel to utilize the temperature difference between the hot gas and cold water. The thermoelectric modules convert the temperature difference of the hot gas and cold water into power using the Seebeck effect [20]. The heat exchanger and the cold fluid channels are constructed with aluminum material, whereas the thermoelectric modules are constructed with the skutterudite material. The heat exchanger is comprised of the frame with straight fins and guide fins in the inlet and outlet diffuser sections to enable the uniform distribution of the hot gas. The cold fluid channels are provided with the internal fins structure to enable the uniformity of water. Four cold fluid channels with two at the top of the heat exchanger and two at the bottom of the heat exchanger are arranged with three modules between each channel and heat exchanger. The thermoelectric generator system for waste heat recovery is installed in the airtight chamber filled with the argon gas at the pressure of 1×10^5 Pa. The electric heater supplies the hot gas at the required temperature using the thermostat controller. The mass flow rate of the hot gas is measured by the mass flow indicator with an accuracy of ±0.5% installed near the thermostat controller. The airtight vacuum chamber provides the constant temperature and pressure controlled by the chamber pressure regulator. In addition, the chamber pressure regulator indicates the inlet and outlet temperatures of the hot gas. The constant temperature chiller supplies the cold water to the cold fluid channels at the required temperature and pressure. The mass flow rate of the water is measured by the mass flow indicator with an accuracy of ±0.5% installed on a tube which transfers the cold water from chiller to the cold fluid channels. The temperatures of the hot gas at the inlet and outlet of the heat exchanger, temperatures of the cold water at the inlet and outlet of the cold fluid channels, temperatures of the thermoelectric modules and the chamber are measured using nine K-type thermocouples with an accuracy ±0.1 °C. The thermocouples are connected to a KEYSIGHT 34970A data logger with an accuracy of ±0.1% for monitoring the temperatures continuously. The thermoelectric modules are connected to the KIKUSUI PLZ334L electronic loader to record the current, voltage and power data with time. The accuracy of the electronic loader is ±0.1%, ±0.2% and ±0.6% for the current, voltage and power measurements, respectively.

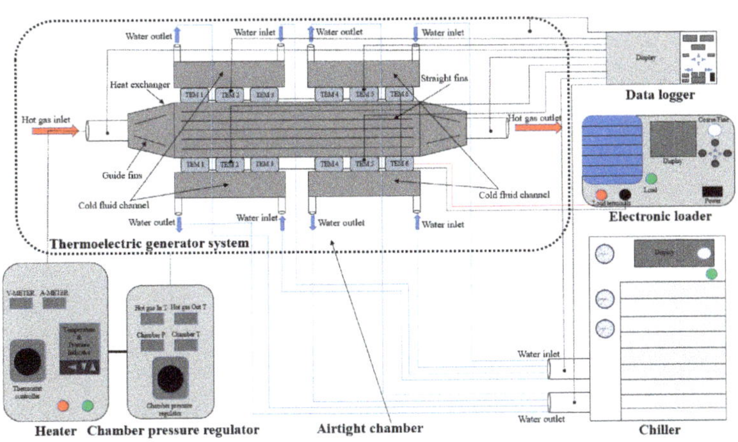

Figure 1. Experimental set-up of the thermoelectric generator system for waste heat recovery.

The uncertainties of various measuring instruments and the measurement errors are considered in the experimental data. The temperature, mass flow rate, voltage, current, power and thermal efficiency are the experimentally predicted output data. Thus, uncertainties of the temperature, mass flow rate, voltage and current as the independent parameters are calculated with errors of the measuring instruments. The uncertainties of the power and thermal efficiency as the dependent parameters are calculated based on the linearized fraction approximation as shown by Equation (1) [21]. The uncertainties in the temperature, mass flow rate, voltage, current, power and thermal efficiency are showed as ±0.24%, ±0.76%, ±1.27%, ±1.19%, ±1.74% and ±1.91%, respectively:

$$w_r = [(\frac{\partial R}{\partial x_1} w_1)^2 + (\frac{\partial R}{\partial x_2} w_2)^2 + \cdots + (\frac{\partial R}{\partial x_n} w_n)^2]^{\frac{1}{2}} \tag{1}$$

Here, R is the dependent parameter, w_r is the uncertainty in the dependent parameter, $x_1, x_2, \ldots x_n$ are the independent parameters and $w_1, w_2, \ldots w_n$ are the uncertainties in the independent parameters.

3. Numerical Method

The numerical analysis is conducted on the physical domain of the thermoelectric generator system for waste heat recovery. The physical domain of the thermoelectric generator system for waste heat recovery is comprised of the heat exchanger, cold fluid channels, thermoelectric modules and fluid domains of the hot gas and water. The coupled numerical approach comprises the CFX and the thermal electric solvers of ANSYS 19.1 commercial software is used to predict the current, power and thermal efficiency of the thermoelectric generator system for waste heat recovery [4]. The continuity, momentum, energy and thermoelectric coupling equations are solved using the coupled numerical approach. The boundary conditions for solving governing equations are shown in Table 1. The experimental voltage conditions with time are used as the high potential voltage conditions for the numerical analysis. The low potential voltage is set to 0 V to ensure the flow of current from the high potential to the low potential of the thermoelectric module. The tetrahedrons mesh structure and fluid domains are used to solve the governing equations because of the complex geometrical structures of the thermoelectric generator system for waste heat recovery with the presence of fins and sharp corners. To verify the convergence of the predicted results, the grid dependency test is carried out for five different grid element numbers. Figure 2 shows the grid dependency test for the simulated power and thermal efficiency of the thermoelectric generator system for waste heat recovery. As shown in Figure 2, the simulated power and thermal efficiency values are converged within ±1% above a grid element number of 11,431,310. Therefore, considering the computational time and the computational cost, the grid element number of 11,431,310 is selected as the final mesh configuration for the thermoelectric generator system for waste heat recovery to predict its current, power and thermal efficiency numerically. In numerical analysis, the density, specific heat, thermal conductivity and dynamic viscosity of the hot gas are set as 1.19 kg/m^3, 1005 J/k·°C, 0.026 W/m·K and 1.8 × 10^{-4} kg/m·s, respectively. The density, specific heat, thermal conductivity and dynamic viscosity of the water are set as 997 kg/m^3, 4182 J/kg·°C, 0.607 W/m·K and 8.9 × 10^{-4} kg/m·s, respectively. For skutterudite, the density, specific heat and thermal conductivity are used as 7598 kg/m^3, 350 J/kg·°C and 3.4 W/m·K, respectively. In addition, the Seebeck coefficient is set as 142.8 μV/K for p-leg and −183.5 μV/K for n-leg [22]. The continuity, momentum and energy equations are expressed with Equations (2) to (5) [23].

Continuity Equation

$$\frac{\partial \rho}{\partial t} + \nabla \cdot (\rho U) = 0 \tag{2}$$

Momentum Equation

$$\frac{\partial (\rho U)}{\partial t} + \nabla \cdot (\rho U \times U) = -\nabla p + \nabla \tau + S_M \tag{3}$$

Stress tensor τ is expressed in terms of strain rate as follows:

$$\tau = \mu(\nabla U + (\nabla U)^T - \frac{2}{3}\delta\nabla\cdot U) \tag{4}$$

Energy equation

$$\frac{\partial(\rho h)}{\partial t} + \nabla\cdot(\rho U h) = \nabla\cdot(\lambda\nabla T) + \tau : \nabla U + S_E \tag{5}$$

where ρ is the density (kg/m^3), U is the average velocity (m/s), ∇ is the nabla operator, p is the static pressure (Pa), τ is the stress tensor, S_M is the momentum source, μ is the dynamic viscosity (Pa·s), h is the enthalpy (J), λ is the thermal conductivity (W/m·K) and S_E is the energy source.

To deal with the turbulence of the hot gas and cold water, the k-ε turbulence model [3] is used with Equations (6) and (7):

$$\frac{\partial(\rho u_i k)}{\partial x_i} = \frac{\partial}{\partial x_i}\left[\left(v + \frac{v_t}{\sigma_k}\right)\frac{\partial k}{\partial x_i}\right] + P_k - \rho\varepsilon \tag{6}$$

$$\frac{\partial(\rho u_i \varepsilon)}{\partial x_i} = \frac{\partial}{\partial x_i}\left[\left(v + \frac{v_t}{\sigma_\varepsilon}\right)\frac{\partial \varepsilon}{\partial x_i}\right] + \rho C_1 S\varepsilon - \rho C_2\frac{\varepsilon^2}{k + \sqrt{v\varepsilon}} \tag{7}$$

where u_i is the velocity component, k is the turbulence kinetic energy, x_i is the cartesian co-ordinates, v_t is the turbulent eddy viscosity, σ_k is 1 based on reference [3], σ_ε is 1.2 based on reference [3], P_k is the shear production of the turbulent kinetic energy, ε is the dissipation rate of the turbulence energy and S is the modulus of the mean rate of the strain tensor.

The thermoelectric coupling equations [24] of Equations (8) and (9) are calculated:

$$\nabla\cdot(p\cdot\vec{J}) - \nabla\cdot(k\cdot\nabla T) = \vec{J}\cdot\vec{E} \tag{8}$$

$$\nabla\cdot(\sigma\cdot\alpha\cdot\nabla T) + \nabla\cdot(\sigma\cdot\nabla\varnothing) = 0 \tag{9}$$

where p is the Peltier coefficient (V), $\vec{J}$ is the electric current intensity (A/m^2), k is the thermal conductivity (W/m·K), ∇T is the temperature gradient, $\vec{E}$ is the electric field intensity (V/m), σ is the electrical conductivity (Ω^{-1}·m^{-1}), α is the Seebeck coefficient (V/K), and $\nabla\varnothing$ is the electric potential (J/C).

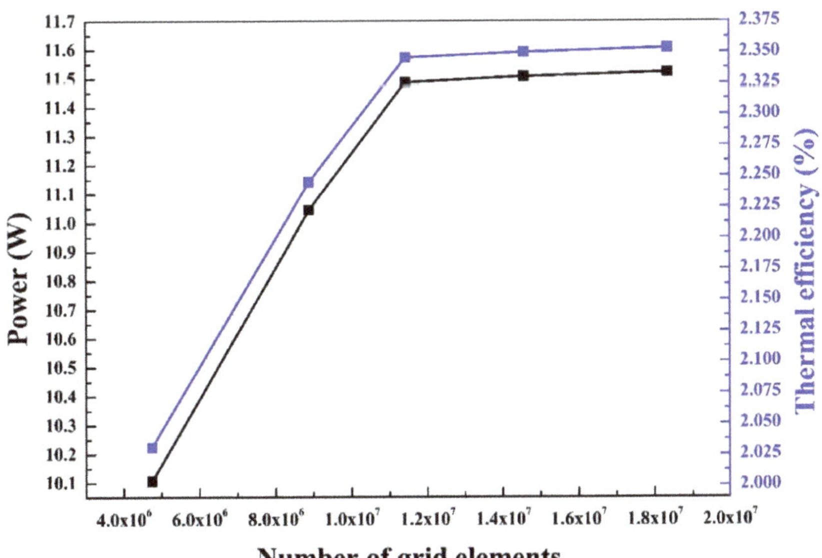

Figure 2. Grid dependency test for the simulated power and thermal efficiency.

Table 1. Boundary conditions.

Parameter	Value
Hot gas (air) inlet temperature (°C)	315.12, 419.26, 521.7, 621.61
Coolant (water) inlet temperature (°C)	30
Hot gas (air) mass flow rate (kg/s)	0.018
Coolant (water) mass flow rate (kg/s)	0.075
High potential voltage (V)	0 to 10 V
Low potential voltage (V)	0

4. Artificial Intelligence Models

4.1. Artificial Neural Network (ANN) Modelling

The artificial neural network is the replica of the biological neural network which could be used for the optimization, simulation, modeling, forecasting and performance prediction of various physical systems [25]. The nonlinear relationship between the input and output variables with larger number of data points could be mapped efficiently using the ANN technique [25]. The ANN consists of three layers with the input layer, output layer and one or more than one hidden layer with a suitable number of neurons in each layer [25,26]. The number of neurons in the input layer is equal to the number of input parameters and number of neurons in the output layer is equal to number of output parameters. The number of hidden layers and hidden neurons are decided based on the training error [27]. The neurons of one layer are connected to the other layer using weights and the single weight value is assigned between two neurons [28]. The ANN structure with three layers and various numbers of neurons in each layer is trained using suitable training algorithm [29]. The training algorithm consists of the back-propagation algorithm, training variants and transfer functions [26]. For the training, the maximum training error and the maximum number of epochs are decided. During the training, the weight values get adjusted to predict the desired values of output parameters. If the error between the predicted output and the actual output is lower than the decided training error, the training is stopped; otherwise, further training is done to achieve the desired output [30]. The neural network structure with the desired prediction accuracy is selected as the optimum neural network structure.

In this study, six ANN models are developed to predict the performances of the thermoelectric generator system for waste heat recovery. Figure 3 shows the formulated ANN structure to predict the performance of thermoelectric generator system for waste heat recovery. The hot gas inlet temperature and voltage load conditions are considered as the inputs to the ANN models for predicting the current, power and thermal efficiency of the thermoelectric generator system for waste heat recovery. The back-propagation training algorithm is used to train the six ANN models. The six models are the combinations of three training variants of Levenberg–Marquardt (LM), Scaled Conjugate Gradient (SCG) and Pola–Ribiere Conjugate Gradient (CGP), and two transfer functions of Tan-Sigmoidal and Log-Sigmoidal and number of hidden neurons (N) of 10, 15, 20, and 25. The maximum number of epochs is set to 1000 and the maximum training error is set to 10^{-6} for training to confirm the prediction accuracy of the tested model for the thermoelectric generator system for waste heat recovery. The experiments are conducted on the thermoelectric generator system for waste heat recovery at the hot gas inlet temperatures and voltage load conditions to collect the data for training. A total of 931 data points of the input and output parameters are used to train the six models. For each ANN model, the training is done until the error becomes steady and the outputs predicted by that trained model are recorded. The predicted output values of the current, power and thermal efficiency are compared with the corresponding experimental values based on three statistical parameters of the coefficient of determination (R^2), root mean square error (RMSE), and coefficient of variance (COV). The ANN model with the highest value of R^2 and the lowest values of RMSE and COV, respectively, is selected as the optimum ANN model to predict the current, power and thermal efficiency of the thermoelectric generator system for waste heat recovery for the hot gas inlet temperatures range of 315.12 to 621.61 °C and voltage load conditions range of 0 to 10 V.

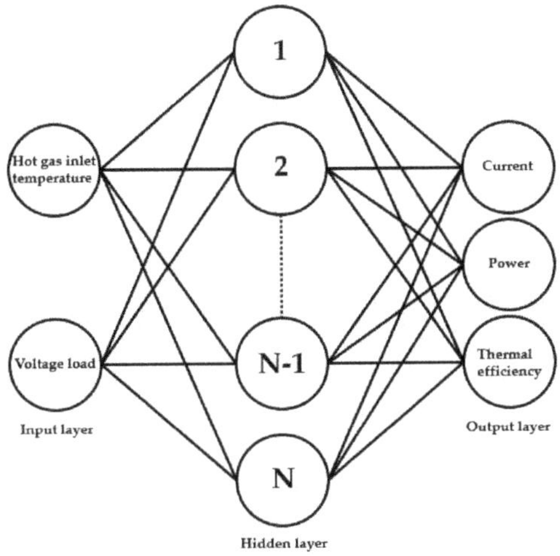

Figure 3. Formulated ANN structure to predict the performances of the thermoelectric generator system for waste heat recovery.

4.2. *Adaptive Neuro-Fuzzy Interface System Modelling (ANFIS)*

The ANFIS is one of the artificial intelligence techniques which is the combination of ANN and fuzzy logic [30]. The nonlinear relationship between the input and output parameters with a larger number of data points could be established accurately using ANFIS [30]. Like ANN, the ANFIS is also used to predict and optimize the performances of the various physical systems [31]. The input and output parameters needed to be related are imported in the ANFIS model in the form of neurons. The data of each input and output parameter are shown in the form of various membership functions [31]. The type of membership function and number of membership functions are decided based on the variation trend of the input and output data. The types of membership functions in the ANFIS model are triangular, trapezoidal, gbell, gauss, gauss2, pi, dsig and psig [32]. In the ANFIS model, the input and output data are connected by rules with the statements by showing the relationship between the input and output data. The prediction of the output values for the various input conditions are decided based on the rules. The ANFIS structure is trained using two algorithms of the back-propagation and hybrid [33]. The maximum number of epochs and maximum error are set for the training of an ANFIS model. During the training of ANFIS, the rules get adjusted to predict the desired output of the various physical systems like solar systems. The training is continued until the desired accuracy is achieved. The ANFIS model with the prediction value closest to the actual output is selected as the optimum model [33]. The output variables are predicted from the optimum ANFIS model by importing the input variables into the rule viewer [34].

Figure 4 shows the structure of formulated ANFIS model to predict the performance of the thermoelectric generator system for waste heat recovery. The selected ANFIS model type is a Takagi–Sugeno which has n number of inputs with only one output prediction [35]. Hence, in the present ANFIS model, two input parameters of voltage and temperature are connected to one output parameter of current, power and thermal efficiency. Seven ANFIS models are formulated to predict the performances of the thermoelectric generator system for waste heat recovery and predict the current, power and thermal efficiency of the thermoelectric generator system for waste heat recovery with the hot gas temperatures and high potential voltage conditions. Seven membership functions of the triangular, trapezoidal, gauss, gauss2, gbell, pi and dsig are used to formulate the ANFIS models, and each ANFIS model is formulated with one type of the membership function. In each ANFIS model, each type of membership function is used with the number of sets of 2, 3, 4 and 5. The ANFIS models are trained for the same experimental data with sets of 931 data points used to train the ANN models.

All ANFIS models are trained using the back-propagation algorithm for the maximum epochs of 1000 and maximum error of 10^{-6}. The ANFIS models are trained until the training error becomes steady. Once the training error converges, the output values are predicted in the rule viewer by importing the input conditions of hot gas temperature and voltage conditions. The predicted values of the current, power and thermal efficiency by each ANFIS model are compared with the corresponding experimental values using three statistical parameters of R^2, RMSE and COV. The ANFIS model with the optimum values of three statistical parameters is considered as the best model to predict the current, power and thermal efficiency of the thermoelectric generator system for waste heat recovery under the influence of various hot gas inlet temperatures and voltage conditions.

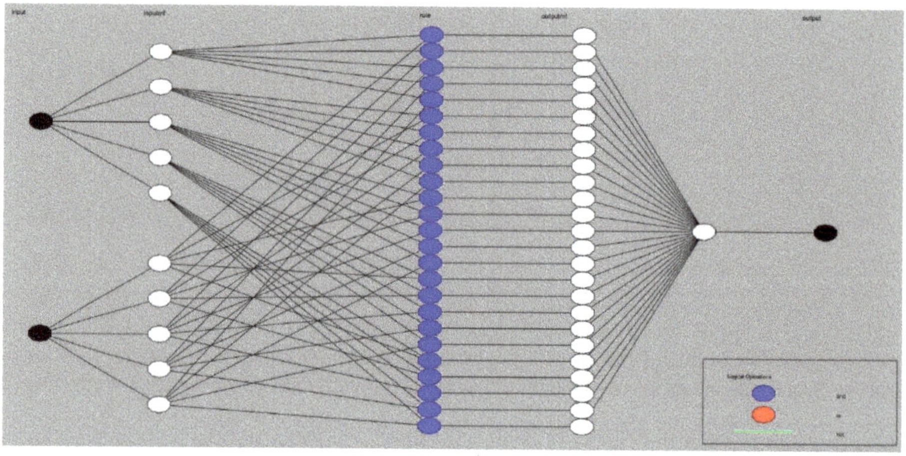

Figure 4. Formulated ANFIS model to predict the performances of the thermoelectric generator system for waste heat recovery.

5. Data Reduction

The power generated by the thermoelectric modules [36] is expressed with Equation (10):

$$P = VI \tag{10}$$

where P is the power (W) of the thermoelectric modules, V is the voltage (V) and I is the current (A).

The thermal efficiency of the thermoelectric generator system for waste heat recovery could be calculated with Equation (11) as the ratio of the power generated by the thermoelectric modules (P) to the heat transfer through the thermoelectric modules ($\dot{Q}$) [36]:

$$\eta_{th} = \frac{P}{\dot{Q}} \times 100\% \tag{11}$$

where η_{th} is the thermal efficiency (%) of the thermoelectric generator system for waste heat recovery and $\dot{Q}$ is the heat transfer (W) through the thermoelectric modules. The heat transfer through the thermoelectric modules [37] is calculated using the Fourier's law of heat conduction as shown by Equation (12):

$$\dot{Q} = \frac{KA}{t}\Delta T \tag{12}$$

where K is the thermal conductivity (W/m·K), A is the surface area (m^2), t is the module thickness (m) and ΔT is the temperature difference (°C) of the thermoelectric module.

The coefficient of determination (R^2), root mean square error (RMSE) and coefficient of variance (COV) are calculated using Equations (13)–(15), respectively [38]:

$$R^2 = 1 - \frac{\sum_{m=1}^{n} \left(X_{pre,m} - Y_{mea,m}\right)^2}{\sum_{m=1}^{n} \left(Y_{mea,m}\right)^2} \tag{13}$$

$$RMSE = \sqrt{\frac{\sum_{m=1}^{n} \left(X_{pre,m} - Y_{mea,m}\right)^2}{n}} \tag{14}$$

$$COV = \frac{RMSE}{\left|\overline{Y}_{mea}\right|} \times 100 \tag{15}$$

where R^2 is the coefficient of determination, $RMSE$ is the root mean square error, COV is the coefficient of variance, n is the number of data points, $X_{pre,m}$ is predicted the value of the output parameter at data point m, $Y_{mea,m}$ is the experimental (actual) value of output parameter at data point m and $\overline{Y}_{mea}$ is the average value at all experimental data points.

6. Results and Discussion

6.1. Experimental Outputs of Current, Power and Thermal Efficiency

The current, power and thermal efficiency as the performance parameters of the thermoelectric generator system for waste heat recovery are experimentally tested with the hot gas inlet temperatures of 315.12 °C, 419.26 °C, 521.70 °C and 621.61 °C and the voltage load range of 0 to 10 V. During the experiments, the voltage load is varied with time for each hot gas inlet temperature. Two experimental data sets for the development of a numerical method, ANN models and ANFIS models are considered as the training data set and testing data set of the thermoelectric generator system for waste heat recovery. The training data set (first) with variations of the current, power and thermal efficiency of thermoelectric generator system for waste heat recovery for hot gas inlet temperatures of 315.12 °C, 419.26 °C, 521.70 °C and 621.61 °C and voltage load range of 0 to 10 V is selected and the testing data set (second) with the variation of current, power and thermal efficiency of thermoelectric generator system for waste heat recovery for hot gas inlet temperature of 419.26 °C and voltage load range of 0 to 5.5 V is selected based on the experiments.

Figure 5 shows the variations of the current, power and thermal efficiency for the training and testing data sets. For all hot gas inlet temperatures, the current of the thermoelectric generator system for waste heat recovery is linearly decreased and the power and thermal efficiency of the thermoelectric generator system for waste heat recovery show the parabolic variations with the voltage load of range 0 to 10 V for the training data set and 0 to 5.5 V for the testing data set, respectively. The current, power and thermal efficiency of the thermoelectric generator system for waste heat recovery increase with the increase of the hot gas inlet temperature from 315.12 °C to 621.61 °C. Therefore, the maximum and average values of the current, power and thermal efficiency of the thermoelectric generator system for waste heat recovery are increased with the increase of the hot gas inlet temperature. For the training data set, the maximum current of 4.1, 7.13, 9.42 and 10.95 A and average current of 2.28, 3.90, 4.99 and 5.95 A, the maximum power of 3.40, 9.75, 17.55 and 24.8 W and average power of 1.94, 5.93, 11.10 and 15.22 W and the maximum efficiency of 1.28, 2.16, 2.87 and 3.39% and average efficiency of 0.72, 1.30, 1.81 and 2.07% are selected experimentally at the hot gas inlet temperatures of 315.12 °C, 419.26 °C, 521.70 °C and 621.61 °C, respectively. For the testing data set, the hot gas inlet temperature of 419.26 °C is the same as the training data set, but the voltage load condition is different with time as shown in Figure 5. Thus, for the testing data set at the hot gas inlet temperatures of 419.26 °C, the maximum current is 8.1 A and the average current is 4.41 A. The maximum power is 11.4 W and the average power is 6.95 W. The maximum efficiency is 2.35% and the average efficiency is 1.43%. As a result, the current, power and thermal efficiency of the thermoelectric generator system for waste heat recovery at the hot gas inlet temperature of 419.26 °C are different for the training and testing data sets because of the different voltage loads.

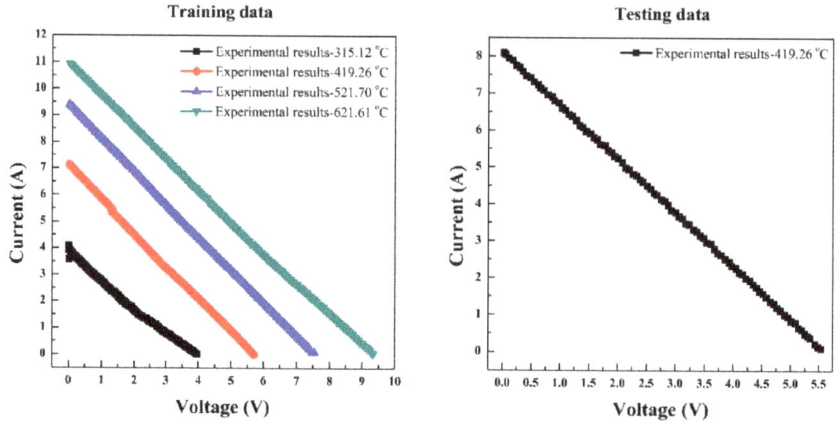

(**a**)

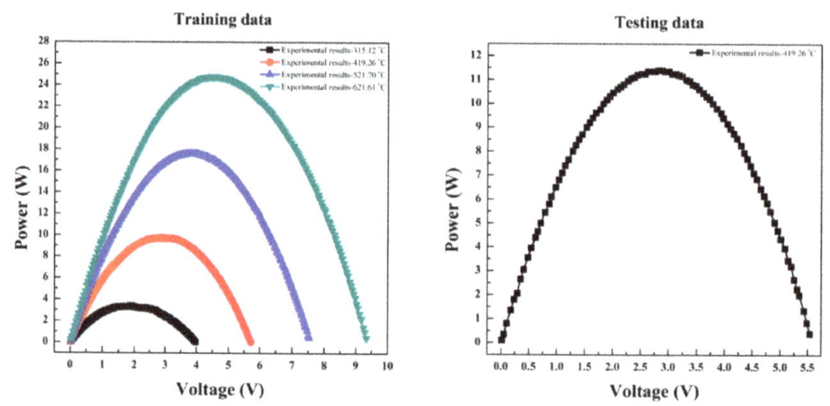

(**b**)

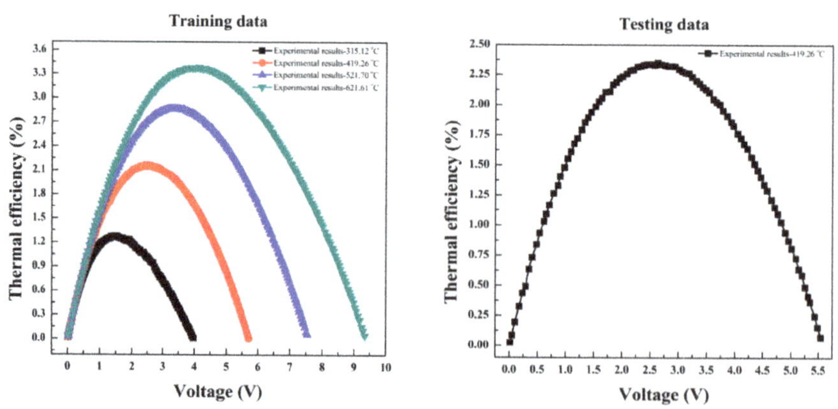

(**c**)

Figure 5. Variation of (**a**) current, (**b**) power, and (**c**) thermal efficiency for training and testing data sets.

6.2. Prediction Results from the Numerical Method

The numerical simulation of the thermoelectric generator system for waste heat recovery at the hot gas inlet temperatures of 419.26 °C, cold water temperature of 30 °C, hot gas mass flow rate of 0.018 kg/s and cold-water mass flow rate of 0.075 kg/s is performed. From the numerical simulation of the thermoelectric generator system for waste heat recovery with various boundary conditions of the hot gas and cold water, the hot and cold surfaces of the thermoelectric modules are simulated.

The temperature of the hot gas decreases with the direction from the inlet to the outlet of the heat exchanger, but the temperature of the cold water increases as the cold water flows from the inlet to the outlet of the cold-water channel. This is because the hot gas transfers the heat and the cold water absorbs the heat from the thermoelectric modules. Therefore, the hot surface and cold surface temperatures of the thermoelectric module are varied with locations because the temperature distributions of the hot gas and cold water depend on the locations.

The temperature distributions of the hot and cold surfaces of the thermoelectric modules with locations (x and y coordinates) at the hot gas inlet temperature of 419.26 °C are showed in Figure 6. Figure 6 shows the temperature distributions of the hot and cold surfaces of the top six thermoelectric modules and the corresponding bottom six thermoelectric modules. In addition, the hot surface temperatures of the thermoelectric module near the inlet of the heat exchanger show higher than those of the thermoelectric modules near the outlet of the heat exchanger. Thus, the current, power and thermal efficiency results of the thermoelectric generator system for waste heat recovery are simulated using the temperature distributions of the hot and cold surfaces of the thermoelectric modules and the voltage load conditions of the testing data set.

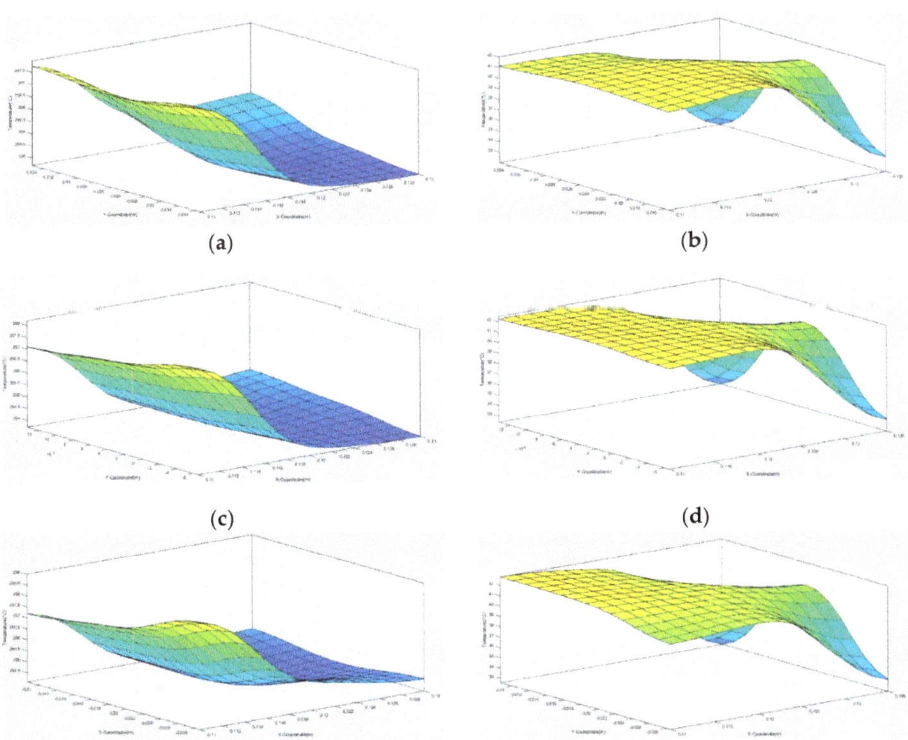

Figure 6. *Cont.*

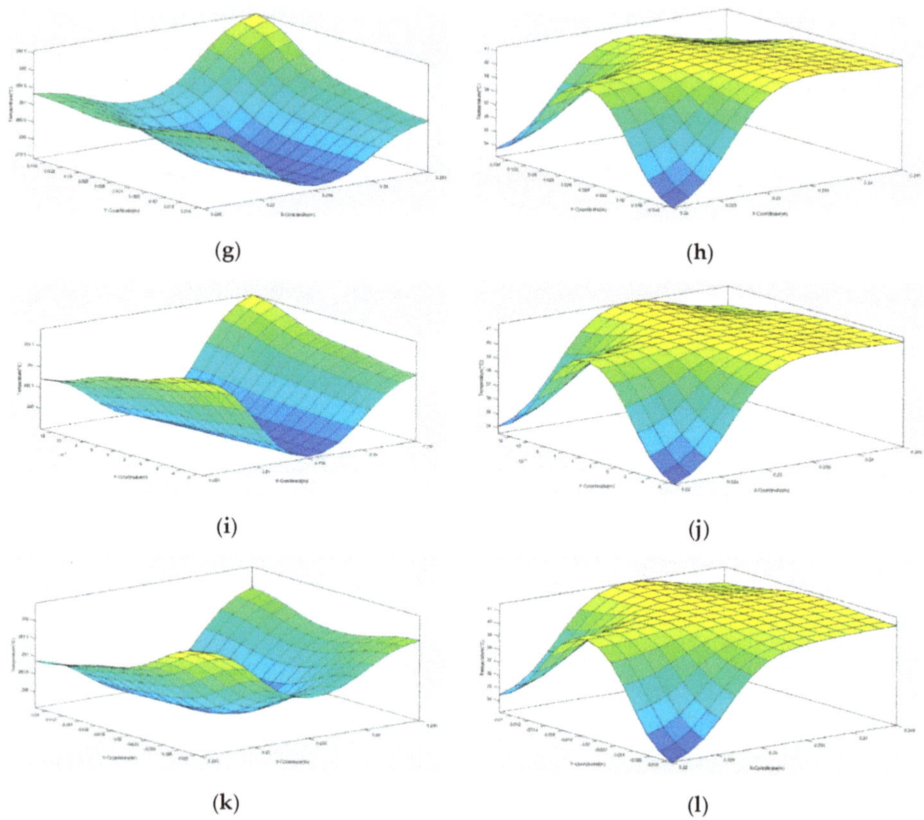

Figure 6. Temperature distributions (**a**) Module 1-hot surface, (**b**) Module 1-cold surface, (**c**) Module 2-hot surface, (**d**) Module 2-cold surface, (**e**) Module 3-hot surface, (**f**) Module 3-cold surface, (**g**) Module 4-hot surface, (**h**) Module 4-cold surface, (**i**) Module 5-hot surface, (**j**) Module 5-cold surface, (**k**) Module 6-hot surface, (**l**) Module 6-cold surface with locations (x and y coordinates) at 419.26 °C.

The comparisons of experimental and numerical results of the current, power and thermal efficiency of the thermoelectric generator system for waste heat recovery for the testing data set are shown in Figure 7. The error between the experimental and numerical values for the current of the thermoelectric generator system for waste heat recovery is validated within 2% except for the initial and end voltage conditions. In addition, the error between the experimental and numerical results for the power and thermal efficiency of the thermoelectric generator system for waste heat recovery is validated within 4% except for the initial and end voltage conditions.

The accuracy of numerical method for the current, power and thermal efficiency of the thermoelectric generator system for waste heat recovery is shown in Table 2. The numerical results of the current, power and thermal efficiency of the thermoelectric generator system for waste heat recovery show a good agreement with the corresponding experimental results [3]. The selection of the accurate boundary condition, meshing configuration with conduction and inflation effects, discretization method and suitable solver result in closer agreement between the numerical and experimental results of the thermoelectric generator system for waste heat recovery. Therefore, the experimental approach of the thermoelectric generator system for waste heat recovery with high manufacturing and installation costs, higher complexity and higher level of efforts could be replaced with a numerical approach of the thermoelectric generator system for waste heat recovery.

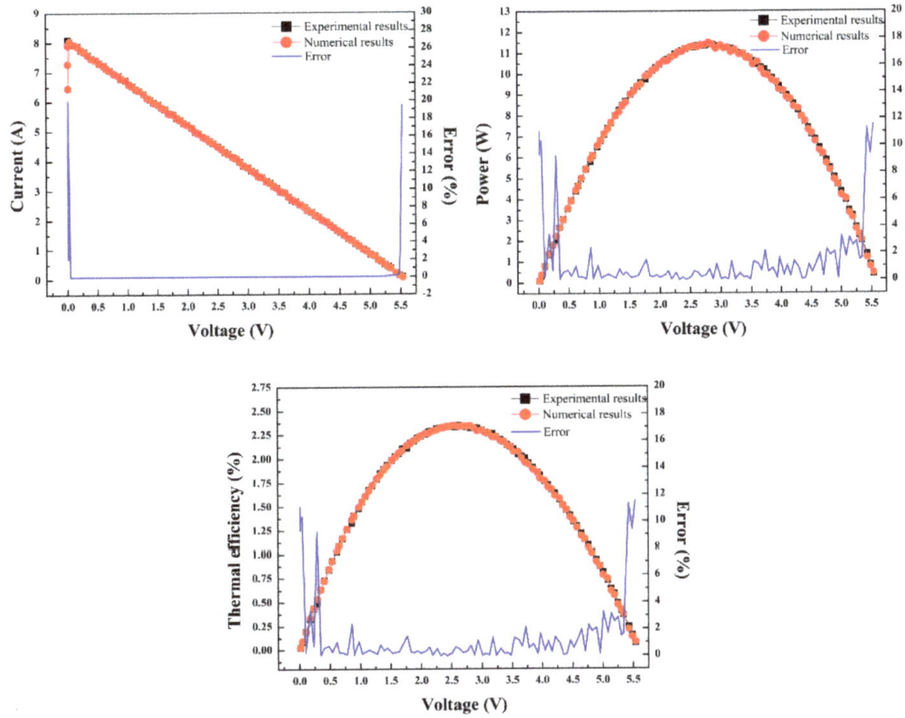

Figure 7. The comparisons of experimental and numerical results of the current, power and thermal efficiency for testing data set.

Table 2. The accuracy of numerical method for the current, power and thermal efficiency.

Parameter	R^2	RMSE	COV
Current	0.99865	0.18633	4.22614
Power	0.99992	0.07032	1.01243
Thermal efficiency	0.99992	0.01422	0.99102

6.3. Training and Testing Data Sets for ANN and ANFIS Models

Figure 5 shows the training and testing data sets used to develop the ANN and ANFIS models of the thermoelectric generator system for waste heat recovery. Using these experimental data of both the hot gas inlet temperature and voltage condition as the input parameters and the current, power and thermal efficiency as the output parameters of the thermoelectric generator system for waste heat recovery, six ANN models and seven ANFIS models are formulated. As the training data set for training the ANN and ANFIS models of the thermoelectric generator system for waste heat recovery, a total of 931 data points of the mixtures of 225 data points at the hot gas inlet temperature of 315.12 °C, 234 data points at the hot gas inlet temperature of 419.26 °C, 236 data points at the hot gas inlet temperature of 521.70 °C and 236 data points at the hot gas inlet temperature of 621.61 °C are deducted from the experimental study. The training for the considered ANN and ANFIS models is stopped when the training error converges. The converged training errors for the considered ANN and ANFIS models are shown in Figure 8. To check the reliability and accuracy of the trained ANN and ANFIS models of the thermoelectric generator system for waste heat recovery, the additional experiment for obtaining the testing data set of 100 data is conducted at the hot gas inlet temperature of 419.26 °C with the different voltage loads.

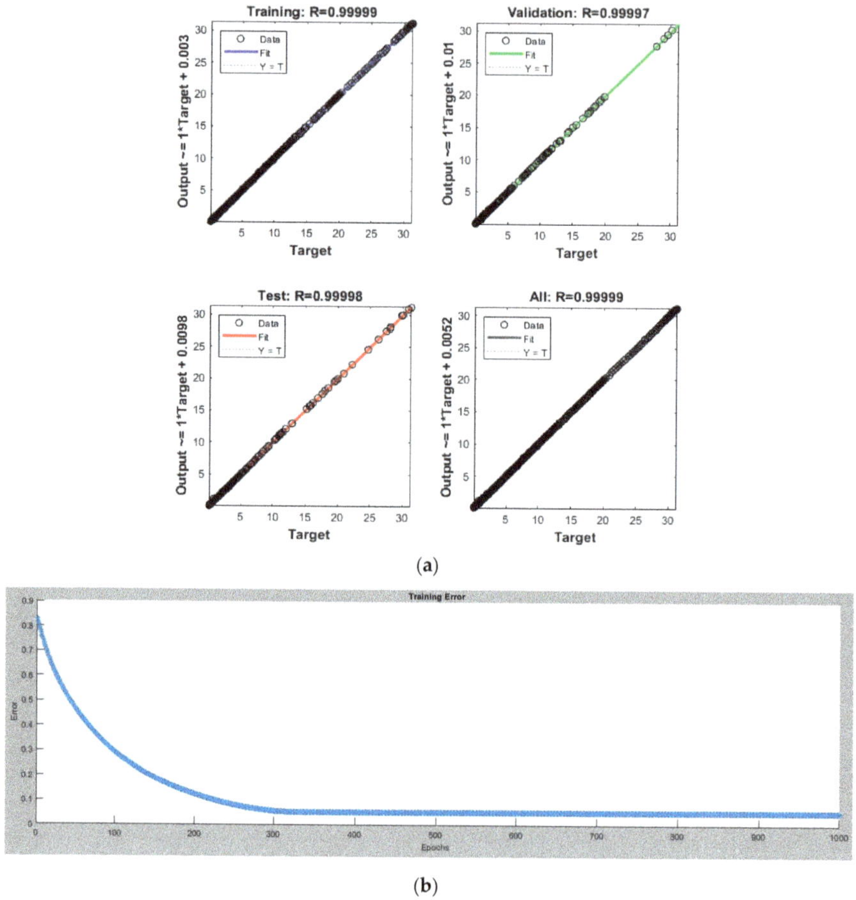

Figure 8. The converged training errors for (**a**) ANN and (**b**) ANFIS models of thermoelectric generator system for waste heat recovery.

The current, power and thermal efficiency of the thermoelectric generator system for waste heat recovery are predicted by ANN and ANFIS models for the hot gas inlet temperature of 419.26 °C and voltage loads of the testing data set. The predicted current, power and thermal efficiency of the thermoelectric generator system for waste heat recovery from the ANN and ANFIS models for the testing data set are compared with the corresponding experimental data of the testing data set. Based on the degree of closeness between the experimental and predicted results of the current, power and thermal efficiency of the thermoelectric generator system for waste heat recovery with ANN and ANFIS models, the optimum ANN and ANFIS models with higher prediction accuracy are decided.

6.4. Prediction Results from ANN Models

The comparison of experimental and ANN predicted results of the current, power and thermal efficiency of the thermoelectric generator system for waste heat recovery using an LM-TanSig algorithm with the various numbers of the hidden neurons is shown in Figure 9a. The increase of the hidden neurons number from 10 to 25 increases the prediction accuracy of the ANN model with an LM-TanSig algorithm. The values of R^2, RMSE and COV of LM-TanSig algorithm with 25 hidden neurons are 0.99998, 0.02163 and 0.49061, respectively for the current, 0.99997, 0.04111 and 0.59192, respectively, for the power and 0.99996, 0.01050 and 0.73183, respectively, for the thermal efficiency.

The comparison of experimental and ANN predicted results of the current, power and thermal efficiency of the thermoelectric generator system for waste heat recovery using an LM-LogSig algorithm with the various numbers of the hidden neurons is shown in Figure 9b. The ANN model for the current and thermal efficiency of the thermoelectric generator system for waste heat recovery with LM-LogSig algorithm and 25 hidden neurons shows the peak prediction accuracy and this prediction accuracy decreases in an order with LM-LogSig algorithm of 20, 15 and 10 hidden neurons, respectively. The values of R^2, RMSE and COV for LM-LogSig algorithm and 25 hidden neurons are 0.99998, 0.02370 and 0.53755, respectively for the current and 0.99994, 0.01225 and 0.85347, respectively for the thermal efficiency. For the power of the thermoelectric generator system for waste heat recovery, LM-LogSig algorithm with 20 hidden neurons shows higher prediction accuracy than that with 25, 15 and 10 hidden neurons, respectively. The values of R^2, RMSE and COV for LM-LogSig algorithm with 20 hidden neurons are 0.99997, 0.04632 and 0.66686, respectively for the power.

The comparison of experimental and ANN predicted results of the current, power and thermal efficiency of the thermoelectric generator system for waste heat recovery using SCG-TanSig algorithm with the various numbers of the hidden neurons is shown in Figure 9c. The ANN model for the current of the thermoelectric generator system for waste heat recovery with SCG-TanSig algorithm and 25 hidden neurons shows the peak prediction accuracy and this prediction accuracy decreases in an order with SCG-TanSig algorithm of 10, 25 and 15 hidden neurons, respectively. The values of R^2, RMSE and COV for SCG-TanSig algorithm with 25 hidden neurons are 0.99992, 0.04524, 1.02613, respectively for the current of the thermoelectric generator system for waste heat recovery. The power and thermal efficiency of the thermoelectric generator system for waste heat recovery using the SCG-TanSig algorithm with 20 hidden neurons shows higher prediction accuracy than that with 10, 25 and 15 hidden neurons, respectively. The values of R^2, RMSE and COV for SCG-TanSig algorithm with 20 hidden neurons are 0.99971, 0.13652 and 1.96554, respectively, for the power and 0.99929, 0.04377 and 3.05105, respectively, for the thermal efficiency.

The comparison of experimental and ANN predicted results of the current, power and thermal efficiency of the thermoelectric generator system for waste heat recovery using an SCG-LogSig algorithm with the various numbers of the hidden neurons is shown in Figure 9d. The prediction accuracy for the current of the thermoelectric generator system for waste heat recovery with the SCG-LogSig algorithm decreases with 25, 15, 20 and 10 hidden neurons. The values of R^2, RMSE and COV for SCG-LogSig algorithm with 25 hidden neurons are 0.99996, 0.03138 and 0.71178, respectively, for the current. The prediction accuracy for the power of the thermoelectric generator system for waste heat recovery with SCG-LogSig algorithm decreases with 15, 10, 25 and 20 hidden neurons but prediction accuracy for the thermal efficiency of the thermoelectric generator system for waste heat recovery with SCG-LogSig algorithm decreases with 15, 25, 10 and 20 hidden neurons. The values of SCG-LogSig with 15 hidden neurons are 0.99980, 0.11376 and 1.63783, respectively, for the power and 0.99958, 0.03359 and 2.34133, respectively, for the thermal efficiency.

The comparison of experimental and ANN predicted results of current, power and thermal efficiency of the thermoelectric generator system for waste heat recovery using the CGP-TanSig algorithm with various numbers of hidden neurons is shown in Figure 9e. The prediction accuracy of the thermoelectric generator system for waste heat recovery with the CGP-TanSig algorithm decreases with 20, 25, 10, and 15 hidden neurons for the current but 20, 10, 15 and 25 hidden neurons for the power, respectively. The values of R^2, RMSE and COV for CGP-TanSig algorithm with 20 hidden neurons are 0.99989, 0.05377 and 1.21965, respectively, for the current and 0.99945, 0.18629 and 2.68213, respectively, for the power. In addition, the prediction accuracy for the thermal efficiency of the thermoelectric generator system for waste heat recovery using CGP-TanSig algorithm with 25 hidden neurons is the most accurate and decreases with 15, 20 and 10 hidden neurons, respectively. The values of CGP-TanSig algorithm with 25 hidden neurons are 0.99875, 0.05805 and 4.04596, respectively, for the thermal efficiency.

The comparison of experimental and ANN predicted results of current, power and thermal efficiency of the thermoelectric generator system for waste heat recovery using the CGP-LogSig algorithm with various numbers of hidden neurons is shown in Figure 9f. The CGP-LogSig algorithm with 25 hidden neurons predicts current values of the thermoelectric generator system for waste heat recovery closer to the corresponding experimental current values of the thermoelectric generator system for waste heat recovery with R^2, RMSE and COV values of 0.99989, 0.05354 and 1.21437, respectively. The CGP-LogSig algorithm with 20, 15, and 10 hidden neurons shows the decreasing order of prediction accuracy for the current of the thermoelectric generator system for waste heat recovery. The CGP-LogSig algorithm with 15, 20, 25 and 10 hidden neurons, respectively, shows the decreasing order of prediction accuracy for the power of the thermoelectric generator system for waste heat recovery and CGP-LogSig algorithm with 15, 25, 10 and 20 hidden neurons, respectively, shows the decreasing order of prediction accuracy for the thermal efficiency of the thermoelectric generator system for waste heat recovery. The R^2, RMSE and COV values for CGP-LogSig algorithm with 15 hidden neurons are 0.99953, 0.17188 and 2.47463, respectively, for power and 0.99848, 0.06391 and 4.45470, respectively, for the thermal efficiency.

The comparison of ANN models with various combinations of the training variants, transfer functions and number of the hidden neurons is shown. The combination of LM training variant with TanSig and LogSig transfer functions and all numbers of the hidden neurons show better accuracy than that of SCG and CGP training variants with TanSig and LogSig transfer functions and all numbers of hidden neurons to predict current, power and thermal efficiency of the thermoelectric generator system for waste heat recovery. In particular, the ANN model with LM-TanSig training algorithm and 25 hidden neurons shows the best prediction accuracy [29,39] and is suggested as the optimum model for predicting the current, power and thermal efficiency of the thermoelectric generator system for waste heat recovery for the hot gas temperature ranges of 315.12 to 621.61 °C and voltage load ranges of 0 to 10 V. The accuracy of the current, power and thermal efficiency of the thermoelectric generator system for waste heat recovery using the ANN model with the LM-TanSig algorithm and 25 hidden neurons are 0.99998, 0.99997 and 0.99996, respectively, as shown in Table 3. Table 3 shows the prediction accuracy of the optimum ANN model with the LM-TanSig algorithm and various numbers of the hidden neurons for the current, power and thermal efficiency of the thermoelectric generator system for waste heat recovery.

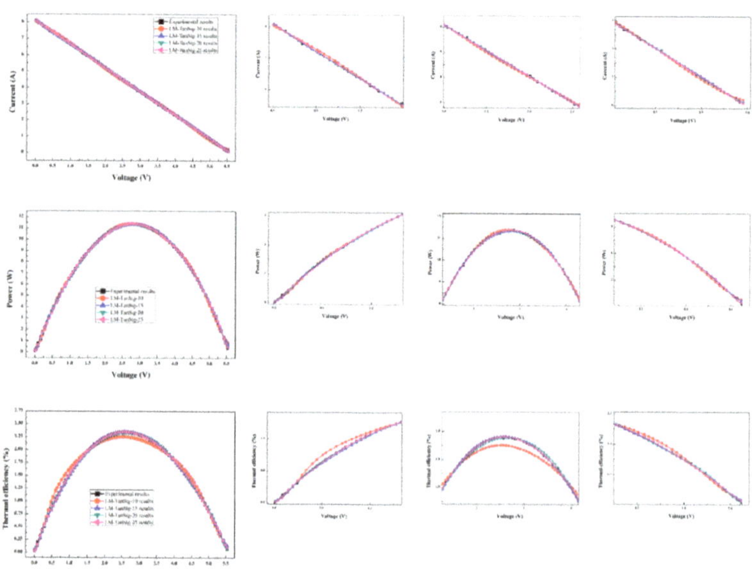

(**a**)

Figure 9. *Cont.*

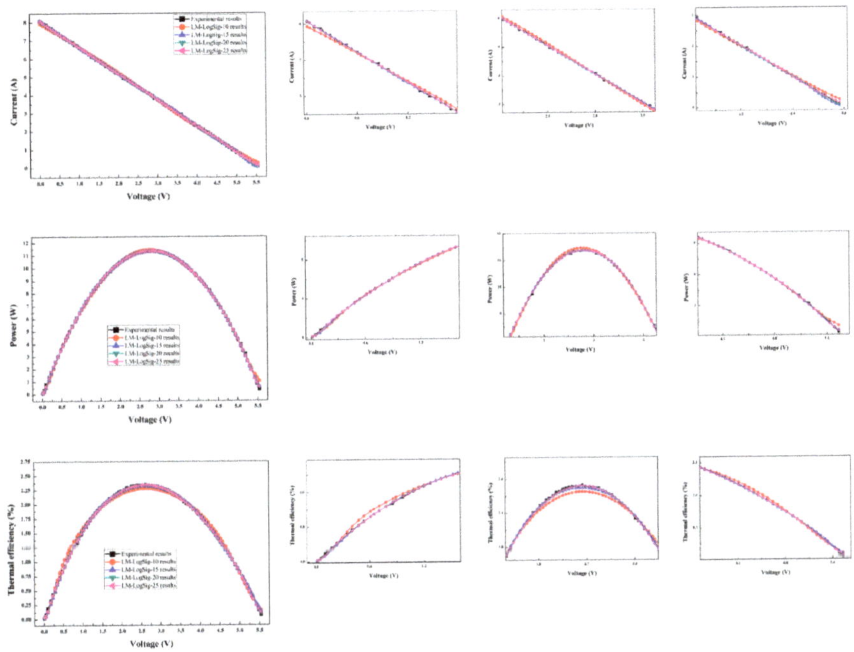

(**b**)

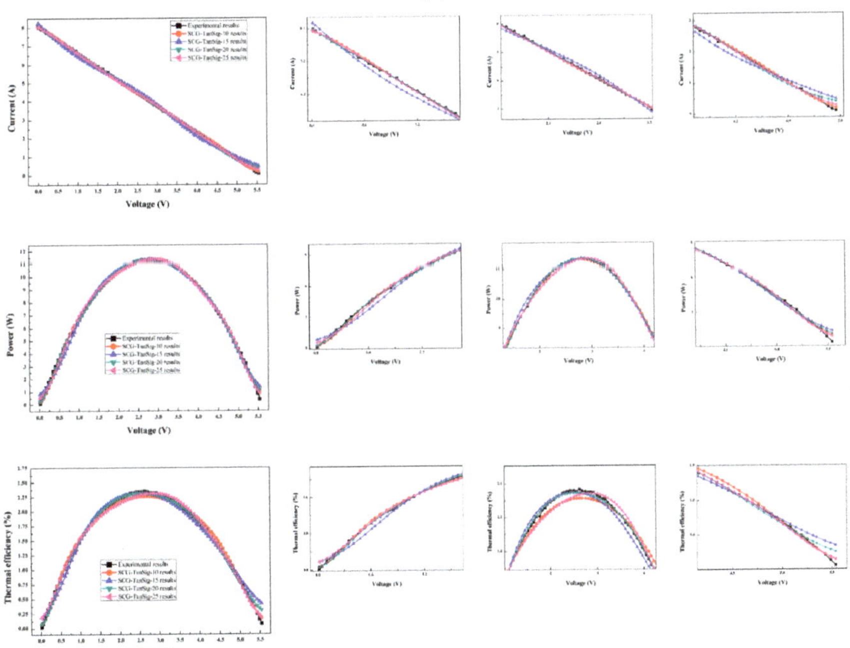

(**c**)

Figure 9. *Cont.*

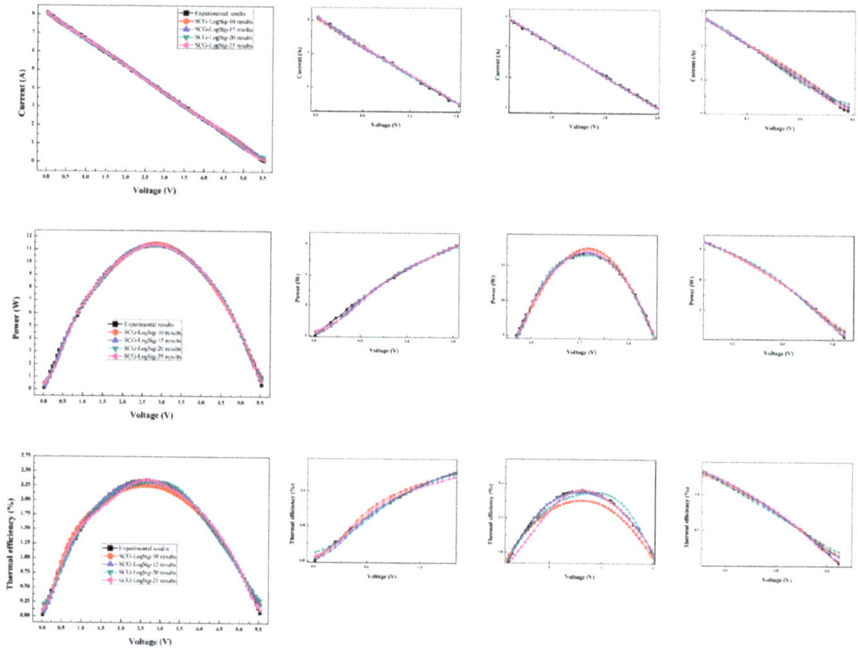

(**d**)

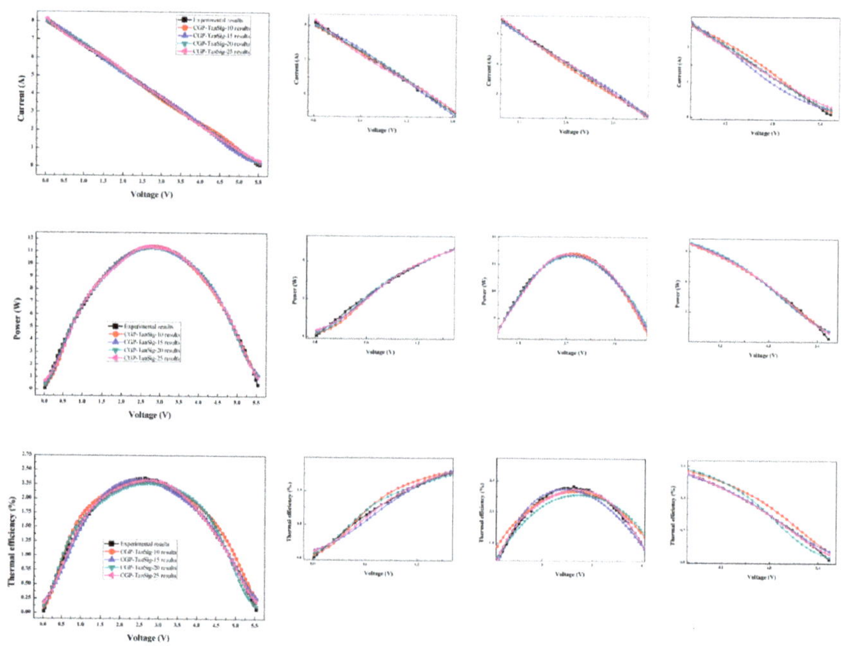

(**e**)

Figure 9. *Cont.*

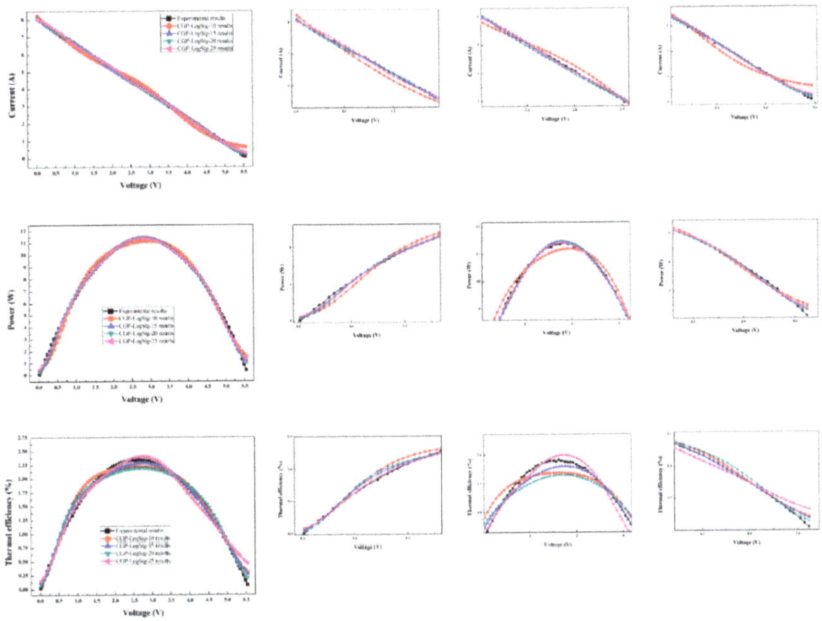

(f)

Figure 9. The comparison of experimental and ANN predicted results of current, power and thermal efficiency for (**a**) LM-TanSig algorithm, (**b**) LM-LogSig algorithm, (**c**) SCG-TanSig algorithm, (**d**) SCG-LogSig algorithm, (**e**) CGP-TanSig algorithm, and (**f**) the CGP-LogSig algorithm with various numbers of hidden neurons.

Table 3. The prediction accuracy of optimum ANN model with LM-TanSig algorithm and various numbers of hidden neurons for current, power and thermal efficiency.

Parameter	Number of Hidden Neurons	R^2	RMSE	COV
Current	10	0.99986	0.06013	1.36373
	15	0.99997	0.02578	0.58476
	20	0.99998	0.02507	0.56860
	25	0.99998	0.02163	0.49061
Power	10	0.99977	0.12180	1.75370
	15	0.99993	0.06769	0.97459
	20	0.99993	0.06609	0.95152
	25	0.99997	0.04111	0.59192
Thermal efficiency	10	0.99809	0.07170	4.99773
	15	0.99983	0.02126	1.48165
	20	0.99981	0.02277	1.58688
	25	0.99996	0.01050	0.73183

6.5. Prediction Results from ANFIS Models

The comparison of experimental and ANFIS predicted results of the current, power and thermal efficiency of the thermoelectric generator system for waste heat recovery using the triangular membership function is shown in Figure 10a. The triangular with 4-membership function shows higher prediction accuracy with R^2, RMSE and COV of 0.99998, 0.02209 and 0.50106, respectively and the prediction accuracy decreases in order of triangular with 5-, 2- and 3-membership functions for the current of the thermoelectric generator system for waste heat recovery. For the power of the thermoelectric generator system for waste heat recovery, the triangular with 4-membership function shows a good agreement with the experimental results with R^2, RMSE and COV of 0.99973, 0.13024 and 1.87522, respectively. The prediction accuracy of the triangular with 5- and 3-membership functions shows a good agreement within ±5%, but the prediction accuracy of the triangular with 2-membership function shows over ±15% from the corresponding experimental which are not a permissible limit. In the case of the thermal efficiency of the thermoelectric generator system for waste heat recovery using the triangular with 4-membership function shows the peak prediction accuracy and this prediction accuracy decreases in an order with the triangular with 5- and 3-membership functions. The values of R^2, RMSE and COV for the thermal efficiency of the thermoelectric generator system for waste heat recovery using the triangular with 4-membership function are 0.99968, 0.02955 and 2.05980, respectively. The thermal efficiency of the thermoelectric generator system for waste heat recovery using the triangular with a 2-membership function shows the errors over ±15% from the corresponding experimental thermal efficiency as shown in Figure 10a.

The comparison of experimental and ANFIS predicted results of the current, power and thermal efficiency of the thermoelectric generator system for waste heat recovery using trapezoidal membership function is shown in Figure 10b. The prediction accuracy for the current and power of the thermoelectric generator system for waste heat recovery using the trapezoidal with 5-membership function shows the best. The values of R^2, RMSE and COV for the trapezoidal with 5-membership function are 0.99998, 0.02333 and 0.52922, respectively, for the current and 0.99982, 0.10528 and 1.51577, respectively, for the power. In the case of the thermal efficiency of the thermoelectric generator system for waste heat recovery, the trapezoidal with a 4-membership function shows higher prediction accuracy than the trapezoidal with 5-membership function and the values of R^2, RMSE and COV for the trapezoidal with 4-membership function are 0.99978, 0.02424 and 1.68948, respectively, for thermal efficiency. However, the current, power and thermal efficiency predicted by the trapezoidal with 2- and 3-membership functions show the errors above ±15% from the experimental which are not within permissible limit.

The comparison of experimental and ANFIS predicted results of the current, power and thermal efficiency of the thermoelectric generator system for waste heat recovery using a gbell membership function is shown in Figure 10c. The prediction accuracy of the gbell with a 4-membership function for the current of the thermoelectric generator system for waste heat recovery shows the best and decreases with 5-, 3- and 2-membership functions. The values of R^2, RMSE and COV for gbell with 4-membership function are 0.99998, 0.02266 and 0.51393, respectively, for the current. For the power and thermal efficiency of the thermoelectric generator system for waste heat recovery, the gbell with 3-membership function shows a better agreement than gbell with 5- and 4-membership functions. The values of R^2, RMSE and COV for gbell with 3-membership function are 0.99996, 0.04812 and 0.69281, respectively, for the power but 0.99994, 0.01241 and 0.86506, respectively for the thermal efficiency. However, the prediction accuracy of the power and thermal efficiency of the thermoelectric generator system for waste heat recovery using gbell with a 2-membership function show the errors above ±15% from the corresponding experimental which are not the permissible limit.

The comparison of experimental and ANFIS predicted results of the current, power and thermal efficiency of the thermoelectric generator system for waste heat recovery using a gauss membership function is shown in Figure 10d. For the current, power and thermal efficiency of the thermoelectric generator system for waste heat recovery, the prediction accuracy of the gauss with a 5-membership function shows the best and decreases with 4-, 3- and 2-membership functions. The values of R^2, RMSE and COV for the gauss with 5-membership function are 0.99998, 0.02165 and 0.49110, respectively for the current, 0.99997, 0.04429 and 0.63770, respectively, for the power and 0.99997, 0.00911 and 0.63516, respectively, for the thermal efficiency. However, the current, power and thermal efficiency of the thermoelectric generator system for waste heat recovery using gauss with 2-membership function show the errors above ±15% from the corresponding experimental which are not a permissible limit.

The comparison of experimental and ANFIS predicted results of the current, power and thermal efficiency of the thermoelectric generator system for waste heat recovery using a gauss 2-membership function is shown in Figure 10e. For the current and thermal efficiency of the thermoelectric generator system for waste heat recovery, the gauss2 with 4-membership function shows higher prediction accuracy than gauss2 with a 5-membership function, but the gauss2 with 5-membership function shows higher prediction accuracy than gauss2 with 4-membership function for the power. The values of R^2, RMSE and COV for the gauss2 with 4-membership function are 0.99998, 0.02377 and 0.53902, respectively, for the current and 0.99992, 0.01437 and 1.0012, respectively, for the thermal efficiency. In addition, the values of R^2, RMSE and COV for gauss2 with a 5-membership function are 0.99992, 0.06965 and 1.00285, respectively, for the power. However, the current, power and thermal efficiency of the thermoelectric generator system for waste heat recovery using gauss2 with 2- and 3-membership function show the errors above ±15% from the corresponding experimental which are not the permissible limit.

The comparison of experimental and ANFIS predicted results of current, power and thermal efficiency of the thermoelectric generator system for waste heat recovery using a pi membership function is shown in Figure 10f. For the current, power and thermal efficiency of the thermoelectric generator system for waste heat recovery, the pi with 5-membership function shows higher prediction accuracy than pi with a 4-membership function. The values of R^2, RMSE and COV for pi with a 5-membership function are 0.99998, 0.02469 and 0.55991, respectively, for the current, 0.99997, 0.04029, and 0.58006, respectively, for the power and 0.99997, 0.00931, and 0.64890, respectively, for the thermal efficiency. However, the current, power and thermal efficiency of the thermoelectric generator system for waste heat recovery using pi with 2 and 3-membership functions show the errors above ±15% from the corresponding experimental which are not a permissible limit.

The comparison of experimental and ANFIS predicted results of current, power and thermal efficiency of the thermoelectric generator system for waste heat recovery using a dsig membership function is shown in Figure 10g. For the current of the thermoelectric generator system for waste heat recovery, dsig with a 4-membership function shows higher prediction accuracy than dsig with a 5-membership function. In addition, for the power and thermal efficiency of the thermoelectric generator system for waste heat recovery, dsig with a 5-membership function shows higher prediction accuracy than dsig with 4-membership function. The values of R^2, RMSE and COV for dsig with a 4-membership function are 0.99998, 0.02360 and 0.53534, respectively, for the current. In addition, the values of R^2, RMSE and COV for dsig with a 5-membership function are 0.99990, 0.07853 and 1.13067, respectively, for the power and 0.99989, 0.01704 and 1.18732, respectively, for the thermal efficiency. However, the current, power and thermal efficiency of the thermoelectric generator system for waste heat recovery using dsig with 2 and 3 membership functions show the errors above ±15% from the corresponding experimental, which are not a permissible limit.

The same number of the membership functions results in almost the same prediction cost. As the number of the membership functions increases, it results in a higher prediction cost. When the prediction accuracy plays a crucial role, the ANFIS model with a pi-5-membership function or a gauss-5-membership function could be recommended to predict the current, power and thermal efficiency of the thermoelectric generator system for waste heat recovery [40]. The prediction accuracy of the ANFIS model for the current, power and thermal efficiency is shown in Table 4a for pi membership function and Table 4b for gauss membership function, respectively. When the prediction cost plays a crucial role, the ANFIS model with gbell-3-membership function could be suggested to predict the current, power and thermal efficiency of the thermoelectric generator system for waste heat recovery as shown in Table 4c. Table 4c shows the prediction accuracy of an ANFIS model with a gbell membership function for the current, power and thermal efficiency of the thermoelectric generator system for waste heat recovery. The proposed ANFIS model with pi-5 or gauss-5 and gbell-3 show better prediction accuracy than the coupled numerical approach for the current, power and thermal efficiency of the thermoelectric generator system for waste heat recovery [41].

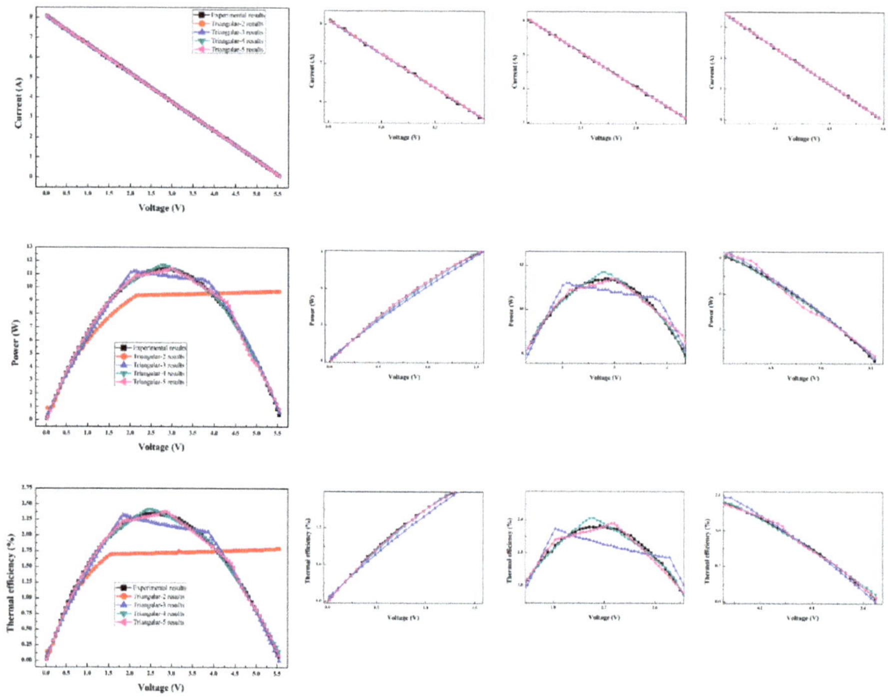

(a)

Figure 10. *Cont.*

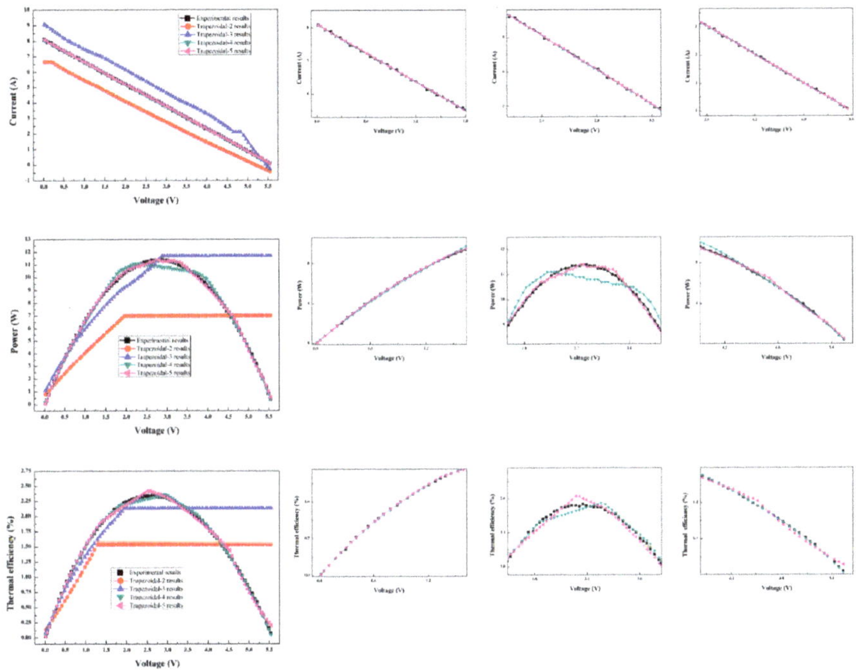

(**b**)

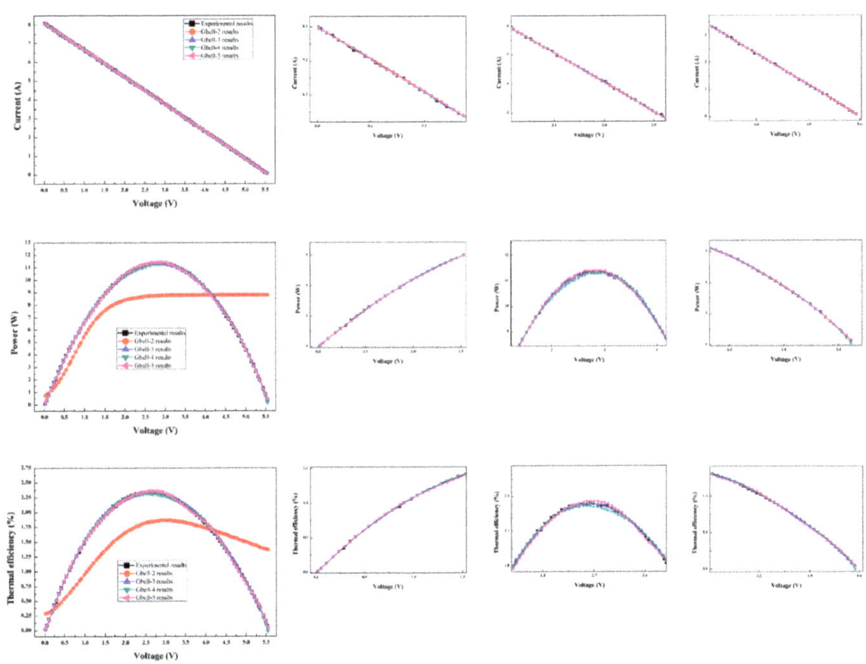

(**c**)

Figure 10. *Cont.*

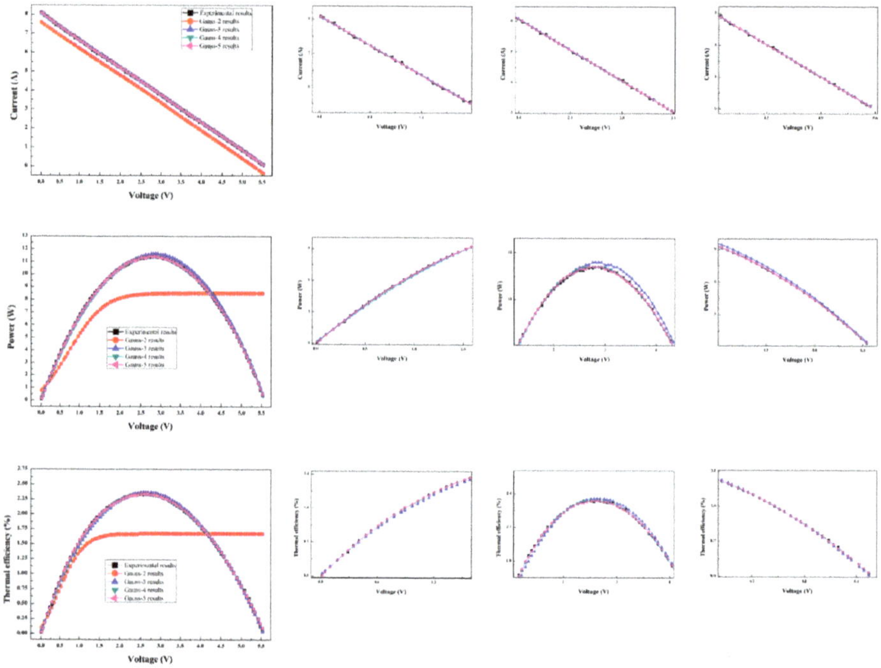

(d)

Figure 10. *Cont.*

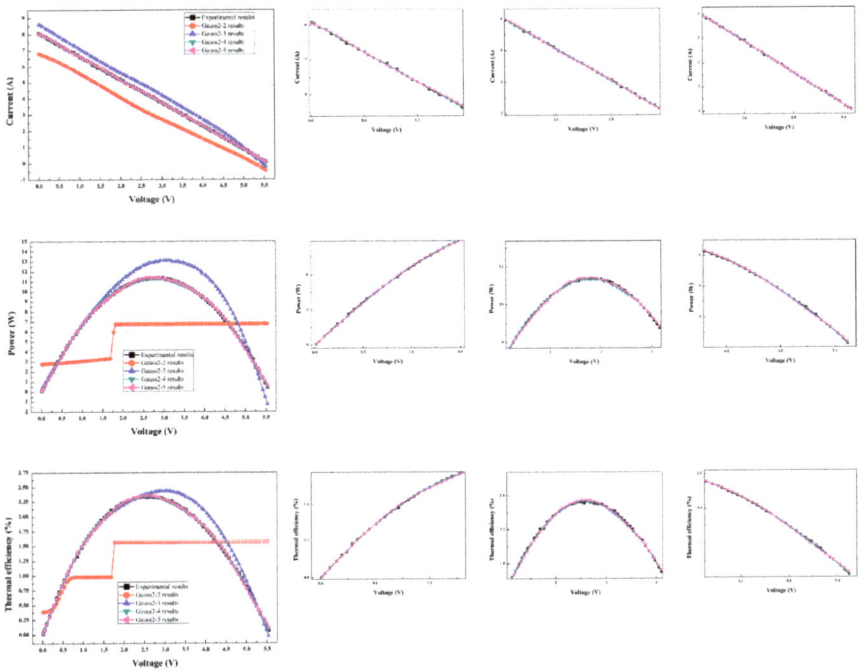

(**e**)

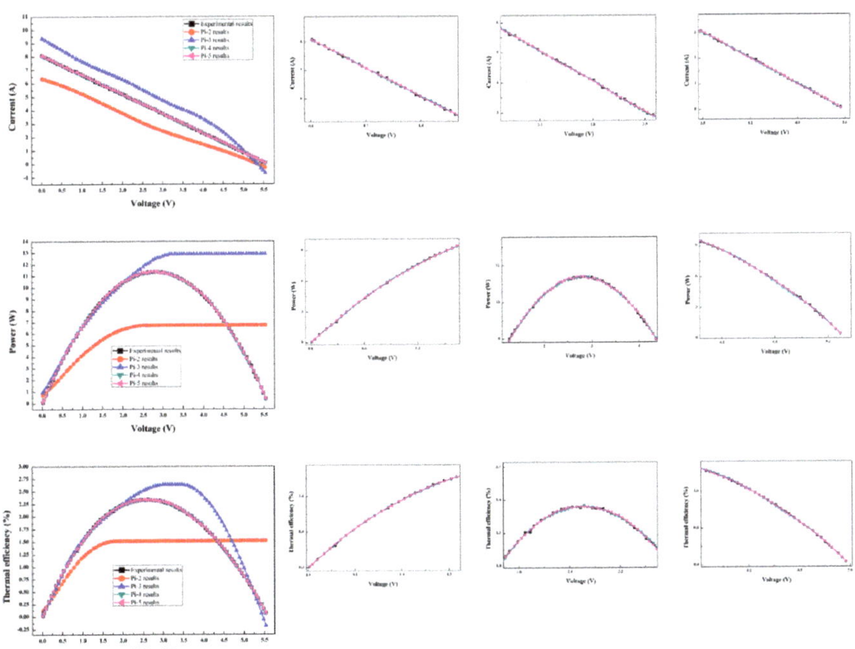

(**f**)

Figure 10. *Cont.*

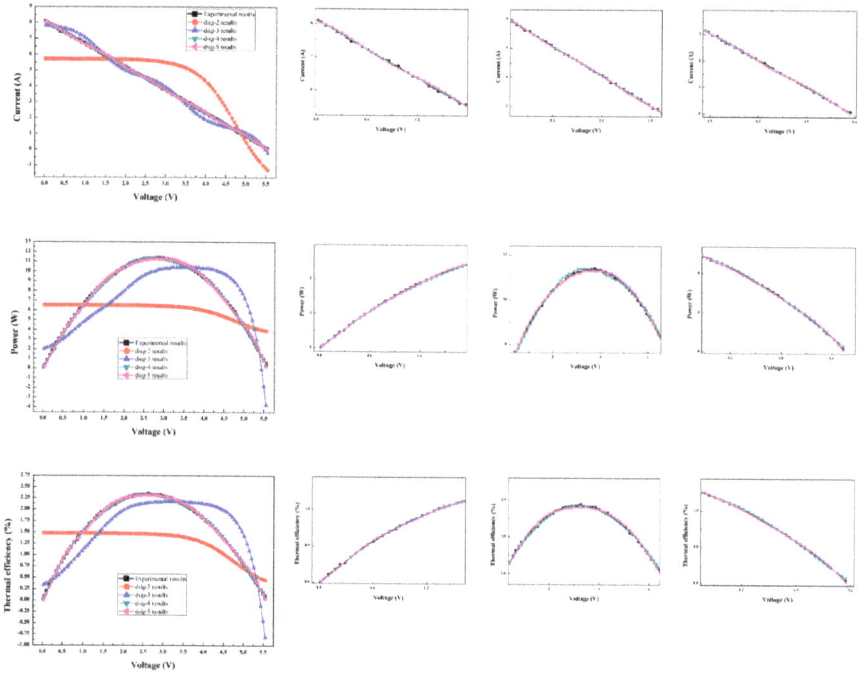

(g)

Figure 10. The comparison of experimental and ANFIS predicted results of current, power and thermal efficiency for (**a**) triangular membership function, (**b**) trapezoidal membership function, (**c**) gbell membership function, (**d**) gauss membership function, (**e**) gauss2-membership function, (**f**) pi membership function, and (**g**) dsig membership function.

Table 4. (a) The prediction accuracy of ANFIS model with pi membership function for current, power and thermal efficiency; (b) The prediction accuracy of ANFIS model with a gauss membership function for current, power and thermal efficiency; (c) The prediction accuracy of ANFIS model with a gbell membership function for current, power and thermal efficiency.

(a)

Parameter	Number of Membership Functions	R^2	RMSE	COV
Current	2	0.94053	1.23658	28.0468
	3	0.96078	1.00416	22.7752
	4	0.99997	0.02852	0.64682
	5	0.99998	0.02469	0.55991
Power	2	0.83830	3.19468	45.9964
	3	0.71781	4.22023	60.7620
	4	0.99994	0.05948	0.85635
	5	0.99997	0.04029	0.58006
Thermal efficiency	2	0.87063	0.58791	41.1021
	3	0.96815	0.29262	20.3952
	4	0.99994	0.01257	0.87619
	5	0.99997	0.00931	0.64890

Table 4. *Cont.*

(b)

Parameter	Number of Membership Functions	R^2	RMSE	COV
Current	2	0.99250	0.43925	9.96247
	3	0.99998	0.02366	0.53671
	4	0.99998	0.02254	0.51116
	5	0.99998	0.02165	0.49110
Power	2	0.88660	2.67538	38.5195
	3	0.99957	0.16457	2.36942
	4	0.99988	0.08667	1.24784
	5	0.99997	0.04429	0.63770
Thermal efficiency	2	0.88430	0.55769	38.8710
	3	0.99967	0.03002	2.09204
	4	0.99996	0.01056	0.73603
	5	0.99997	0.00911	0.63516

(c)

Parameter	Number of Membership Functions	R^2	RMSE	COV
Current	2	0.99997	0.02820	0.63965
	3	0.99998	0.02432	0.55166
	4	0.99998	0.02266	0.51393
	5	0.99998	0.02325	0.52728
Power	2	0.88847	2.65317	38.1999
	3	0.99996	0.04812	0.69281
	4	0.99990	0.08003	1.15220
	5	0.99996	0.04865	0.70043
Thermal efficiency	2	0.90510	0.50507	35.2033
	3	0.99994	0.01241	0.86506
	4	0.99984	0.02086	1.45405
	5	0.99987	0.01848	1.28804

Additionally, the developed ANN and ANFIS models could accurately predict the performances like the current, power and thermal efficiency of the thermoelectric generator system for waste heat recovery with less computational time and cost because the experimental and coupled numerical approaches could be expensive and time consuming. Therefore, the proposed methodology to develop the ANN and ANFIS models could be applicable to accurately predict the performances of the various physical systems like solar based systems, refrigeration systems, heat exchanger systems, thermoelectric coolers, etc.

7. Conclusions

The coupled numerical approach is investigated to predict the current, power and thermal efficiency of the thermoelectric generator system for waste heat recovery. Six ANN and seven ANFIS models are developed to predict the current, power and thermal efficiency of the thermoelectric generator system for waste heat recovery using the hot gas inlet temperatures and the voltage conditions as the inputs. Six ANN models with combinations of three training variants of Levenberg–Marquardt (LM), Scaled Conjugate Gradient (SCG) and Pola–Ribiere Conjugate Gradient (CGP), two transfer functions of Tan-Sigmodal and Log-Sigmoidal and the number of hidden neurons of 10, 15, 20 and 25 are compared. Seven ANFIS models are compared with seven types of the membership functions of triangular, trapezoidal, gauss, gauss2, gbell, pi and dsig, and the number of the membership functions of 2, 3, 4 and 5. The optimum ANN and ANFIS models are proposed from the comparison with experimental data using three statistical parameters of the coefficient of determination (R^2), root mean square error (RMSE) and coefficient of variance (COV). The ANN model with back-propagation algorithm, Levenberg–Marquardt training variant, Tan-Sigmoidal transfer function, and 25 hidden neurons is suggested as the optimum model based on optimum values of statistical parameters for the prediction of the current, power and thermal efficiency of the thermoelectric generator system for waste heat recovery. The ANFIS model with gbell membership function in a number of sets of 3 is suggested as the optimum model based on optimum values of statistical parameters to predict the current, power and thermal efficiency of the thermoelectric generator system for waste heat recovery with low prediction cost and acceptable prediction accuracy. The ANFIS model with pi or gauss membership function in the number of sets of 5 is suggested as the optimum model based on optimum values of statistical parameters to predict the current, power and thermal efficiency of the thermoelectric generator system for waste heat recovery with higher prediction accuracy. The optimum ANN and ANFIS models show better prediction of the current, power and thermal efficiency of the thermoelectric generator system for waste heat recovery with low computational time and cost than the coupled numerical approach.

Author Contributions: Conceptualization, K.S.G.; J.-H.S., and M.-Y.L.; methodology, K.S.G.; J.-H.S.; C.-P.C., and M.-Y.L.; software, K.S.G. and J.-H.S.; validation, K.S.G. and J.-H.S.; formal analysis, K.S.G.; J.-H.S.; C.-P.C., and M.-Y.L.; investigation, K.S.G.; J.-H.S.; C.-P.C., and M.-Y.L.; resources, K.S.G. and J.-H.S.; data curation, K.S.G.; writing—original draft preparation, K.S.G.; J.-H.S. and M.-Y.L.; writing—review and editing, K.S.G.; J.-H.S. and M.-Y.L.; visualization, K.S.G.; supervision, M.-Y.L.; project administration, M.-Y.L.; funding acquisition, M.-Y.L. All authors have read and agreed to the published version of the manuscript.

Funding: This research received no external funding.

Acknowledgments: This work was supported by the Dong-A University research fund.

Conflicts of Interest: The authors declare no conflict of interest.

References

1. Demirel, Y. *Energy: Production, Conversion, Storage, Conservation, and Coupling*; Springer Science & Business Media: Berlin/Heidelberg, Germany, 2012.
2. Riffat, S.B.; Ma, X. Thermoelectrics: A review of present and potential applications. *Appl. Therm. Eng.* **2003**, *23*, 913–935. [CrossRef]
3. Wang, Y.; Li, S.; Xie, X.; Deng, Y.; Liu, X.; Su, C. Performance evaluation of an automotive thermoelectric generator with inserted fins or dimpled-surface hot heat exchanger. *Appl. Energy* **2018**, *218*, 391–401. [CrossRef]
4. Niu, Z.; Diao, H.; Yu, S.; Jiao, K.; Du, Q.; Shu, G. Investigation and design optimization of exhaust-based thermoelectric generator system for internal combustion engine. *Energy Convers. Manag.* **2014**, *85*, 85–101. [CrossRef]
5. Liu, X.; Deng, Y.D.; Zhang, K.; Xu, M.; Xu, Y.; Su, C.Q. Experiments and simulations on heat exchangers in thermoelectric generator for automotive application. *Appl. Therm. Eng.* **2014**, *71*, 364–370. [CrossRef]
6. Quan, R.; Liu, G.; Wang, C.; Zhou, W.; Huang, L.; Deng, Y. Performance investigation of an exhaust thermoelectric generator for military SUV application. *Coatings* **2018**, *8*, 45. [CrossRef]
7. Luo, D.; Wang, R.; Yu, W.; Sun, Z.; Meng, X. Modelling and simulation study of a converging thermoelectric generator for engine waste heat recovery. *Appl. Therm. Eng.* **2019**, *153*, 837–847. [CrossRef]
8. Nithyanandam, K.; Mahajan, R.L. Evaluation of metal foam based thermoelectric generators for automobile waste heat recovery. *Int. J. Heat Mass Transf.* **2018**, *122*, 877–883. [CrossRef]
9. Cao, Q.; Luan, W.; Wang, T. Performance enhancement of heat pipes assisted thermoelectric generator for automobile exhaust heat recovery. *Appl. Therm. Eng.* **2018**, *130*, 1472–1479. [CrossRef]
10. He, W.; Wang, S. Performance Comparison of Different Exhaust Exchanger Types Considering Peak Net Power and Optimal Dimension in a Thermoelectric Generator System. *Adv. Theory Simul.* **2018**, *1*, 1800012. [CrossRef]
11. Lu, X.; Yu, X.; Qu, Z.; Wang, Q.; Ma, T. Experimental investigation on thermoelectric generator with non-uniform hot-side heat exchanger for waste heat recovery. *Energy Convers. Manag.* **2017**, *150*, 403–414. [CrossRef]
12. Rana, S.; Date, A.; Iqbal, A.; Akbarzadeh, A. Optimization model for power generation using thermoelectric generator. *Energy Procedia* **2019**, *160*, 723–730. [CrossRef]

13. Suter, C.; Jovanovic, Z.R.; Steinfeld, A. A 1 kWe thermoelectric stack for geothermal power generation–Modeling and geometrical optimization. *Appl. Energy* **2012**, *99*, 379–385. [CrossRef]
14. Zhao, Y.; Wang, S.; Ge, M.; Liang, Z.; Liang, Y.; Li, Y. Performance analysis of automobile exhaust thermoelectric generator system with media fluid. *Energy Convers. Manag.* **2018**, *171*, 427–437. [CrossRef]
15. Zhao, Y.; Wang, S.; Ge, M.; Liang, Z.; Liang, Y.; Li, Y. Performance investigation of an intermediate fluid thermoelectric generator for automobile exhaust waste heat recovery. *Appl. Energy* **2019**, *239*, 425–433. [CrossRef]
16. Lu, H.; Wu, T.; Bai, S.; Xu, K.; Huang, Y.; Gao, W.; Chen, L. Experiment on thermal uniformity and pressure drop of exhaust heat exchanger for automotive thermoelectric generator. *Energy* **2013**, *54*, 372–377. [CrossRef]
17. Dheenamma, M.; Soman, D.P.; Muthamizhi, K.; Kalaichelvi, P. In pursuit of the best artificial neural network configuration for the prediction of output parameters of corrugated plate heat exchanger. *Fuel* **2019**, *239*, 461–470. [CrossRef]
18. Angeline, A.A. Power generation from combusted "Syngas" using hybrid thermoelectric generator and forecasting the performance with ANN technique. *J. Therm. Eng.* **2018**, *4*, 2149–2168. [CrossRef]
19. Angeline, A.A.; Asirvatham, L.G.; Hemanth, D.J.; Jayakumar, J.; Wongwises, S. Performance prediction of hybrid thermoelectric generator with high accuracy using artificial neural networks. *Sustain. Energy Technol. Assess.* **2019**, *33*, 53–60. [CrossRef]
20. Liu, X.; Deng, Y.D.; Li, Z.; Su, C.Q. Performance analysis of a waste heat recovery thermoelectric generation system for automotive application. *Energy Convers. Manag.* **2015**, *90*, 121–127. [CrossRef]
21. Raj, A.K.; Kunal, G.; Srinivas, M.; Jayaraj, S. Performance analysis of a double-pass solar air heater system with asymmetric channel flow passages. *J. Therm. Anal. Calorim.* **2019**, *136*, 21–38. [CrossRef]
22. Park, S.H.; Jin, Y.; Cha, J.; Hong, K.; Kim, Y.; Yoon, H.; Chung, I. High-Power-Density Skutterudite-Based Thermoelectric Modules with Ultralow Contact Resistivity Using Fe–Ni Metallization Layers. *ACS Appl. Energy Mater.* **2018**, *1*, 1603–1611. [CrossRef]
23. Seo, J.H.; Patil, M.S.; Cho, C.P.; Lee, M.Y. Heat transfer characteristics of the integrated heating system for cabin and battery of an electric vehicle under cold weather conditions. *Int. J. Heat Mass Transf.* **2018**, *117*, 80–94. [CrossRef]
24. Ma, Q.; Fang, H.; Zhang, M. Theoretical analysis and design optimization of thermoelectric generator. *Appl. Therm. Eng.* **2017**, *127*, 758–764. [CrossRef]
25. Yazdani-Chamzini, A.; Zavadskas, E.K.; Antucheviciene, J.; Bausys, R. A model for shovel capital cost estimation, using a hybrid model of multivariate regression and neural networks. *Symmetry* **2017**, *9*, 298. [CrossRef]
26. Mohanraj, M.; Jayaraj, S.; Muraleedharan, C. Performance prediction of a direct expansion solar assisted heat pump using artificial neural networks. *Appl. Energy* **2009**, *86*, 1442–1449. [CrossRef]
27. Islam, K.T.; Raj, R.G.; Mujtaba, G. Recognition of traffic sign based on bag-of-words and artificial neural network. *Symmetry* **2017**, *9*, 138. [CrossRef]
28. Moya-Rico, J.D.; Molina, A.E.; Belmonte, J.F.; Tendero, J.C.; Almendros-Ibanez, J.A. Characterization of a triple concentric-tube heat exchanger with corrugated tubes using Artificial Neural Networks (ANN). *Appl. Therm. Eng.* **2019**, *147*, 1036–1046. [CrossRef]
29. Ullah, I.; Fayaz, M.; Kim, D. Improving accuracy of the kalman filter algorithm in dynamic conditions using ANN-based learning module. *Symmetry* **2019**, *11*, 94. [CrossRef]
30. Şahin, A.Ş. Performance analysis of single-stage refrigeration system with internal heat exchanger using neural network and neuro-fuzzy. *Renew. Energy* **2011**, *36*, 2747–2752. [CrossRef]
31. Bogiatzis, A.; Papadopoulos, B. Global Image Thresholding Adaptive Neuro-Fuzzy Inference System Trained with Fuzzy Inclusion and Entropy Measures. *Symmetry* **2019**, *11*, 286. [CrossRef]
32. Liu, Y.; Wang, Q.; Zhang, X.; Song, S.; Niu, C.; Shangguan, Y. Using ANFIS and BPNN Methods to Predict the Unfrozen Water Content of Saline Soil in Western Jilin, China. *Symmetry* **2019**, *11*, 16. [CrossRef]
33. Mehrabi, M.; Pesteei, S.M. Modeling of heat transfer and fluid flow characteristics of helicoidal double-pipe heat exchangers using adaptive neuro-fuzzy inference system (ANFIS). *Int. Commun. Heat Mass Transf.* **2011**, *38*, 525–532. [CrossRef]
34. Sarkar, M.; Julai, S.; Wen Tong, C.; Toha, S.F. Effectiveness of Nature-Inspired Algorithms using ANFIS for Blade Design Optimization and Wind Turbine Efficiency. *Symmetry* **2019**, *11*, 456. [CrossRef]

35. Yeom, C.U.; Kwak, K.C. Performance Comparison of ANFIS Models by Input Space Partitioning Methods. *Symmetry* **2018**, *10*, 700. [CrossRef]
36. Erturun, U.; Mossi, K. A feasibility investigation on improving structural integrity of thermoelectric modules with varying geometry. In Proceedings of the ASME 2012 Conference on Smart Materials, Adaptive Structures and Intelligent Systems, Stone Mountain, GA, USA, 19–21 September 2012; pp. 939–945.
37. Phillips, S.S. Characterizing the Thermal Efficiency of Thermoelectric Modules. Ph.D. Thesis, Massachusetts Institute of Technology, Cambridge, MA, USA, 2009.
38. Gunasekar, N.; Mohanraj, M.; Velmurugan, V. Artificial neural network modeling of a photovoltaic-thermal evaporator of solar assisted heat pumps. *Energy* **2015**, *93*, 908–922. [CrossRef]
39. Esen, H.; Inalli, M. ANN and ANFIS models for performance evaluation of a vertical ground source heat pump system. *Expert Syst. Appl.* **2010**, *37*, 8134–8147. [CrossRef]
40. Belaout, A.; Krim, F.; Mellit, A.; Talbi, B.; Arabi, A. Multiclass adaptive neuro-fuzzy classifier and feature selection techniques for photovoltaic array fault detection and classification. *Renew. Energy* **2018**, *127*, 548–558. [CrossRef]
41. Yaïci, W.; Entchev, E. Adaptive neuro-fuzzy inference system modelling for performance prediction of solar thermal energy system. *Renew. Energy* **2016**, *86*, 302–315. [CrossRef]

symmetry

MDPI

Article

Thermal Abuse Behavior of the LIR2450 Micro Coin Cell Battery Having Capacity of 120 mAh with Internal Short Circuit by Penetrating Element

Moo-Yeon Lee [1], Namwon Kim [2], Jae-Hyeong Seo [1] and Mahesh Suresh Patil [1,*]

1 Department of Mechanical Engineering, Dong-A University, 37 Nakdong-Daero 550, Saha-gu, Busan 607-714, Korea; mylee@dau.ac.kr (M.-Y.L.); cheonchw@naver.com (J.-H.S.)

2 Ingram School of Engineering, Texas State University, San Marcos, TX 78666, USA; namwonkim@txstate.edu

* Correspondence: msp692@gmail.com; Tel.: +82-51-200-5560

Received: 8 January 2020; Accepted: 2 February 2020; Published: 5 February 2020

check for updates

Abstract: Internal short circuit in lithium-ion battery by penetrating element leads to exothermic behavior due to accumulated heat. In the present study, investigations are conducted on the thermal behavior of the LIR2450 micro coin cell haivng capacity of 120 mAh, with internal short circuit by penetrating element. The experimental coin cell discharge study was conducted and validated with numerical study within ±5.0%. The effect of penetrating element size, location of penetrating element, state of charge, discharge rate, short-circuit resistance, and heat transfer co-efficient on maximum coin cell temperature and heat generation rate are analyzed. The penetrating element diameters of 0.5, 1.0, 1.5, 2.0, 2.5, 3.0, and 3.5 mm are considered. The effect of initial state of charge (SOC) is considered with 100%, 80%, 60%, and 40%. Three locations for penetrating element are considered with the center, the middle of the radius, and on the edge of the coin cell radius. The different discharge rates of 1C, 2C, 3C, and 4C are considered. The higher-penetrating element size of 3.5 mm with location at the center of the coin cell with 100% SOC showed maximum heat generation rate and maximum temperature of the coin cell. In addition, the optimum value of the dimensionless heat generation rate is obtained at dimensionless short-circuit resistance. The study provides comprehensive insights on the thermal behavior of the lithium-ion cell during thermal abuse condition with internal short circuit by penetrating element.

Keywords: heat generation; internal short; lithium-ion; nail penetration; temperature; thermal abuse

1. Introduction

Lithium-ion batteries (LIBs) are extensively used in various applications, including from small-scale portable electronics to large-scale application in electric vehicles, along with other applications in the field of drones, airplanes, and robots [1]. LIBs have shown efficient and practical applications in domestic as well as hand-held devices [2]. The LIBs are preferred over other chemistries owing to their strong advantages of high-energy density, low self-discharge, fast charging ability, enhanced cycle life, among many other advantages [3].

However, LIBs are likely to fail in case of over-charging and over-discharging with the possibility of a severe accident in case of overheating or internal-external short circuit [4]. The thermal abuse of LIBs can be caused by penetration, external-short, overheating, or over-charge [5]. In the case of a thermal abuse incident, the available chemical energy in LIBs rapidly converts to heat energy, generating a large amount of heat with the possibility of thermal runaway—fire accompanied by the occasional explosion [6]. Owing to high-energy density and flammable components, LIBs are prone to safety issues. Physical abuse of LIBs could lead to a temperature rise to the levels of 100 to 150 °C, which could trigger cascading exothermic electrochemical reactions along with chemical disintegration, resulting in the rise of temperature to the levels of 500 °C in a few seconds. Although many safety methods, including effective cooling, have been installed, several cases of hazards have been reported over the years [7,8]. The recent events of LIB failure events leading to fire and explosion accidents have focused research attention on LIB safety [9]. For example: iPod caught fire due to overheated LIB [10] and three fire accidents of Boeing 747 at different locations [11]. Although, recently, LIB safety issues have received attention due to increased instances of hazards in electrical vehicle application, the safety concerns were admitted for many years [12,13].

In the case of continuous increase of temperature, rapid reactions start to occur. At around 110 °C, the graphite anode reacts with the electrolyte solvent. At around 165 °C, cathode material decomposes and releases oxygen. At around 180 °C, electrolyte solvent decomposes, and flammable components undergo combustion leading to thermal runaway [14]. The mechanical abuse in terms of penetration, or crush leading to thermal abuse, is an important safety concern and has received increased attention in recent years [15]. In many applications, the overheating and overcharge is controlled as a safe temperature and cut-off voltage are maintained. However, mechanical abuse leading to a short circuit is a challenging threat, as LIB can be subjected to crushing or metal penetration in unpredictable situations, such as accidents. The penetration provides a low-resistance path, drawing extremely high levels of currents with a rapid discharge of available chemical energy and simultaneously converting it into thermal energy. In the absence of resistive load, with the short circuit providing a low-resistance path, the energy is dissipated as joule heat, resulting in a rapid temperature increase [16].

Chen et al. developed an electro-thermal model to investigate the thermal performance during normal discharge and internal short circuit. The authors investigated the effect of an internal short circuit at the center of the battery on the temperature rise of the battery, and suggested that the increasing thermal conductivity of a separator can be an effective method to reduce the heat accumulation at the location of the internal short circuit [17]. Shi et al. conducted an experimental study on the exothermic behavior of LIB under mechanical abuse and suggested that dibenzylamine (DBA) could be used to prevent thermal runaway. The authors suggested that under normal conditions DBA does not affect the battery performance, and during mechanical abuse, DBA is released, which increases electrolyte resistivity preventing thermal runaway [18]. Vyroubal et al. presented a finite element model of nail penetration into the lithium-ion battery and showed that shorting resistance has a significant influence on the cell electrochemical-thermal process of batteries [19]. Noelle investigated internal short circuit on LIB by conducting direct current internal resistance, extremal shorting, and nail penetration experiment, and presented an electrolyte resistance model with experimental validation [20]. Fang et al. developed a 3D electrochemical-thermal model for internal short in Li-ion cell, and predicted that discharge rate was higher in Anode-Aluminum short as compared to Anode-Cathode owing to lower short resistance [21]. Mao et al. investigated the failure mechanism of the lithium-ion battery during nail penetration tests and presented the influence of penetration position and depth on temperature of the jelly-roll-type cylindrical cell. The authors pointed out that maximum temperature was observed when the nail was penetrated at the center of the cylindrical cell with the region of thermal runaway covering the entire cell. According to the micro short-circuit cell model presented by authors, if the temperature reaches between 90–120 °C, heat can spread to whole battery. However, reactions cannot lead to thermal runaway due to the unavailability of oxygen. But if the temperature reaches up to more than 233 °C, then the separator shrinks, thus leading to a larger short-circuit area, with the start of a reaction between cathode and electrolyte releasing a large amount heat and combustion byproducts in terms of gases [22]. Zao et al. conducted external and internal short circuit tests on batteries with different capacities. The authors showed that for external short test, with lower internal resistance, heat was accumulated between clamps and battery tabs, whereas, in case of nail penetration, heat accumulation increases with battery capacity and may catch fire. The authors suggested hydrogel-based thermal management to prevent thermal runaway [23]. There are few studies on how to mitigate the effect of internal short circuit and prevent thermal runaway. For example, Wang et al. suggested a modified current collector with surface-notch, instead of a flat current collector, leading to a negligible temperature increase associated with internal sorting [24]. However, safety concerns still remain for the accumulation of large amounts of heat leading to thermal runaway in LIB during internal short circuit, and more understanding is still needed of the thermal behavior of LIB during internal shorting.

The goal of the present study is to investigate the thermal abuse behavior of the LIR2450 coin cell with internal short circuit by penetrating element. The experimental coin cell discharge study is conducted and validated with numerical study within ±5.0%. The effect of penetrating element size, location of penetrating element, state of charge (SOC), discharge rate, short-circuit resistance, and heat transfer coefficient on maximum temperature and heat generation rate are presented. The study provides comprehensive insights on the thermal behavior of LIBs during thermal abuse condition with internal short circuit by penetrating element.

2. Experimental Study

Figure 1 shows the schematic for the experimental study. The LIB under consideration is the LIR2450 coin cell with 120 mAh capacity, and the specifications of the coin cell are presented in Table 1. The coin cell was connected to KIKUSUI electronic load PLU-150 for evaluating and maintaining current, voltage, and power. The discharge test of the coin cell was conducted under constant current condition. The coin cell was fully charged to 4.2 V and discharge tests were conducted with a cut-off voltage of 2.75 V. The thermocouple was attached to a coin cell positive tab for measuring the surface temperature. The second thermocouple was attached to measure the ambient temperature. All the experiments were conducted in a room set at a constant temperature. The temperature of the room was controlled and kept constant at 25 °C. The details of the equipment used in the experimental study are presented in Table 2.

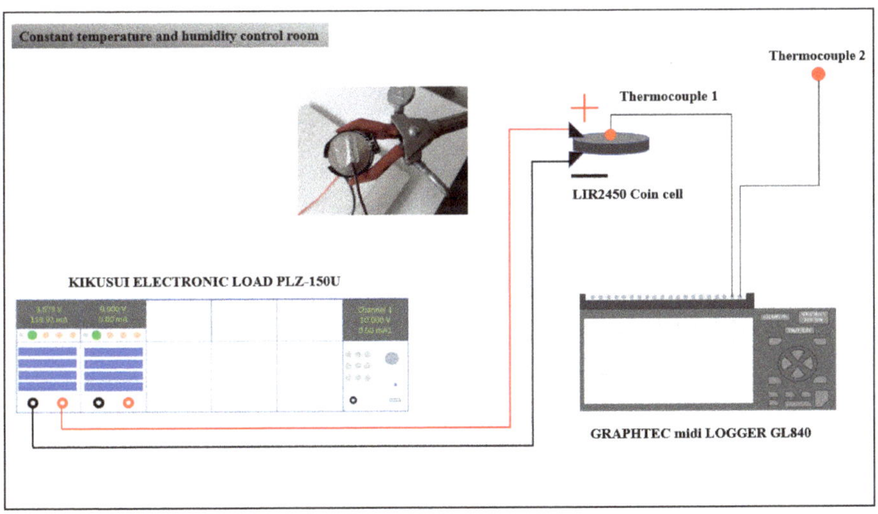

Figure 1. Experimental setup.

Table 1. The thermal and electrical properties of the 120 mAh LIR2450 coin cell.

Specifications	Values
Cathode material	$LiCoO_2$
Anode material	Graphite
Nominal capacity (mAh)	120
Mass (g)	5
Specific heat capacity (J/kg·K)	1000
Internal resistance (mΩ)	≤400
Density (kg/m^3)	1940
Thermal conductivity (W/m-K)	18.2
Geometry specifications	
Diameter (mm)	24.5
Height (mm)	5

Table 2. Equipment details.

Equipment	Parameter	Specifications
KIKUSUI electronic load PLU-150	Operating voltage	1.5 to 150 V
	Current range	300 mA to 30 A
	Maximum power	150 W
GL820 Data logger	Operating range	−200 °C *leq* TS *leq* 400 °C
Constant temperature and humidity control room	Temperature range	– 30 TO 60 °C
	Humidity range	30% to 95% RH

The measured parameter uncertainty is presented based on the accuracy of the instrument. The measured parameters were voltage, current, and temperature. The accuracy for current measurement was calculated based on Equation (1). The maximum uncertainty in measuring current was 0.45%. The accuracy of the voltage measurement was calculated based on Equation (2). The maximum uncertainty in voltage measurement was 0.08%. The accuracy in temperature measurement was calculated based on Equation (3). The maximum uncertainty in temperature measurement was 1.25%.

$$I_U = \pm(0.2\% \text{ of set} + 0.2\% \text{ of full scale}) + \text{V}_{in}/500\text{k}\Omega \quad (1)$$

$$V_U = \pm(0.2\% \text{ of set} + 0.2\% \text{ of full scale}) \quad (2)$$

$$T_U = \pm(0.1\% \text{ of reading} + 0.5^\circ\text{C}) \quad (3)$$

3. Numerical Model

The numerical model was developed using the MSMD (multi-scale multi-dimensional) model of Ansys Fluent [25]. The LIR2450 coin cell was modeled with radius of 24.5 mm and height of 5 mm. The mesh was generated with 107,478 nodes and 99,552 elements. The geometry and meshing images are provided in Figure 2. The current density, j_{Ech}, was calculated from Equation (4), proposed by Newman, Tiedemann, Gu, and Kim (NTGK) [26] as follows:

$$j_{Ech} = \frac{Q_{nominal}}{Q_{ref} Vol} Y(U - V) \quad (4)$$

where V is battery cell voltage; $Q_{nominal}$ is the battery capacity in Ampere hours; Q_{ref} is the battery capacity used in the experiments to obtain Y and U. Y and U are LIB depth of discharge functions. In the present study, discharge tests were conducted at different constant currents to obtain Y and U functions. The obtained Y and U are fitting parameters with functions of depth of discharge in

Equations (5) and (6), as initially suggested by Gu et al. [27]. The values obtained by curve fitting for the 120 mAh LIR2450 coin cell battery are presented in Table 3.

$$U = a_0 + a_1(DOD)^1 + a_2(DOD)^2 + a_3(DOD)^3 + a_4(DOD)^4 + a_5(DOD)^5 \quad (5)$$

$$Y = b_0 + b_1(DOD)^1 + b_2(DOD)^2 + b_3(DOD)^3 + b_4(DOD)^4 + b_5(DOD)^5 \quad (6)$$

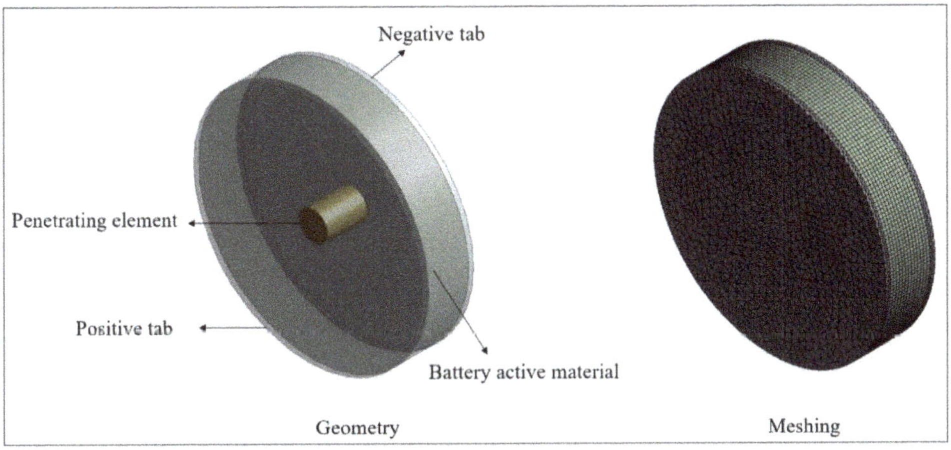

Figure 2. Numerical model geometry and meshing.

Table 3. Experimental values for co-efficient of U and Y functions.

Co-Efficient of U	Values	Co-Efficient of Y	Values
a_0	4.167186	b_0	0.923942
a_1	−1.12224	b_1	−7.07927
a_2	1.522472	b_2	37.43602
a_3	−3.46622	b_3	−87.1731
a_4	5.954965	b_4	90.32512
a_5	−3.55203	b_5	−34.5455

The thermal- and electrical-coupled field equation for battery operation were solved using Equations (7)–(9) [25], where σ_+ and σ_- are electrical conductivities of the positive and negative electrode, respectively; ϕ_+ and ϕ_- phase potentials for the positive and negative electrodes, respectively; j_{ECh} and $\dot{q}_{short}$ represent volumetric current transfer rate and heat due to electrochemical reaction, respectively; j_{short} and $\dot{q}_{short}$ represent current transfer rate and heat generation during internal short circuit, respectively. ρ, k, and T represent density, thermal conductivity, and temperature, respectively.

$$\frac{\partial \rho C_p T}{\partial t} = -\nabla \cdot (k\nabla T) = \sigma_+ |\nabla\phi_+|^2 + \sigma_- |\nabla\phi_-|^2 + \dot{q}_{ECh} + \dot{q}_{short} \quad (7)$$

$$\nabla \cdot (\sigma_+ \nabla\phi_+) = -(j_{ECh} - j_{short}) \quad (8)$$

$$\nabla \cdot (\sigma_- \nabla\phi_-) = j_{ECh} - j_{short} \quad (9)$$

In Equation (10), $Q_{nominal}$ and Q_{ref} are the same as the reference battery parameters and are estimated through experimentation. The electrochemical reaction heat was calculated as given in

Equation (11). The first term $(U - V)$ in Equation (11) [25] represents heat generated due to overpotential, and the second term is related to heat generated due to entropic heating.

$$j_{ECh} = \frac{Q_{nominal}}{Q_{ref} Vol} Y[U - V] \tag{10}$$

$$\dot{q}_{ECh} = j_{Ech}\left[U - V - T\frac{dU}{dT}\right] \tag{11}$$

During the normal operation of LIB battery, the cathode and anode are separated by separators, which is generally thin polymer material. The separator prevents the direct contact of positive and negative electrodes. In the event of short circuiting, as a result of penetration or crash, the separator gets damaged, which results in a secondary current along with regular current flowing through tabs. The transfer short current density is computed from Equation (12). The heat generated by internal short circuit is computed by volumetric contact resistance (r_c/a), where r_c is contact resistance and a is the specific area of the electrode, as shown in Equation (13) [23].

$$j_{short} = a(\phi_+ - \phi_-)/r_c \tag{12}$$

$$\dot{q}_{short} = a(\phi_+ - \phi_-)^2/r_c \tag{13}$$

4. Results and Discussion

The results present the effect of various parameters on the thermal behavior of the coin cell with internal short circuit caused by penetrating element. The numerical model is used to simulate the thermal behavior of the penetrating element before the coin cell reaches 150 °C [17]. Thermal runaway behavior, including material decomposition, is not included in the present study. The focus of the present study is to analyze the thermal abuse behavior due to internal short circuit before the coin cell temperature reaches thermal runaway [17]. The experimental study was conducted at a 1C discharge rate, with a fully-charged coin cell, at a controlled ambient temperature of 25 °C. A battery specific model was developed with U and Y as functions of depth of discharge, using experimental voltage, current, and temperature data with the parameter estimation tool. For all numerical simulations in the present study, the developed model parameters were treated as reference model parameters.

4.1. Validation

The Y and U functions were developed as a function of depth of discharge and these functions were curve fitted using the experimental study. The experimental study was conducted at a 1C discharge rate, with a fully-charged coin cell at 4.2 V, with a discharge cut-off voltage of 2.75 V. The results of voltage and temperature from the developed numerical model were compared with the experimental study, as shown in Figure 3. Similar trends were observed for the experimental and numerical studies for voltage and temperature profiles. The maximum deviation for the voltage and temperature results of numerical models, as compared to the experimental study, were within ±5%, as shown in the Figure 3. Thus, the developed numerical model is considered valid for conducting numerical simulations.

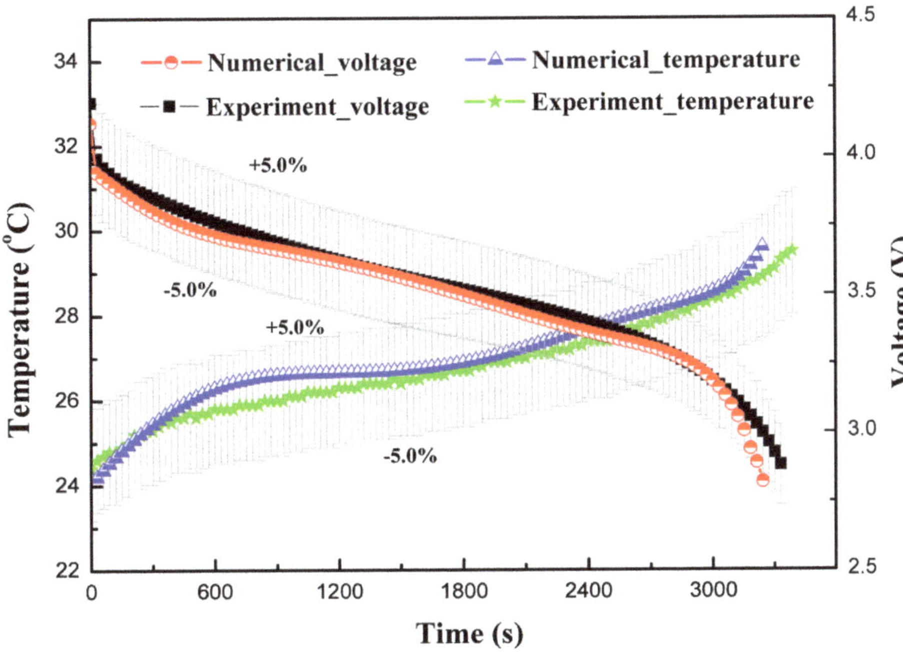

Figure 3. Voltage and temperature results comparison for the experimental study and the developed numerical model.

4.2. Effect of Discharge Rate

The discharge rate of 1C discharges a fully-charged 100% SOC LIB in approximately 1 h. For the safety and longevity of the battery life, the LIBs are advised to charge less than 100% and not to discharge fully to 0% SOC. In simple terms, the continuous discharge rate indicates rate of energy extraction from battery, whereby the higher the discharge rate, the higher the energy extraction rate will be. However, during high-discharge output, LIBs suffer enhanced heat generation, which must be dissipated to safely operate LIBs. Moreover, the batteries suffer performance degradation at extreme temperatures, resulting in lower performance or sometimes malfunctioning. In this section, the effect of different discharge rates during normal operation is compared to coin cell behavior with penetrating element. Figure 4a shows the temperature profiles for different discharge rates and with penetrating element. The trend is similar to the one observed by Vyroubal et al. [19]. As expected for normal discharge tests, the temperature increases as the discharge rate of the coin cell is increased. For safety reasons, the LIR2450 coin cell temperature must not exceed 60 °C during normal operation. Therefore, either the discharge rate should be maintained low, or a high heat transfer co-efficient cooling system needs to be provided. As the discharge rate increases from 1C to 4C, the maximum coin cell temperature increases by 26.2 °C, and with the penetrating element the temperature increases by 84.0 °C. The temperature of the coin cell with penetrating element increases sharply due to development of a secondary current at the short-circuit site. Interestingly, the maximum temperature is observed at the center of the coin cell in both cases—without penetrating element and with penetrating element at center.

The discharge of LIBs at different C-rates is associated with exothermic reactions. Moreover, the heat generation rate is also dependent on the material of construction of electrodes and electrolyte. In the present study, the $LiCoO_2$ chemistry-based LIR2450 coin cell is considered. The in-depth heat generation rate during normal charging–discharging has been previously studied [28]. Figure 4b shows the heat generation rate variation with various discharge rates and penetrating element. For discharge rates of 1C and 2C, the total heat generation rate increases steadily, whereas for higher discharge rates the heat generation rate spikes slightly. During the normal operation of LIBs, the major part of heat is generated due to electrochemical reaction. Moreover, this heat generation is dependent on operating temperature, as electrochemical reactions are very sensitive to temperature. The complexity of electrochemical modeling arises as the continuous increase in temperature of the cell affects the electrochemical reaction. In addition, for the cases involving thermal abuse caused by nail penetration or crash, more complexity is added as LIB may behave abruptly, leading to fire or explosion, if thermal runaway temperature is reached. As shown in Figure 4b, in the case of penetrating element, the heat generation rate increases abruptly initially and then increases steadily. The abrupt increase in the heat generation rate is attributed to the formation of a high-current density area at the site of penetrating element, with very low resistance compared to battery internal resistance. In addition, the small sub-peaks for heat generation observed during the discharge process are related to the phase change influence of electrodes [28].

Figure 4c shows the voltage response of a coin cell when different discharge rates are employed and compared with voltage response, with penetrating element of diameter 3 mm at 1C discharge rate. For the normal operation of the coin cell, the voltage response shows a general trend. With penetrating element, a discharge cut-off voltage of 2.75 V is reached at around 1200 s with 1C discharge. The rapid attainment of a discharge cut-off voltage attributes to the flow of a secondary high-density current due to an internal short circuit at the site of penetrating element. The results of maximum temperature, heat generation rate, and voltage profiles of the coin cell at different discharge rates and with penetrating element show that the discharge rate as well as penetration have a substantial effect on the thermal behavior of the coin cell.

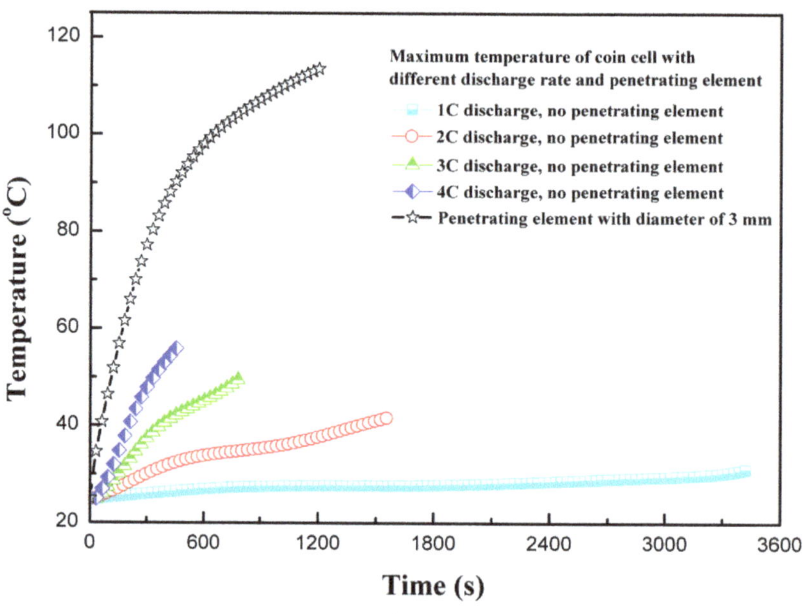

(a)

Figure 4. *Cont.*

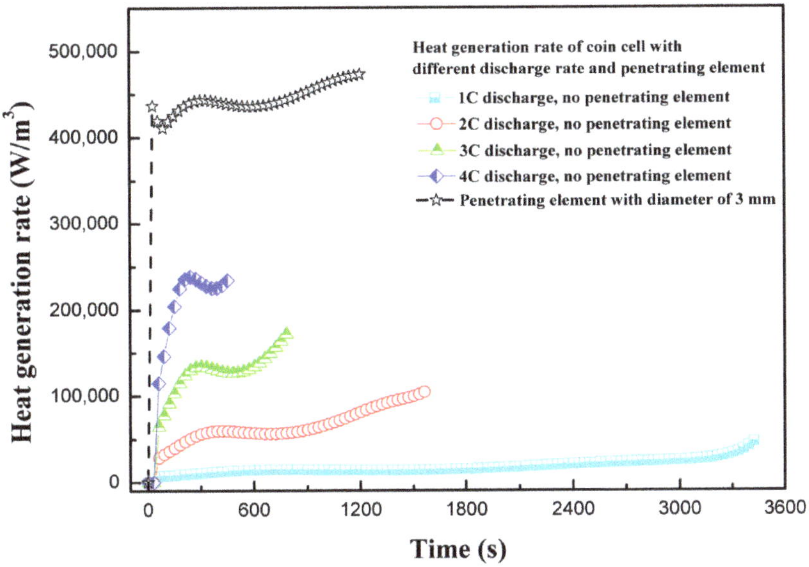

(**b**)

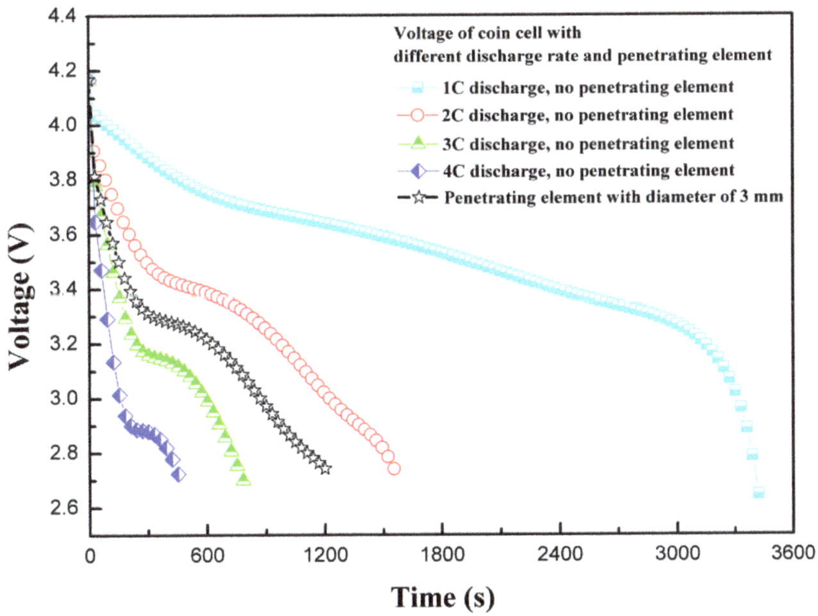

(**c**)

Figure 4. (**a**) Temperature profiles for different discharge rates and with penetrating element. (**b**) Heat generation rate profiles for different discharge rates and with penetrating element. (**c**) Voltage profiles for different discharge rates and with penetrating element.

4.3. *Effect of Penetarting Element Size*

In LIBs, the electrodes are separated by a thin separator. The separator prevents the direct contact of electrodes. The electrodes and electrolytes are tightly packed, as in the case of the LIR2450 coin cell, by positive and negative caps. The generally-used polymer separator has safety issues when combined with high-density and high-capacity LIBs, because, during internal short circuit by penetration or crash, the separator shrinks in volume due to heat, making the direct contact of electrodes possible, which is a dangerous scenario [29]. The effect of different shapes of penetrating element on thermal behavior have been studied previously with ellipsoid, flat, cone, and sphere shapes [6]. In the current study, a cylindrical penetrating element in the battery active material (electrodes and electrolytes) with different sizes were considered. The penetrating element with diameters of 0.5, 1.0, 1.5, 2.0, 2.5, 3.0, and 3.5 mm were considered. During the operation of battery with penetrating element, the electrodes come into contact due to deformation and, subsequently, joule heat is produced at a very high rate. The heat is transferred from the point of penetration to the whole battery by conduction, and then to the outside by convection and radiation. In some cases, the high temperature activates the chemical reactions leading to exothermic behavior, and the cell components, including electrolyte and electrode, explode either from penetrating location or safety valve with flames [5]. The nail penetration tests are characterized by a localized hotspot and the propagation of heat to the whole battery through conduction [30]. Figure 5a shows the effect of the penetrating element size on the maximum coin cell temperature. It is evident from Figure 5a that a large penetrating element produces higher temperatures, and temperature continues increasing as the penetrating element size increases. The maximum temperature of the coin cell increases by 103 °C as the penetrating element diameter was increases from 0.5 to 3.5 mm. This thermal behavior, of a large increase in temperature for higher penetrating element size, is associated with a large short-circuit area occurring during internal short circuit, leading to a large cross-sectional area available for the flow of secondary current that developed due to the short circuit (i.e., the short-circuit current is proportional to the square of the penetrating element radius) [6]. In addition, a large penetrating element leads to a considerable reaction force and a high-buckling displacement [6]. The thermal behavior due to accumulated heat with penetrating element is closely related to battery internal resistance and contact resistance, which is a direct function of penetrating element size (i.e., diameter) [31].

The heat generation rate increases rapidly at the start due to the internal short circuit for all the cases of different diameters of penetrating element. The total heat generation rate is comprised of heat due to ohmic source, heat due to electrochemical reaction source, and heat due to short-circuit source. As the diameter of the penetrating element decreases, the heat due to the short circuit decreases, as shown in Figure 5b. As the diameter of the penetrating element increases, the contribution of heat due to the short circuit increases. The maximum total volumetric heat generation rate of 636,744.6 W/m^3 is recorded for the penetrating element with diameter of 3.5 mm, whereas the maximum total heat generation rate of 57,784.6 W/m^3 is recorded for diameter of 0.5 mm. Figure 5c shows the voltage response of a coin cell with different penetrating element sizes. It is seen from Figure 5c that as the penetrating diameter increases, from 0.5 to 3.5 mm, the time to attain the discharge cutoff voltage of 2.75 V reduces. The coin cell with the penetrating element of 0.5 mm diameter attained a discharge cut-off voltage of 2.75 V in 3270 s, whereas, for the penetrating element diameter of 3.5 mm only 780 s was needed. The results of maximum temperature, heat generation rate, and voltage profiles of the coin cell for different penetrating element diameters show that penetrating element size has a substantial effect on the thermal behavior of a coin cell.

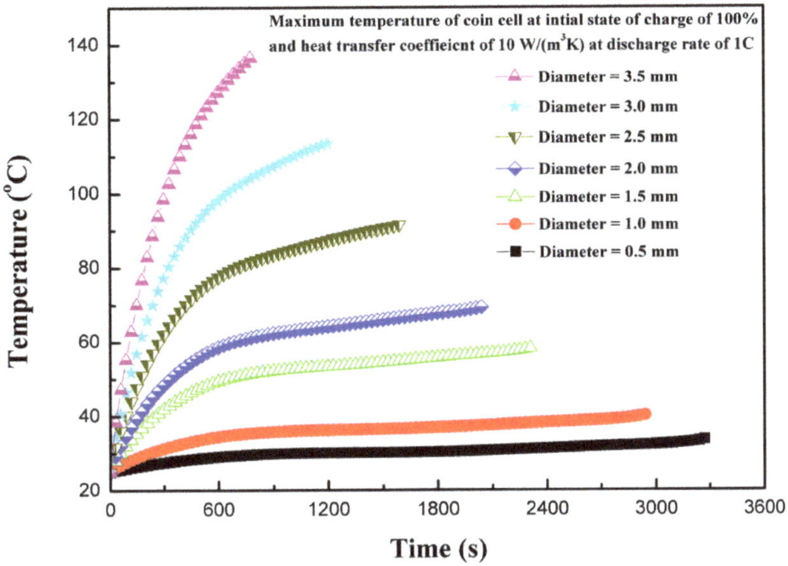

(**a**)

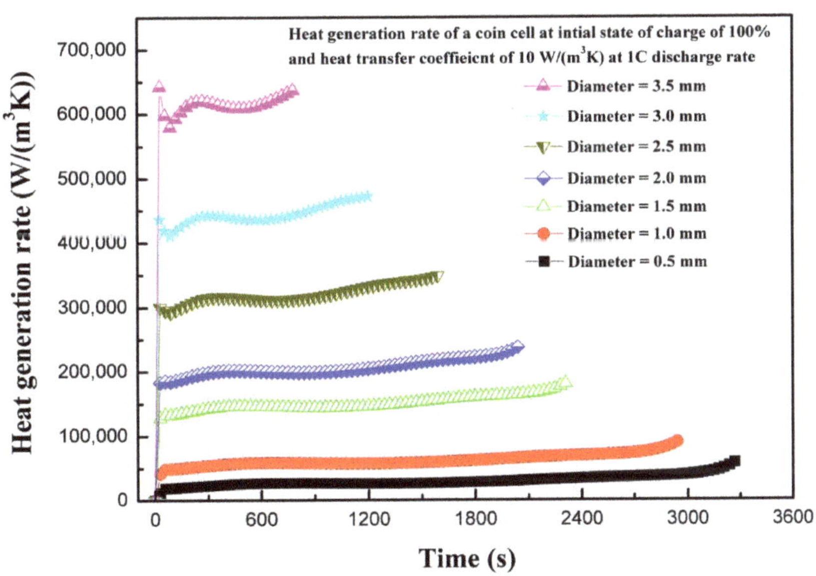

(**b**)

Figure 5. *Cont.*

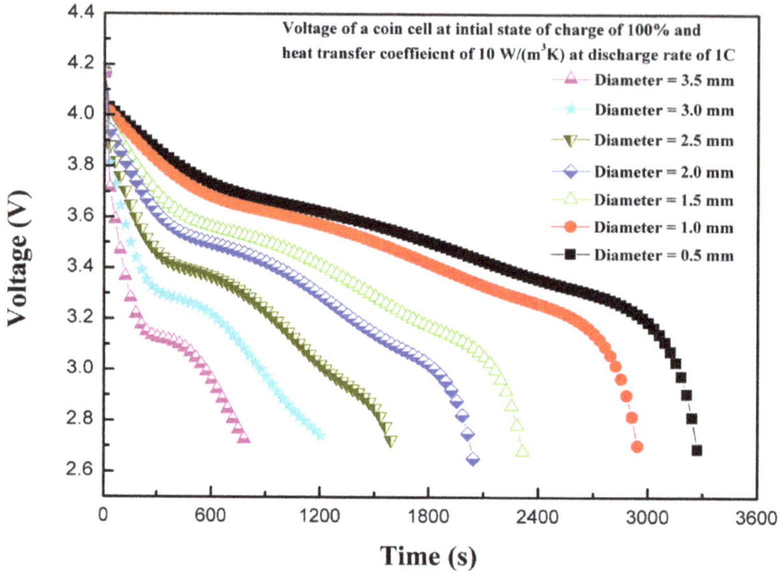

(c)

Figure 5. (**a**) Temperature profiles of the coin cell for different penetrating element sizes. (**b**) Heat generation rate profiles of the coin cell for different penetrating element sizes. (**c**) Voltage profiles of the coin cell for different penetrating element sizes.

4.4. Effect of Initial State of Charge

The state of charge is an indicator of usable energy available compared to maximum rated usable energy. Figure 6a shows the effect of initial state of charge (SOC) on the maximum temperature of the coin cell. The maximum temperature of the coin cell increased with the increase in the initial SOC of the coin cell. Although temperature increased continuously for all SOC levels, the cases with high initial SOC levels showed high temperatures owing to stable high currents for sufficiently long periods [6]. Moreover, the rate of increase of temperature is slightly higher for low SOC level, because the available energy depleted quickly in the case of low SOC levels. The maximum and minimum temperatures of 113.5 and 67.7 °C are observed for 100% and 40% SOC levels, respectively. Similarly, the continuous heat generation rate is maintained for long periods with higher SOCs leading to higher temperatures, as shown in Figure 6b. The characteristic rapid rise of the heat generation rate at the start is observed for all cases of SOCs. The maximum and minimum total heat generation rates of 472,707.2 and 340,656.4 W/m^3 are observed for 100% and 40% SOC levels, respectively. The results of various SOCs show that the thermal runaway temperature attainment is strongly dependent on the SOC level. The results support the findings of Cai et al., which showed that chances of partially-charged LIBs (at 50% SOC or lower) to attain thermal runaway temperatures are rare during mechanical abuse [32]. The maximum temperature decreases as the SOC decreases. On the other hand, more energy is produced for higher SOC levels, as visible for the area below the heat generation curve shown in Figure 6a [31]. The thermal stability of LIB components is strongly dependent on SOC levels [33]. LIBs operate based on the intercalation–deintercalation phenomenon, in which lithium-ion occupies porous structures of cathodes or anodes during charging–discharging. As the level of lithiation in negative electrodes increases, more lithium-ions are available for reactions that are exothermic in nature [34]. The exothermic reactions involving cathodes and electrolytes increase heat generation linearly with the increase in the SOC level, which indicates that SOC levels can play an important role in the event of thermal runaway. This is experimentally shown by Mao et al., where batteries with 100% SOC were burnt owing to thermal runaway, and low SOC level batteries with 0% and 50% were not burnt [22].

Besides this, other parameters, including battery type, battery design, components of active battery material, battery shape, and size of battery, are key important factors that can affect the thermal behavior of LIBs. Therefore, the SOC level effect with penetrating element on the thermal behavior of LIBs is very specific, and may have different results (with above suggested parameters) for different batteries. In addition, the temperature non-uniformity is dependent on the size of the battery under consideration. In the present study, due to the compact size of the coin cell, the temperature non-uniformity study is not considerably relevant. The initial rise of the heat generation rate is maximum for 100% SOC and decreases as the SOC level decreases. Zao et al. discussed practical ways to prevent thermal runaway, and ruled out the decrease of SOC as a practical option to prevent thermal runaway in the case of nail penetration [31]. However, this strategy can be used for aircraft "cargo-only" transportation of batteries, where SOC should be kept at less than 30% to prevent thermal runaway [35]. Figure 6c shows the voltage response of the coin cell for different SOC levels. For different SOC levels, the initial voltage is different [9]. For low SOC, the capacity depletes soon after the stable current is established, with the attainment of a discharge cut-off voltage of 2.75 V. This results in a relatively less heat generation rate, leading to a comparatively low temperature. The voltage curve trends are the same for all the cases of SOC, as all other parameters are kept constant except SOC. The present model did not consider the discussions on thermal runaway behavior for temperatures exceeding 128 °C, as thermal runaway behavior including material decomposition is not included in the present study [6,36]. The results of maximum temperature, heat generation rate, and voltage profile of the coin cell for different SOC levels show that the initial SOC level has a substantial effect on the thermal behavior of the coin cell.

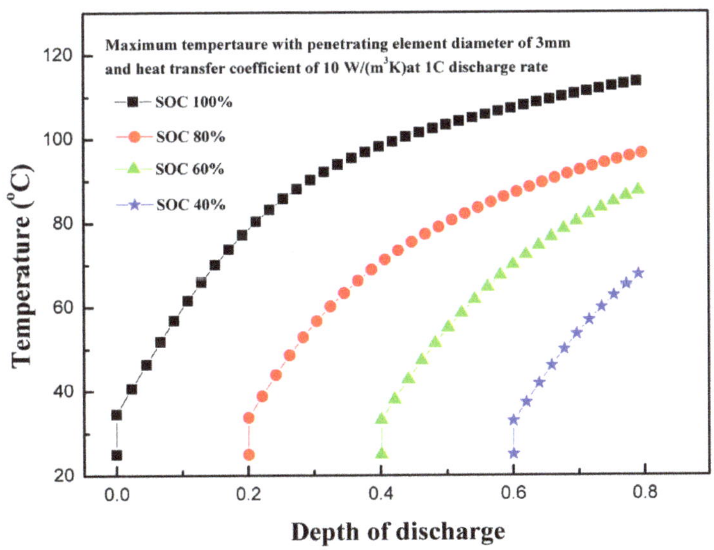

(a)

Figure 6. *Cont.*

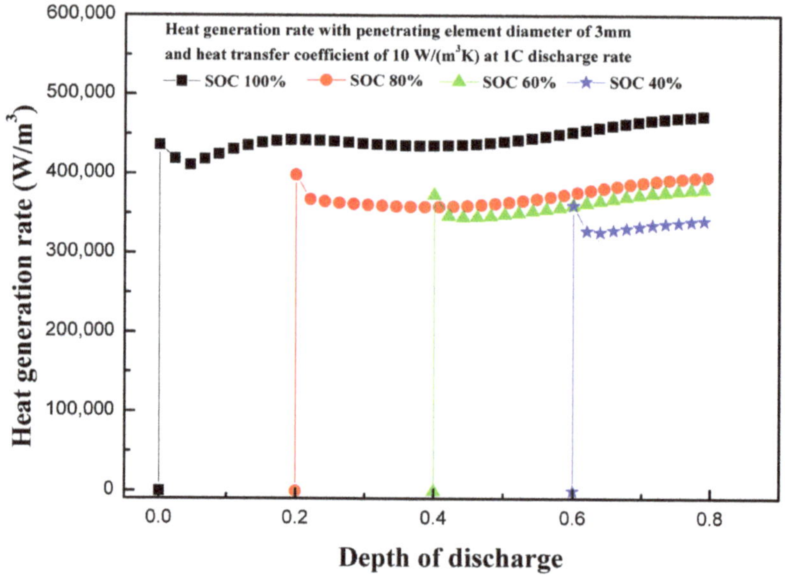

(b)

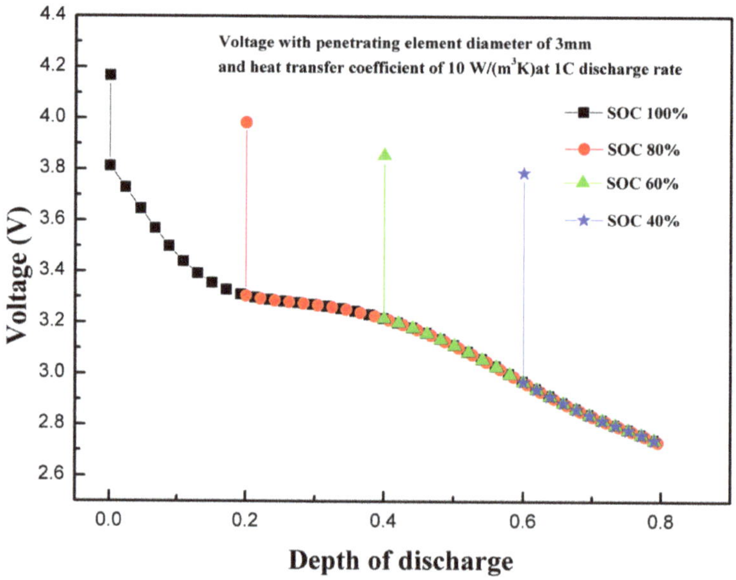

(c)

Figure 6. (**a**) Temperature profiles of the coin cell for different levels of initial state of charge. (**b**) Heat generation rate profiles of the coin cell for different levels of initial state of charge. (**c**) Voltage profiles of the coin cell for different levels of initial state of charge.

4.5. Effect of the Location of the Penetrating Element

The location of the penetrating element can have different effects on different types of batteries. In case of accidents, it is unpredictable as to how and where the battery will be impacted in terms of penetration or crash. As the battery design is different for cylindrical, pouch, or coin cells with different capacities, the penetration or crash during any accident will have varying effects. In the present study, three different locations of the coin cell with penetrating element were considered. These three locations were: Center of the coin cell, middle of the radius, and edge of the coin cell. Figure 7a shows the effect of penetrating element location on the temperature profile of the coin cell. As evident, the temperature increases continuously regardless of penetrating element position. The penetrating element at the center of the coin cell showes slightly higher temperatures than on the middle of the radius and on edge of the coin cell. The results indicate that the penetrating location at the center of the coin cell is the most dangerous, because, with abrupt high heat generation, heat is accumulated at the nail penetration location due to the inability to dissipate heat quickly be means of conduction, convection, or radiation. The location of penetrating element can have varying effects on different types of commercially available LIBs. For example, for the 18650-type cylindrical cell, the penetrating position at the center of the cylinder along the height is the most dangerous case, as thermal runaway can spread to the entire cell. The penetrating element at other locations, than center, may not cover thermal runaway over the entire cell [22]. However, this is dependent on the size, shape, and design of the LIB. For the pouch-type LIB, which has a generally high ampere-hour capacity, the penetrating element at the center of the pouch cell on the flat side is very dangerous as a large amount of heat is accumulated. Additionally, if the penetrating element is near the tabs, that could lead to a large wraparound current source [16]. As in case of the coin cell, due to smaller size and capacity, the position of the penetrating element does not have much differentiating effect, indicating the batteries are prone to thermal runaway or fire hazards, irrespective of the location of the penetrating element. The maximum temperatures of 113.5, 108.7, and 107.5 °C are observed at the center of the coin cell, middle of the radius, and on edge of the coin cell, respectively. Figure 7b shows the voltage response of a coin cell with different penetrating element locations. The penetrating effect has minimal effect on voltage profile. The discharge cut-off voltage for all three cases reached 2.75 V at almost the same time of battery operation.

Figure 8 presents the magnitude of current density and temperature of the coin cell on the middle of the height plane and details of three different cases are presented. The vectors of current density at the central plane of the coin cell for three different cases are presented in Figure 8a,c,e. The maximum current density reaches as high as 8200 A/m^2, which is an extremely high level of current concentrated at the site of penetrating element for a coin cell. The high level of flow of current occurs owing to the low resistance of the penetrating element. The arrow indicates the direction and the color indicates the magnitude of the current density vector at the central plane of the coin cell. The magnitude of the current density decreases with the distance from the penetrating element. Evidently, the temperature also decreases with the distance from the penetrating element. From Figure 8b, it can be seen that the lowest temperature is observed at the edges and this can be attributed to the availability of heat dissipation surface near the edges. The temperature contours for three different cases are presented in Figure 8b,d,f. In addition, the difference between maximum and minimum temperatures for all three cases with penetrating element are below 3.5 °C, owing to the compact size of the coin cell and heat spreading to the entire coin cell. Radial symmetry is observed, as shown in Figure 8a,b, for current density pattern and temperature distribution pattern, owing to the uniform temperature gradient between the center of the coin cell and the edge of the coin cell. The highest temperature is observed at the center due to the penetrating element, and the lowest temperature is observed at the edges with uniform convective heat transfer. The results of the maximum temperature profile, voltage profile, and current density profile for the coin cell with different penetrating locations show minimal variation considering the thermal behavior of the coin cell.

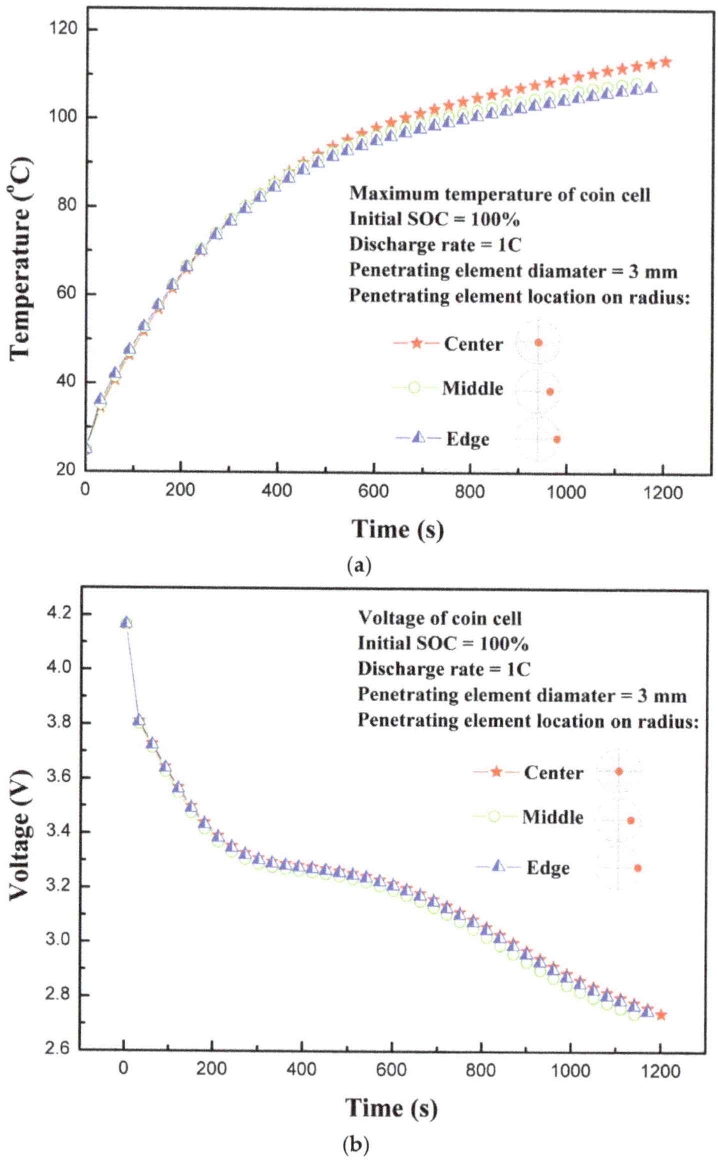

Figure 7. (**a**) Temperature profiles for different locations of the penetrating element. (**b**) Voltage profiles for different locations of the penetrating element.

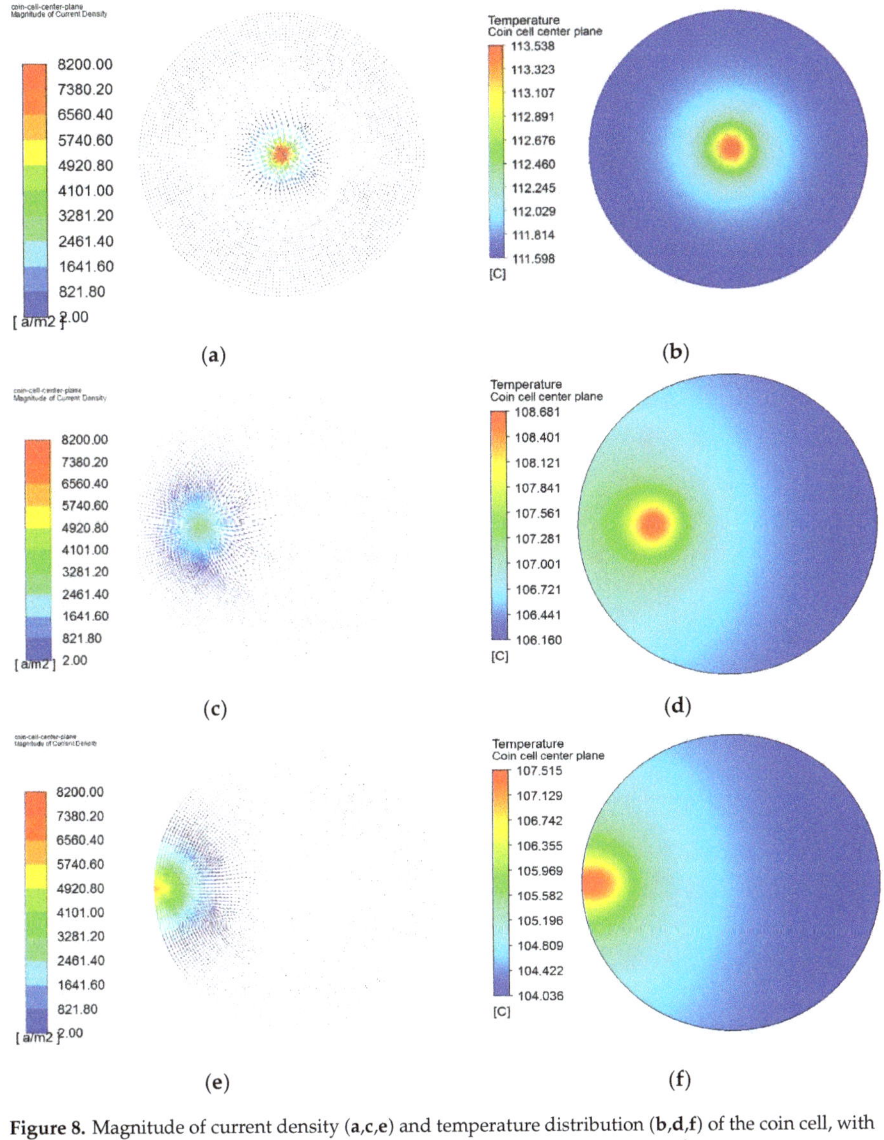

Figure 8. Magnitude of current density (a,c,e) and temperature distribution (b,d,f) of the coin cell, with penetrating element diameter of 3 mm and heat transfer coefficient of 10 W/(m^3K), at a discharge rate of 1C, with initial state of charge (SOC) of 100% at the center of the cell, middle of the radius, and edge of the coin cell.

4.6. Effect of the Heat Transfer Coefficient

Figure 9 shows the effect of the heat transfer co-efficient on the maximum temperature of the coin cell during the internal short circuit. As expected, the heat transfer co-efficient has a consequential effect on the maximum temperature of the coin cell. A cooling convective heat transfer coefficient of 1 W/(m^2K) or less represents a condition similar to a battery packed with insulating materials, whereas 200 W/(m^2K) or more indicates an effective liquid cooling situation [31]. For the case of the coin cell, the heat transfer coefficients ranging from 5 to 25 W/(m^2K) are considered, which are reasonable assumptions for partially-insulated to forced air cooling convection. The heat transfer coefficient of 10 W/(m^{2}K) can be considered as a natural convection case. The heat dissipation term presented in Equation (7) is dependent on temperature variations of LIBs. The maximum temperature of 165.3 °C is observed in the case of a heat transfer coefficient of 5 W/(m^{2}K), whereas the lowest temperature of 59.5 °C is observed for a heat transfer coefficient of 25 W/(m^{2}K). The thickness of the battery plays an important role in transferring heat to the surface. For thicker batteries, the heat transfer coefficient is insignificant [31]; however, for thinner batteries, such as LIR2450 coin cells, the heat transfer coefficient has a substantial effect on the maximum temperature of the coin cell. Providing efficient cooling is useful, especially for cases of thermal abuse, as this is an effective tool to prevent thermal runaway or to reduce the effect of propagation of thermal runaway in LIBs, as previously discussed [23]. The results show that the heat transfer coefficient has a substantial effect on the thermal behavior of coin cells.

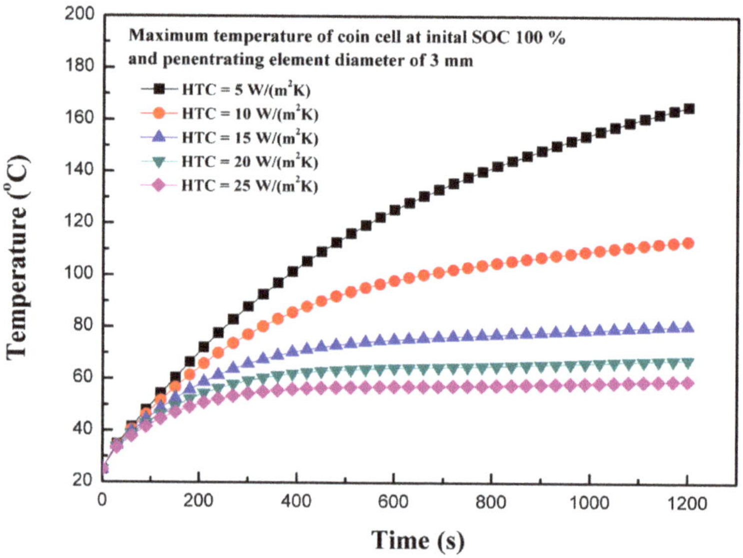

Figure 9. Temperature profiles for different heat transfer co-efficients.

4.7. Effect of Short-Circuit Resistance

Figure 10 shows the variation of dimensionless short-circuit heat generation rates with dimensionless internal short-circuit resistance. The dimensionless short-circuit heat generation rate is represented by Equation (14) and the dimensionless internal short resistance is represented by Equation (15). The internal resistance of the coin cell is 400 mΩ. During the internal short circuit, heat is generated due to a short circuit along with heat generation due to electrochemical reaction source and ohmic source. The total heat generation is the sum of the electrochemical heat source, ohmic heat source, and short-circuit heat source. The results from Figure 10 indicate that there is an optimum point at which the heat generation contribution from the short circuit is maximum. A similar trend was presented with the 1 Ah capacity pouch cell by Fang et al., although the values differ as the battery under consideration is different from previously studied [21]. There are few researches which focus on the increase in charge transfer resistance by using thermal runaway retardant (TRR) such as dibenzylamine [18], by using flexible separators, or by using high-viscosity protection films [31]. With decrease of internal resistance, the contribution of heat source from the short circuit increased due to a higher short-circuit current with constant contact resistance (constant size of penetrating element) [31].

$$\text{Dimensionless short circuit heat generation rate} = \frac{q_{short}}{q_{ECh} + q_{Ohmic} + q_{short}} \quad (14)$$

$$\text{Dimensionless short circuit resistance} = \frac{R_{short}}{R_{cell}} \quad (15)$$

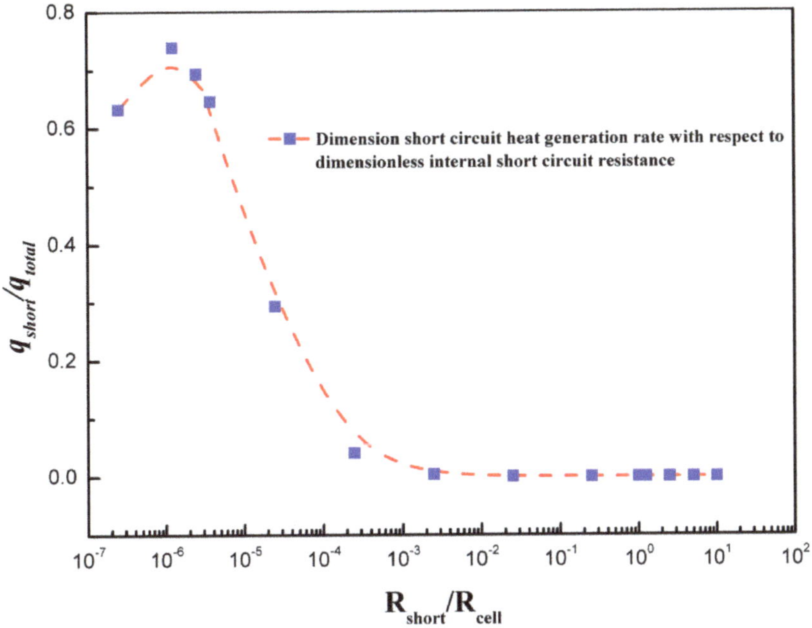

Figure 10. Dimensionless short-circuit heat generation rate variation with dimensionless internal short-circuit resistance.

The short-circuit resistance during the penetration can play an important role, as it can affect the extent of damage to the coin cell in the case of thermal abuse leading to thermal runaway. The short-circuit resistance varies from 0.001 to 0.0000005 Ω and effects on the thermal behavior of coin cell is reported. A high value of short-circuit resistance can contain thermal runaway, whereas a very low-resistance value could facilitate the exothermic reactions leading to thermal runaway. Figure 11a shows the effect of various short-circuit resistances on the maximum coin cell temperature. It is evident from Figure 11a that a lower short-circuit resistance produces a high temperature, and the temperature continues increasing as the short-circuit resistance decreases. The maximum temperature of the coin cell increases to 82.9 °C as the short-circuit resistance decreases from 0.001 to 0.0000005 Ω. This thermal behavior of a large increase in temperature for lower short-circuit resistance is associated with the large flow of a short-circuit current with a low-resistance path. The variation in heat generation rates is considerably large for different short-circuit resistances, as shown in Figure 11b. The low heat generation for higher short-circuit resistance is reported, as the contribution from the short-circuit heat source is very low. In such cases, the electrochemical heat source is a dominating heat source. However, as the short-circuit resistance decreases, the short-circuit heat source dominates over the electrochemical heat source. This is also supported from the results of Figure 10, which shows that the contribution from the short-circuit heat source to the total heat source is relatively large for low short-circuit resistances. Figure 11c shows the voltage response of a coin cell with penetrating element diameter of 3 mm and different short-circuit resistances. It is seen from the Figure 11c that as the short-circuit resistance decreases, the time to attain the discharge cutoff voltage of 2.75 V reduces. The coin cell with a short-circuit resistance of 0.001 Ω attained a discharge cut-off voltage of 2.75 V in 3420 s, whereas the coin cell with a short-circuit resistance of 0.0000005 Ω only needed

1200 s. The results of maximum temperature, heat generation rate, and voltage profiles of the coin cell for different short-circuit resistances show that short-circuit resistance has a substantial effect on the thermal behavior of the coin cell. In addition, the findings from the present studies are useful, especially focusing on the thermal-runaway retardant (TRR) of lithium-ion batteries, as heat generation, temperature distribution, and effect of resistance could be used to evaluate the effect of TRR.

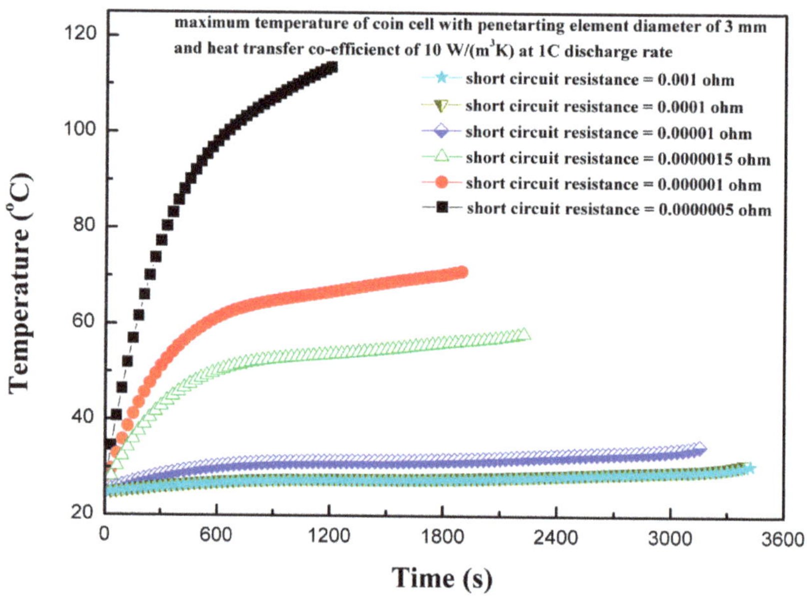

(a)

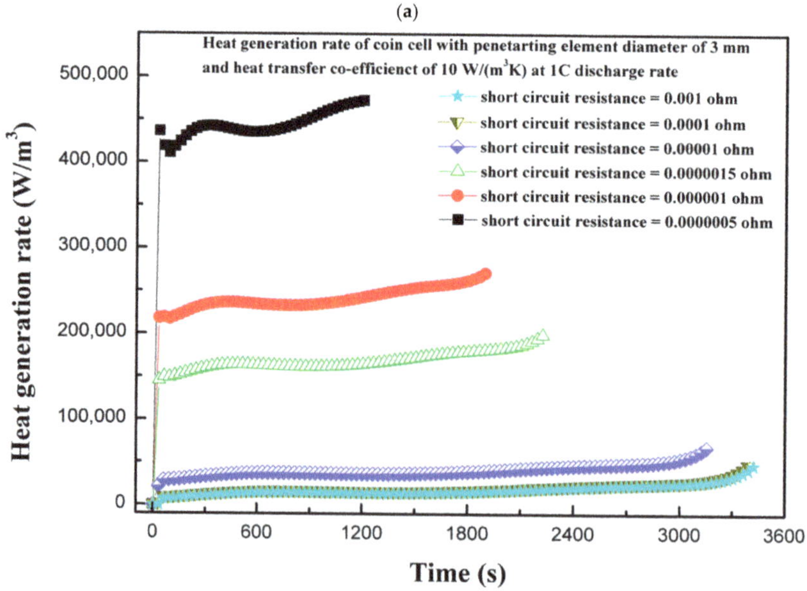

(b)

Figure 11. *Cont.*

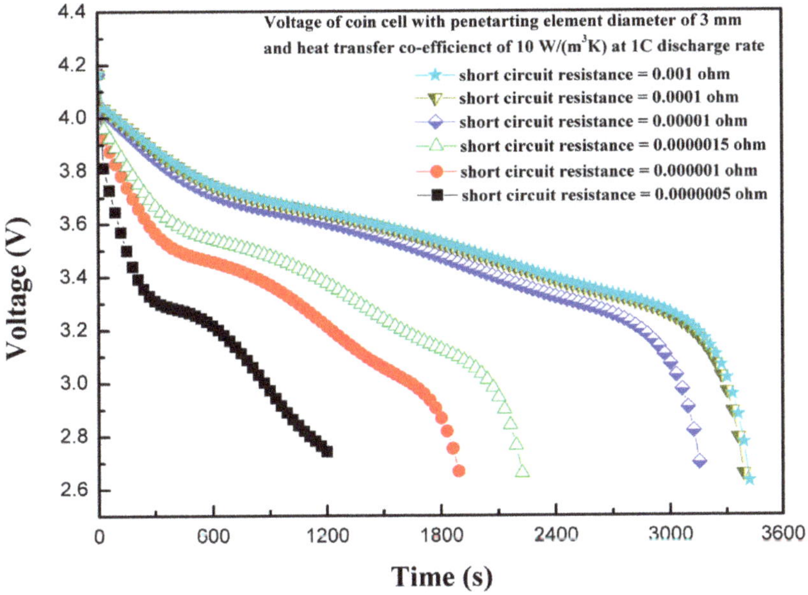

(c)

Figure 11. (**a**) Temperature profiles of the coin cell for different internal short-circuit resistances. (**b**) Heat generation rate profiles of the coin cell for different internal short-circuit resistances. (**c**) Voltage profiles of the coin cell for different internal short-circuit resistances.

5. Conclusions

This study presents the thermal behavior of the LIR2450 micro coin cell battery, with capacity of 120 mAh, with internal short circuit by penetrating element. The numerical model is developed using voltage, temperature, and current characteristics from experimental study and validated within ±5.0%. The effect of the penetrating element size, the location of the penetrating element, initial state of charge, discharge rate, short-circuit resistance, and heat transfer coefficient on the maximum temperature and the heat generation rate of the coin cell are investigated. The maximum temperature and heat generation rate increased with the increase of the penetrating element size and initial state of charge, whereas it decreased with the increase of the heat transfer coefficient. The penetrating element at the center of the coin cell reached the highest temperature and heat generation rate, as compared to the penetrating element at middle of the radius or on the edge of the coin cell. The variation of dimensionless short-circuit heat generation rates with dimensionless short-circuit resistances showed an optimum point. The study provides comprehensive insights on the thermal behavior of lithium-ion cells during thermal abuse condition, with internal short circuit by penetrating element, and can be used for enhancing the design of safe lithium-ion batteries.

Author Contributions: Conceptualization, M.-Y.L. and M.S.P.; methodology, M.-Y.L.; software, M.S.P.; validation, M.-Y.L., M.S.P. and N.K.; experimental study, M.-Y.L., M.S.P. and J.-H.S.; numerical investigation, M.-Y.L. and M.S.P.; resources, M.-Y.L.; data reduction, M.-Y.L., M.S.P. and J.-H.S.; writing—original draft preparation, M.-Y.L. and M.S.P.; writing—review and editing, M.-Y.L., M.S.P., J.-H.S. and N.K.; visualization, M.S.P.; supervision, M.-Y.L.; project administration, M.-Y.L.; funding acquisition, M.-Y.L. All authors have read and agreed to the published version of the manuscript.

Funding: This research received no external funding.

Acknowledgments: This work was supported by LG Yonam Foundation (of Korea).

Conflicts of Interest: The authors declare no conflict of interest.

References

1. Zubi, G.; López, R.D.; Carvalho, M.; Pasaoglu, G. The lithium-ion battery: State of the art and future perspectives. *Renew. Sustain. Energy Rev.* **2018**, *89*, 292–308. [CrossRef]
2. Wang, Y.; Liu, B.; Li, Q.; Cartmell, S.; Ferrara, S.; Deng, Z.D.; Xiao, J. Lithium and lithium ion batteries for applications in microelectronic devices: A review. *J. Power Sources* **2015**, *286*, 330–345. [CrossRef]
3. Patil, M.S.; Panchal, S.; Kim, N.; Lee, M.-Y. Cooling Performance Characteristics of 20 Ah Lithium-Ion Pouch Cell with Cold Plates along Both Surfaces. *Energies* **2018**, *11*, 2550. [CrossRef]
4. Chen, Z.; Xiong, R.; Lu, J.; Li, X. Temperature rise prediction of lithium-ion battery suffering external short circuit for all-climate electric vehicles application. *Appl. Energy* **2018**, *213*, 375–383. [CrossRef]
5. Wang, Q.; Mao, B.; Stoliarov, S.I.; Sun, J. A review of lithium ion battery failure mechanisms and fire prevention strategies. *Prog. Energy Combust. Sci.* **2019**, *73*, 95–131. [CrossRef]
6. Liu, B.; Yin, S.; Xu, J. Integrated computation model of lithium-ion battery subject to nail penetration. *Appl. Energy* **2016**, *183*, 278–289. [CrossRef]
7. Liao, Z.; Zhang, S.; Li, K.; Zhao, M.; Qiu, Z.; Han, D.; Zhang, G.; Habetler, T.G. Hazard analysis of thermally abused lithium-ion batteries at different state of charges. *J. Energy Storage* **2020**, *27*, 101065. [CrossRef]
8. Chen, M.; Liu, J.; He, Y.; Yuen, R.; Wang, J. Study of the fire hazards of lithium-ion batteries at different pressures. *Appl. Therm. Eng.* **2017**, *125*, 1061–1074. [CrossRef]
9. Liu, B.; Jia, Y.; Yuan, C.; Wang, L.; Gao, X.; Yin, S.; Xu, J. Safety issues and mechanisms of lithium-ion battery cell upon mechanical abusive loading: A review. *Energy Storage Mater.* **2020**, *24*, 85–112. [CrossRef]
10. Wang, Q.; Ping, P.; Zhao, X.; Chu, G.; Sun, J.; Chen, C. Thermal runaway caused fire and explosion of lithium ion battery. *J. Power Sources* **2012**, *208*, 210–224. [CrossRef]
11. Williard, N.; He, W.; Hendricks, C.; Pecht, M. Lessons Learned from the 787 Dreamliner Issue on Lithium-Ion Battery Reliability. *Energies* **2013**, *6*, 4682–4695. [CrossRef]
12. Qin, T.; Zeng, S.; Guo, J.; Skaf, Z. State of Health Estimation of Li-ion Batteries with Regeneration Phenomena: A Similar Rest Time-Based Prognostic Framework. *Symmetry* **2017**, *9*, 4. [CrossRef]
13. Feng, X.; Sun, J.; Ouyang, M.; Wang, F.; He, X.; Lu, L.; Peng, H. Characterization of penetration induced thermal runaway propagation process within a large format lithium ion battery module. *J. Power Sources* **2015**, *275*, 261–273. [CrossRef]
14. Ping, P.; Wang, Q.; Huang, P.; Sun, J.; Chen, C. Thermal behaviour analysis of lithium-ion battery at elevated temperature using deconvolution method. *Appl. Energy* **2014**, *129*, 261–273. [CrossRef]
15. Xia, Y.; Chen, G.; Zhou, Q.; Shi, X.; Shi, F. Failure behaviours of 100% SOC lithium-ion battery modules under different impact loading conditions. *Eng. Fail. Anal.* **2017**, *82*, 149–160. [CrossRef]
16. Yamanaka, T.; Takagishi, Y.; Tozuka, Y.; Yamaue, T. Modeling lithium ion battery nail penetration tests and quantitative evaluation of the degree of combustion risk. *J. Power Sources* **2019**, *416*, 132–140. [CrossRef]
17. Chen, M.; Bai, F.; Song, W.; Lv, J.; Lin, S.; Feng, Z.; Li, Y.; Ding, Y. A multilayer electro-thermal model of pouch battery during normal discharge and internal short circuit process. *Appl. Therm. Eng.* **2017**, *120*, 506–516. [CrossRef]
18. Shi, Y.; Noelle, D.J.; Wang, M.; Le, A.V.; Yoon, H.; Zhang, M.; Meng, Y.S.; Qiao, Y. Exothermic behaviors of mechanically abused lithium-ion batteries with dibenzylamine. *J. Power Sources* **2016**, *326*, 514–521. [CrossRef]
19. Vyroubal, P.; Kazda, T. Finite element model of nail penetration into lithium ion battery. *J. Energy Storage* **2018**, *20*, 451–458. [CrossRef]
20. Noelle, D.J. Investigating Internal Short Circuit Heating to Inform Novel Lithium-Ion Battery Safety Strategies. Ph.D. Thesis, University of California, San Diego, CA, USA, 2018.
21. Fang, W.; Ramadass, P.; Zhang, Z. Study of internal short in a Li-ion cell-II. Numerical investigation using a 3D electrochemical-thermal model. *J. Power Sources* **2014**, *248*, 1090–1098. [CrossRef]
22. Mao, B.; Chen, H.; Cui, Z.; Wu, T.; Wang, Q. Failure mechanism of the lithium ion battery during nail penetration. *Int. J. Heat Mass Transf.* **2018**, *122*, 1103–1115. [CrossRef]
23. Zhao, R.; Liu, J.; Gu, J. Simulation and experimental study on lithium ion battery short circuit. *Appl. Energy* **2016**, *173*, 29–39. [CrossRef]
24. Wang, M.; Le, A.V.; Noelle, D.J.; Shi, Y.; Meng, Y.S.; Qiao, Y. Internal-short-mitigating current collector for lithium-ion battery. *J. Power Sources* **2017**, *349*, 84–93. [CrossRef]

25. Ansys Inc. "ANSYS Fluent theory guide 2019 release," ANSYS, Inc., December 2019. [Online]. Available online: http://www.ansys.com (accessed on 1 December 2019).
26. Kim, U.S.; Yi, J.; Shin, C.B.; Han, T.; Park, S. Modeling the Dependence of the Discharge Behavior of a Lithium-Ion Battery on the Environmental Temperature. *J. Electrochem. Soc.* **2011**, *158*, A611–A618.
27. Gu, H. Mathematical Analysis of a Zn/NiOOH Cell. *J. Electrochem. Soc.* **1983**, *130*, 1459–1464. [CrossRef]
28. Giel, H.; Henriques, D.; Bourne, G.; Markus, T. Investigation of the heat generation of a commercial 2032 (LiCoO2) coin cell with a novel differential scanning battery calorimeter. *J. Power Sources* **2018**, *390*, 116–126. [CrossRef]
29. Kim, C.-S.; Yoo, J.-S.; Jeong, K.-M.; Kim, K.; Yi, C.-W. Investigation on internal short circuits of lithium polymer batteries with a ceramic-coated separator during nail penetration. *J. Power Sources* **2015**, *289*, 41–49. [CrossRef]
30. Maleki, H.; Howard, J.N. Internal short circuit in Li-ion cells. *J. Power Sources* **2009**, *191*, 568–574. [CrossRef]
31. Zhao, R.; Liu, J.; Gu, J. A comprehensive study on Li-ion battery nail penetrations and the possible solutions. *Energy* **2017**, *123*, 392–401. [CrossRef]
32. Cai, W.; Wang, H.; Maleki, H.; Howard, J.; Curzio, E.L. Experimental simulation of internal short circuit in Li-ion and Li-ion-polymer cells. *J. Power Sources* **2011**, *196*, 7779–7783. [CrossRef]
33. Wang, Q.; Sun, J.; Yao, X.; Chen, C. Thermal Behavior of Lithiated Graphite with Electrolyte in Lithium-Ion Batteries. *J. Electrochem. Soc.* **2006**, *153*, A329–A333. [CrossRef]
34. Maleki, H.; Deng, G.; Haller, I.K.; Anani, A.; Howard, J.N. Thermal Stability Studies of Binder Materials in Anodes for Lithium-Ion Batteries. *J. Electrochem. Soc.* **2000**, *147*, 4470–4475. [CrossRef]
35. IATA. 2020 Lithium Battery Guidance Document. 2020. Available online: https://www.iata.org/contentassets/05e6d8742b0047259bf3a700bc9d42b9/lithium-battery-guidance-document-2020.pdf (accessed on 1 January 2020).
36. Chiu, K.-C.; Lin, C.-H.; Yeh, S.-F.; Lin, Y.-H.; Chen, K.-C. An electrochemical modeling of lithium-ion battery nail penetration. *J. Power Sources* **2014**, *251*, 254–263. [CrossRef]

MDPI
St. Alban-Anlage 66
4052 Basel
Switzerland
Tel. +41 61 683 77 34
Fax +41 61 302 89 18
www.mdpi.com

Symmetry Editorial Office
E-mail: symmetry@mdpi.com
www.mdpi.com/journal/symmetry

www.ingramcontent.com/pod-product-compliance
Lightning Source LLC
Chambersburg PA
CBHW061827210926
51850CB00006B/830